The Future of Hyperspectral Imaging

The Future of Hyperspectral Imaging

Special Issue Editor

Stefano Selci

MDPI • Basel • Beijing • Wuhan • Barcelona • Belgrade

MDPI

Special Issue Editor
Stefano Selci
Institute of Photonics and
Nanotechnologies, IFN-CNR,
Roma, Italy

Editorial Office
MDPI
St. Alban-Anlage 66
4052 Basel, Switzerland

This is a reprint of articles from the Special Issue published online in the open access journal *Journal of Imaging* (ISSN 2313-433X) from 2018 to 2019 (available at: https://www.mdpi.com/journal/jimaging/special issues/future Hyperspectral Imaging).

- - -

For citation purposes, cite each article independently as indicated on the article page online and as indicated below:

LastName, A.A.; LastName, B.B.; LastName, C.C. Article Title. *Journal Name* **Year**, *Article Number*, Page Range.

ISBN 978-3-03921-822-6 (Pbk)
ISBN 978-3-03921-823-3 (PDF)

Cover image courtesy of Stefano Selci.

Contents

About the Special Issue Editor

Stefano Selci is a senior scientist at IFN-CNR. After many years of activity in optical spectroscopies and scanning probes, he realized the first Italian STM (Scanning Tunneling Microscope). He has produced some fundamental papers on the exploration, analysis, and modeling of surface states, in particular about the dielectric function concept applied to surface states. Within his STM activity, he has also participated in the CNR's "Genetic Engineering" Project, within the International "Genoma project" lead by Nobel laureate Prof. Dulbecco. He is now the principal investigator of the hyperspectral confocal microscopy activity at IFN-CNR, responsible for its design and implementation, leading the effort of merging concepts from optical physics with biologic exploration.

Journal of
Imaging

MDPI

Editorial

The Future of Hyperspectral Imaging

Stefano Selci

Institute of Photonics and Nanotechnologies, IFN-CNR, via Cineto Romano 42, 00156 Roma, Italy;
Stefano.Selci@cnr.it

Received: 22 October 2019; Accepted: 23 October 2019; Published: 25 October 2019

Abstract: The Special Issue on hyperspectral imaging (HSI), entitled "The Future of Hyperspectral Imaging", has published 12 papers. Nine papers are related to specific current research and three more are review contributions: In both cases, the request is to propose those methods or instruments so as to show the future trends of HSI. Some contributions also update specific methodological or mathematical tools. In particular, the review papers address deep learning methods for HSI analysis, while HSI data compression is reviewed by using liquid crystals spectral multiplexing as well as DMD-based Raman spectroscopy. Specific topics explored by using data obtained by HSI include alert on the sprouting of potato tubers, the investigation on the stability of painting samples, the prediction of healing diabetic foot ulcers, and age determination of blood-stained fingerprints. Papers showing advances on more general topics include video approach for HSI dynamic scenes, localization of plant diseases, new methods for the lossless compression of HSI data, the fusing of multiple multiband images, and mixed modes of laser HSI imaging for sorting and quality controls.

Keywords: hyperspectral imaging; medical imaging by HSI; HSI for biology; remote sensing; hyperspectral microscopy; fluorescence hyperspectral imaging; Raman hyperspectral imaging; infrared hyperspectral imaging; statistical methods for HSI; hyperspectral data mining and compression; statistical methods for HSI; hyperspectral data mining and compression

---

1. Introduction to This Special Issue

Hyperspectral imaging (HSI) manages to gather images where any single pixel is associated a full spectrum in a given range. Unlike multispectral techniques that are already capable of acquiring similar data within a few spectral bands, a continuous spectrum is available through HSI. As research on specific materials and their properties is commonly identified by spectral signatures, HSI exploits the same objective down to the pixel level with the further advantage of possible classification and segmentation of the overall spectral and imaging information, like for land analysis by satellite observation showing distinct regions with water, mineral resources, or plant extensions. No prior knowledge is usually needed except for what can be obtained exclusively from the dataset.

The remarkable mix of information is often represented by "hypercubes", a multidimensional representation of the obtained data along multiple axes providing a picture of the spatial distribution of the observed information: Spectroscopic (one axis, with signal coming from reflectance, fluorescence, Raman, or any other spectroscopic probe), structural (three axes), and also time (a possible further axis).

A single image pixel can range down to microscopic detail and, for HSI-based microscopes, up to meters for satellite data. It should be noted that the spectral component does not have to be limited to the visible range, for instance, the infrared spectral region can be a common choice.

The rapid increase of possible HSI applications and the availability of better spectral hardware requires a much higher speed in acquisition, smart data elaboration, new set-ups, and new ideas, in order to make HSI tools compatible with real time needs, for instance looking for food contamination or for obtaining prompt cancer detection along routine medical checks.

Mathematical tools and algorithms are also needed for HSI data across many different applications, facilitating chemo-physical identification of materials on the basis of their unique spectral signature that is usually buried under a huge amount of data.

The aim of this Special Issue is to offer a view of current topics that are believed to become more significant in the future, hoping that the methods of acquisition and analysis shown in the articles can be used as template cases or for a better understanding of a large variety of problems.

2. Contributions

Hyperspectral imaging offers great opportunities, but it requires careful handling of information obtained in order to have a clear vision of the spectroscopic signatures that might become hidden within a huge amount of data. Moreover, different sources of signals can be obtained simultaneously to each other and a problem arises in how to meaningfully merge the different contributions.

Therefore, there are contributions related to fuse different spectral bands or deal with signals originating from diverse decay channels (e.g., luminescence and Raman).

Likewise, in order to reduce the effort, various compression strategies in the spectroscopic realm are well illustrated.

More theoretical smart tools are then introduced and reviewed that are sophisticated and at the same time versatile so as to better investigate the intricacies of HIS hypercube in searching for specific spectroscopic signatures.

Various specific applications are presented as well. Aside from their specific application areas, their practical realization requires a high skill level from the technical set-up to the data analysis, to make the presentation more widely interesting.

In particular, there are experimental and theoretical contributions for time series observations, compression, and specific signatures, which are a very firm step toward 4D representation and usage of HSI data.

In Radi et al. [1], data spectral fusion is shown as coming from a Visible and Near Infrared VIS/NIR spectroscopic system and from a VIS/NIR hyperspectral imaging systems used respectively in interactance mode and in reflection mode with the aim of acquiring the spectral information of whole tubers for predicting the primordial leaf count of potatoes. In particular, the HSI unit uses an optical fiber to illuminate a large portion of the sampled potato with the imaging spectrograph directly attached to a CCD camera, acquiring spectral information along a line of the potato that is held by a scanner system, therefore in push broom modality.

Results inferred from this study, with the aid of partial least squares (IPLS), initiate the possibility of developing a portable or stationary electronic system aimed at obtaining a rapid and accurate prediction of the sprouting activity of stored potatoes. Future steps could be in testing more growing seasons for many cultivated varieties (cultivars), as well as improving the robustness and reproducibility of the prediction models. Furthermore, an increase in the productivity of the method could be achieved through improved data elaboration, for instance reducing the number of selected wavelengths, and other IPLS models, such as moving average IPLS, synergy IPLS, backward/forward IPLS, or else a genetic algorithm.

In Bonifazi et al. [2], a HSI set-up is used to evaluate the stability against light and UV ageing of several painting materials, in particular powder pigments and commercial watercolors to be used in retouching. In order to obtain HSI data, a pushbroom commercial system acquires hyperspectral images in the short-wave infrared region, in the spectral wavelength interval between 1000 nm and 2500 nm. The pixel resolution by scanning the sample is between 30 micron and 300 micron, building the image one line at a time while the sample is scanned on a moving sample tray in front of the camera. The paper shows the possibility of evaluating minute spectral variations due to ageing times well before the eyes can detect changes: The classification techniques based on principal component analysis (PCA) and k-nearest neighbor (KNN) of hyperspectral data are effectively capable of monitoring the changes occurring in the painting layers. Therefore, HSI coupled with a chemometric approach allows one to monitor ageing modifications of paint layers, showing the possibility of detecting damages before its irreversibility, a result of great relevance in the field of cultural heritage and particularly useful in monitoring artworks and restoration interventions over time at a lower cost compared to similar methods. Future research lines are suggested, for instance in studying other restoration materials, like synthetic resins, or the classification and forecast of material behavior involved in cultural heritage artifacts.

Yang et al. [3], explore the medical implications of predicting the healing of diabetic foot ulcers by a HSI instrument. Such a tool is presented as highly helpful because foot ulcers are indeed a major complication of diabetes. Previously, the biomedical application of HSI has been able to anticipate wound healing based on SpO_2 values due to the different absorption spectra of oxy- and deoxyhemoglobin. Here, PCA is addressed as an alternative approach to improving the prediction of wound healing. A comparison is also made with the performance of SpO_2 mapping. It is found that the PCA second principal component elaborated on hyperspectral images appears superior to analysis in comparison to SpO_2 values in predicting the healing of wounds, taken at a baseline of 12 weeks. A HSI camera operating in pushbroom mode, for which the images are taken one line at a time from the scene, is part of the set-up that include a CCD camera coupled to an imaging spectrograph. Each 3D data cube taken in this study, after a sweep from heel to toe, is reduced to simpler regions of interest of 50 pixels × 50 pixels, for which the authors found enough information to characterize the wound and surrounding tissue for all obtained images. The PC analysis is deeply discussed, pointing to PC2 discrimination in the oxy- and deoxyhemoglobin spectra with superior performance to SpO_2 measurement strategies.

In Cadd et al. [4], a novel application of visible wavelength reflectance HSI is shown for both detection and age determination of blood-stained fingerprints on white ceramic tiles based on the signature of hemoglobin in the visible absorption spectrum between 400 and 680 nm and for the presence of a Soret peak at 415 nm. Blood-stained fingerprints were aged for over 30 days and analyzed using HSI. Data produced results organized on a 24 h scale and a 30-day scale. A clear age estimation of deposited blood-stained fingerprints has been shown to be therefore possible, while a similar visual examination was not possible using a standard digital photographic camera. The HSI system used in this study had the same setup as a previous one shown by the same authors consisting in a liquid crystal tunable filter (LCTF), coupled to a digital camera, and a scene illumination with two LED light sources, for VIS and UV. Control of the LCTF and the entire process by means of a custom software takes approximately 30 s to acquire and process each image, demonstrating that HSI could be used for the detection and identification of both blood stains and blood-stained fingerprints together with determining their age. More rugged and portable instruments for use at crime scenes could be realized in the future, which would be particularly beneficial for criminal investigations.

In Gruber et al. [5], a fast line-scanning hyperspectral imaging system is described. The experiments carried out, shown by four different applications, demonstrate that a Laser-excited HSI system makes possible the acquisition of Raman and fluorescence spectra on relatively large sample areas, with fast and high spatial resolution scans. As it is clear and well known, it is not easy to detect the weaker Raman signals compared to the higher fluorescence background as the two are competitive and simultaneous. However, the observation is made that HSI only requires the highest possible spectral variance data to work well for evaluation or classification purposes, so the exact knowledge of the origin or a localized distinction of the signals is of secondary importance. The set-up presented opens up interesting application possibilities in many areas. In fact, the modular design of the system makes it possible to adapt the measuring range and spatial resolution to many different application areas, from quality control in the food industry to surface inspection and recycling. The Laser-HSI system, as it is called in the paper, operates as a pushbroom imager with the hypercube generated line by line while the sample is scanned using a linear motion unit over tens of centimeters to millimeter ranges, with a ad-hoc alignment of the dichroic mirror to achieve a balanced Raman and fluorescence intensities. Classification of the measurements using machine learning algorithms was also demonstrated, after a careful spectral calibration against reference substances. Results support the idea that Laser-HSI can be used for various applications in the field of process and food monitoring or sorting tasks just where conventional HSI systems fail.

Behmann et al. [6], address the problem of characterization of plant disease. A method is presented to spatially reference the time series of a close range of hyperspectral images, overcoming the inability of previous studies of symptoms over time to detect early stages of a disease when they are still invisible. Automatically tracked hyperspectral information could be a promising approach to overcome this limitation.

Results have been shown for the first symptoms and their development in the presence of septoria tritici blotch (STB) and brown rust as well, by using a VISNIR camera. The tracking of symptoms in real space is obtained measuring reflectivity from plants between 400 nm and 1000 nm in pushbroom mode, and, then, making a space reference on multiple hyperspectral image cubes along the time series direction. Those spatially referenced images form therefore a new 4D data type with two spatial axis, one spectral axis, and a fourth temporal axis (x, y, λ, t). Within this new data type, disease symptoms can easily be traced back in time, even to the point when no symptom is visible to the human eye. The point correspondence problem is solved through referencing by including the RANSAC algorithms and multiple 2D geometric transformations in combination with a well-defined set of control points.

The possibility to annotate invisible symptoms by tracing visible symptoms back in time to the invisible phase of pathogenesis shows that automated referencing of hyperspectral images is possible. The training of machine learning models will also allow higher sensitivity even at the very early symptom stages. The claim is that this is a really new approach in hyperspectral series imaging, moving the focus from mature symptoms and their appearing in the visible bands to very early and invisible stages of plant diseases.

In Bachmann et al. [7], hyperspectral image sequences are considered for the detection and tracking of vehicles realizing low-rate video hyperspectral imaging systems. The vehicles are assumed to be driving through the parking lot and passing behind various occlusions within the scene, such as trees in the background and parked cars, at a specified maximum speed. The set-up includes an integration of a high rate data acquisition hyperspectral line scanner with a high-speed maritime pan-tilt unit that contains position and pointing information. Then, the realization of digital elevation models (DEM) is devised: HSI time series imagery after integration onto a telescopic pole system from multiple vantage points, stereo hyperspectral views, and the definition of bi-directional reflectance distribution function can all be derived. Two examples of the low-rate hyperspectral video approach are shown, that is the imaging of the dynamics of the surf zone in a coastal setting and moving vehicle imaging in the presence of many occlusions. For future, the authors propose further improvements in hyperspectral image acquisition rates by reducing the size of the across-track spatial dimension or the adoption of already commercially available sensors with on-chip spectral binning.

In Arablouei [8], a new algorithm is proposed capable of simultaneously fusing multiple multiband HSI images. The used method relies on a forward observation model together with a linear mixture model. The low rank images produced will have a sparse representation in the spectral domain, while preserving the edges and discontinuities in the spatial domain in agreement with the fact that HSI image data are generally known to have a low-rank structure. It is noted, in fact, that, due to correlations among the spectral bands and the fact that the spectrum of each pixel can often be represented as a linear combination of relatively few spectral signatures, the images reside in a subspace that usually has a much smaller dimension than the number of the spectral bands. As a result, it is possible to decompose linearly a hyperspectral image into its constituent endmembers, spectral signatures of the material present at the scene, down to fractional abundances of the endmembers for each pixel. This linear decomposition is what the authors define as spectral unmixing and the corresponding data model the linear mixture model. A comparison with the state-of-the-art fusion methods is made in this paper, demonstrating the advantages of the proposed algorithm. In particular, they show results from experiments with five real hyperspectral images that were done following the Wald's protocol, a general paradigm for quality assessment of fused images regarding their consistency and synthesis properties.

Shen et al. [9], have proposed a new predictive lossless compression algorithm for multiple time series of time-lapse hyperspectral image data using a low-complexity sign algorithm with an expanded prediction context. It is noted that HSI technology has been used for various remote sensing applications due to its excellent capability of monitoring regions-of-interest over a period of time. Simulation results have demonstrated the outstanding capability of this algorithm to compress a temporal series of HSI data through spectral and temporal decorrelation. The actual compression results are congruent with the information theoretic analysis and estimation based on conditional entropy. The paper is about an information theoretical analysis to estimate the potential compression performance gain with varying configurations of context vectors. In fact, it shows how compression performance varies as a function of the initial set of spectral bands for prediction by exploiting the spectral and temporal correlations in the datasets. As examples of future work, the authors propose a full integration of the proposed algorithm and the analytic framework to achieve real-time compression on streaming hyperspectral images, including an adaptive selection of bands to build up an optimal context vector data. By extending a lossless compression of regions-of-interest in hyperspectral images, it is also possible to gain a much higher compression than compressing the entire hyperspectral image dataset.

A review by Oiknine et al. [10], shows the advances of a specific HSI system, the compressive sensing miniature ultra-spectral imaging (CS-MUSI) camera. This article provides an evaluation of the CS-MUSI camera, its evolution, and its different applications. The CS-MUSI camera has been designed for using a liquid crystal (LC) phase retarder in order to modulate the spectral domain, realizing therefore a spectral compression. The outstanding advantage of the CS-MUSI camera is that at least one order of magnitude of fewer measurements are needed for the entire HSI image results in comparison with conventional HSI images, as a consequence that the scene's spectral properties are often redundant in nature. This paper shows the reconstruction of HSI images for both cases when the camera and scene are stationary as well as for when the camera is moving in the along-track direction, demonstrating the ability to use the CS-MUSI camera for 4D spectral-volumetric imaging. Experiments in these scenarios and applications have provided a spectral uncertainty of less than one nanometer. Alternatively, this method can also be realized with other spectral modulators. Other compressive methods, like Fabry-Perot resonator (mFPR), which has a much faster response time than LC cells, and snapshot HS camera, by using parallel spectral multiplexing and including an array of mFPRs and a lens array, are presented in the extensive references of the same authors, thus completing the topic.

Compressing Raman HSI data is reviewed in Cebeci et al. [11]. As noted by the authors, a fast Raman analysis for real-time monitoring and hyperspectral imaging has a key bottleneck in the time required to acquire and post-process HSI data. Multichannel detectors (CCD) are commonly used for this task, although they are generally more expensive and less sensitive than single channel detectors. The CCDs also require cooling because of the usual need for long integration times and low dark counts. Consequently, a CCD-based Raman spectrometer cannot operate fast enough to be applicable to dynamic system measurements. Using ordinary Raman HSI spectroscopy, where thousands to millions of different spatial points are measured, would imply the collection of a one-megapixel image over 12 days using 1 s of acquisition rate per spectrum, clearly an impossible task. Instead, a compressive spectrometer with spatial light modulator (SLM) technology offers higher sensitivity and speed, together with a lower-cost alternative to CCDs because of the single detector adoption. Most notably, the compressive detection allows chemical imaging information in the very low signal limit, which is simply impossible to achieve by a conventional CCD-based Raman spectroscope. The chosen key technology is an optimized binary compressive detection strategy (OBCD) adopting a reflective light modulator DMD (by Texas Instruments), widely used in standard computer projection systems. A DMD is a semiconductor-based "light switch" array of hundreds of thousands of individually switchable mirror pixels, which have switch speed, contrast ratio, and broad spectral capability outperforming analog-based SLMs. The light switching speeds in the order of kHz at which each mirror can modulate between "on" and "off" states enable CD measurements at kHz frequencies. Compressive digital (CD) Raman systems can reproduce the functionality of conventional array-based Raman spectroscopy to collect full spectral information by raster-scanning each array column. The full speed advantage of CD Raman is fully discussed in relation to the compressive detection modes, with filter functions, and a low signal regime or high-speed conditions, all conditions for which a CCD cannot work, while a single channel detector, such as a photomultiplier, instead increases dramatically data SNR.

Finally, the merge of HSI and deep learning (DL) technologies is the subject of a review made by Signoroni et al. [12]. HSI data constitutes an undeniable advantage for any research that benefits from computer-assisted spectroscopic analysis. In fact, HSI images have plenty of information coded that can be thought as a high-dimensional vector in a space and spectral dimension with much more information than any RGB or multispectral data. Each pixel can keep measurements in relatively wide spectral intervals, from VIS to NIR, resolved in hundreds of contiguous narrow band spectral channels down to a few nm of spectral resolution. However, as any industrial or scientific technology, HSI requires cost-benefit evaluations and any method to unlock its deployment potentialities is important and needs careful consideration. As explained by the authors, the advantages introduced by DL solutions in the HSI arena are in the automatic and hierarchical learning process of data itself. A model with increasingly higher semantic layers can then be built, until the searched analysis, e.g., classification, regression, segmentation, detection, or other indexes, has a useful representation. Some caution is however needed to exploit the gain potentially detained by DL when it is applied to hyperspectral data, as pointed out in this review. In particular, there is a need for a reasonably large dataset in HSI data (e.g., hundreds of thousands of examples) that has to be congruent with the large amount of parameters of DL models (typically in the tens of millions), to avoid overfitting, while an HSI dataset composed of hundreds of examples can be considered too small. In conclusion, the deployment of HSI technologies by means of DL solutions can be a possible driver enabling HSI for a wider spectrum of small-scale applications in industry, biology and medicine, cultural heritage, and other professional fields.

3. Conclusions

Hopefully it is clear, from the lecture of the various contributions, that all authors have realized a very serious effort in devising the perspective of their work in a broader realm than the proper single activity.

Various compression and fusion methods, coupled with time series HSI acquisitions, embedded in several real-world examples and theoretical tools, make this Special Issue "The Future of Hyperspectral Imaging" a very interesting instrument to better understand the possible future of this versatile technique.

Funding: This research received no external funding.

Acknowledgments: The Guest Editor would like to acknowledge the time and contributions of the authors (both successful and unsuccessful) who prepared papers for this Special Issue. Special thanks go to all the reviewers who provided constructive reviews of the papers in a timely manner; your analysis and feedback has ensured the quality of the papers selected. It is also necessary to acknowledge the assistance given by the MDPI editorial team, in particular Managing Editors Alicia Wang and Veronica Wang, who made my task as Guest Editor much easier.

Conflicts of Interest: The author declares no conflict of interest.

References

1. Rady, A.; Guyer, D.; Kirk, W.; Donis-González, I.R. Prediction of the Leaf Primordia of Potato Tubers Using Sensor Fusion and Wavelength Selection. *J. Imaging* **2019**, *5*, 10. [CrossRef]
2. Bonifazi, G.; Capobianco, G.; Pelosi, C.; Serranti, S. Hyperspectral Imaging as Powerful Technique for Investigating the Stability of Painting Samples. *J. Imaging* **2019**, *5*, 8. [CrossRef]

3. Yang, Q.; Sun, S.; Jeffcoate, W.J.; Clark, D.J.; Musgove, A.; Game, F.L.; Morgan, S.P. Investigation of the Performance of Hyperspectral Imaging by Principal Component Analysis in the Prediction of Healing of Diabetic Foot Ulcers. *J. Imaging* **2018**, *4*, 144. [CrossRef]
4. Cadd, S.; Li, B.; Beveridge, P.; O'Hare, W.T.; Islam, M. Age Determination of Blood-Stained Fingerprints Using Visible Wavelength Reflectance Hyperspectral Imaging. *J. Imaging* **2018**, *4*, 141. [CrossRef]
5. Gruber, F.; Wollmann, P.; Grählert, W.; Kaskel, S. Hyperspectral Imaging Using Laser Excitation for Fast Raman and Fluorescence Hyperspectral Imaging for Sorting and Quality Control Applications. *J. Imaging* **2018**, *4*, 110. [CrossRef]
6. Behmann, J.; Bohnenkamp, D.; Paulus, S.; Mahlein, A.-K. Spatial Referencing of Hyperspectral Images for Tracing of Plant Disease Symptoms. *J. Imaging* **2018**, *4*, 143. [CrossRef]
7. Bachmann, C.M.; Eon, R.S.; Lapszynski, C.S.; Badura, G.P.; Vodacek, A.; Hoffman, M.J.; McKeown, D.; Kremens, R.L.; Richardson, M.; Bauch, T.; et al. A Low-Rate Video Approach to Hyperspectral Imaging of Dynamic Scenes. *J. Imaging* **2019**, *5*, 6. [CrossRef]
8. Arablouei, R. Fusing Multiple Multiband Images. *J. Imaging* **2018**, *4*, 118. [CrossRef]
9. Shen, H.; Jiang, Z.; Pan, W.D. Efficient Lossless Compression of Multitemporal Hyperspectral Image Data. *J. Imaging* **2018**, *4*, 142. [CrossRef]
10. Oiknine, Y.; August, I.; Farber, V.; Gedalin, D.; Stern, A. Compressive Sensing Hyperspectral Imaging by Spectral Multiplexing with Liquid Crystal. *J. Imaging* **2019**, *5*, 3. [CrossRef]
11. Cebeci, D.; Mankani, B.R.; Ben-Amotz, D. Recent Trends in Compressive Raman Spectroscopy Using DMD-Based Binary Detection. *J. Imaging* **2019**, *5*, 1. [CrossRef]
12. Signoroni, A.; Savardi, M.; Baronio, A.; Benini, S. Deep Learning Meets Hyperspectral Image Analysis: A Multidisciplinary Review. *J. Imaging* **2019**, *5*, 52. [CrossRef]

Journal of
Imaging

MDPI

Article

Prediction of the Leaf Primordia of Potato Tubers Using Sensor Fusion and Wavelength Selection

Ahmed Rady [1,*], Daniel Guyer [2], William Kirk [3] and Irwin R Donis-González [4]

1   Department of Agricultural and Biosystems Engineering, Alexandria University, Alexandria 21545, Egypt
2   Department of Biosystems and Agricultural Engineering, Michigan State University,
    East Lansing, MI 48824, USA; guyer@msu.edu
3   Department of Plant, Soil, and Microbial Sciences, Michigan State University, East Lansing, MI 48824, USA;
    kirkw@msu.edu
4   Department of Biological and Agricultural Engineering, University of California, Davis, CA 95616, USA;
    irdonisgon@ucdavis.edu
*   Correspondence: ahmed.ahmed@alexu.edu.eg or radyahme2@gmail.com; Tel.: +20-0109-127-8487

Received: 10 November 2018; Accepted: 3 January 2019; Published: 9 January 2019

Abstract: The sprouting of potato tubers during storage is a significant problem that suppresses obtaining high quality seeds or fried products. In this study, the potential of fusing data obtained from visible (VIS)/near-infrared (NIR) spectroscopic and hyperspectral imaging systems was investigated, to improve the prediction of primordial leaf count as a significant sign for tubers sprouting. Electronic and lab measurements were conducted on whole tubers of Frito Lay 1879 (FL1879) and Russet Norkotah (R.Norkotah) potato cultivars. The interval partial least squares (IPLS) technique was adopted to extract the most effective wavelengths for both systems. Linear regression was utilized using partial least squares regression (PLSR), and the best calibration model was chosen using four-fold cross-validation. Then the prediction models were obtained using separate test data sets. Prediction results were enhanced compared with those obtained from individual systems' models. The values of the correlation coefficient (the ratio between performance to deviation, or r(RPD)) were 0.95(3.01) and 0.9s6(3.55) for FL1879 and R.Norkotah, respectively, which represented a feasible improvement by 6.7%(35.6%) and 24.7%(136.7%) for FL1879 and R.Norkotah, respectively. The proposed study shows the possibility of building a rapid, noninvasive, and accurate system or device that requires minimal or no sample preparation to track the sprouting activity of stored potato tubers.

Keywords: potatoes; sprouting; primordial leaf count; hyperspectral imaging; spectroscopy; fusion; wavelength selection; PLSR; interval partial least squares

1. Introduction

Recent studies have shown various health-promoting nutritional resources in potato tubers including protein, dietary fibers, minerals, ascorbic acids, anthocyanins, and antioxidants. Moreover, phenolic compounds, contained in the tuber or the peel, are known for their anti-inflammatory and anticarcinogenic effects on human health [1]. Due to the rapid change of lifestyles towards fast food and ready-to-cook meals, the consumption of potatoes in the United States, especially frozen French fries and chips, has shown a significant increase during the last four decades [2]. The U.S. per capita French fry consumption jumped from 12.93 Kg in 1970 to 22.89 Kg in 2017 [2]. Hence, maintaining the appropriate degree of tuber quality during handling and storage operations is a major concern for growers and processors, to preserve a high level of marketability.

Storage significantly affects the chemical composition of tubers and subsequent processed products. Potatoes, as other agricultural commodities, continue to perform several postharvest biological processes, among which respiration represents an important metabolic process that needs to be controlled during storage, to extend the shelf life and reduce the accumulation of sugars [3,4]. Dormancy of potato tubers is the duration after harvest during which tubers will not sprout with the presence of the suitable environmental and biochemical conditions. Dormancy usually lasts from several weeks to months, depending on cultivar and storage conditions [5,6]. Following the dormancy period and with warmer temperatures (10–20 °C), sprouts, i.e., the meristematic regions of the tubers (eyes), begin to grow at a low rate that increases until one sprout dominates others [7]. Sprouting is affected by storage conditions, the cultivar, and the presence of damage. Sprouting has shown a significant impact on the physiological status and age of potatoes during storage [7]. Levels of reducing sugars accumulated during low-temperature storage result in after-frying browning, and excess sucrose content causes improper sweetening flavor of fried products [4]. On the other hand, high levels of reducing sugars and sucrose result in an increase in sprouting [4]. Additionally, unrestrained sprouting results in an increase of respiration rate, which leads into an increase of the sprouting, physiological age, weight loss, and the glycoalkaloid levels that are known to be toxic [8]. Thus, uncontrolled storage sprouting causes a considerable decline in the marketability of raw and subsequent processed potato products.

Various techniques have been used to control or inhibit potato sprouting during storage. Low temperature storage is beneficial for minimizing the sprouting of seed tubers [9]. However, sugar accumulation is an expected consequence of storing potatoes at a low temperature [10]. Chemical inhibitors are commonly applied during storage, including isopropyl N-phenylcarbamate (ICP; propham), isopropyl N-(3chlorophenyl) carbamate (CIPC; chloro-IPC, chloropropham), and maleic hydrazide (MH) [10–12]. However, ICP and CIPC cannot be applied on seed potatoes for their irreversible sprouting inhibition [13,14].

Near-infrared (NIR) spectroscopy has been studied for detecting chemical constituents and physical properties of agricultural and food products, in addition to pharmaceutical, textiles, cosmetics, and medicine domains [15]. The utilization of NIR technology in the agricultural domain included the quality evaluation of grains [16,17], fruits, and vegetables [18–20]. More specifically, the possibility of using NIR systems on determining several quality attributes of potatoes showed promising results. Such properties include specific gravity [21,22], dry matter [23,24], and sugars [24–26]. Rady et al. [25] stated that prediction models of leaf primordia for potato tubers had correlation coefficient (r) values of 0.89 and 0.77 for FL1879 and R.Norkotah, respectively, using a VIS/NIR spectroscopic system in the interactance mode. In the case of the VIS/NIR hyperspectral imaging system, the prediction models yielded *r* values of 0.47 and 0.43 for FL1879 and R.Norkotah, respectively. Jeong et al. [27] investigated the application of VIS/NIR diffuse reflectance spectroscopy (400–2500 nm) for estimating the sprouting capacity of Atlantic and Superior potato cultivars. The authors stated that the sprouting capacity could be evaluated by measuring the weight of sprouts grown under a standard sprouting method. Thus, sprouting capacity was measured based on the weight percentage of sprouts for tubers stored for 30 days in the dark at 20 °C and 90% relative humidity. Results showed a good correlation between lab measurements and predicted sprouting capacity, with *r* values falling between 0.87 and 0.97.

The fusion of data acquired from different electronic sensors has been studied for the potential benefits of improving the prediction models of quality attributes of fruits, vegetables, and food products. The data combined from each individual sensor should, however, provide distinguishing and non-redundant information about the measured property. Consequently, the improvement of prediction and classification models can be feasible. Data fusion can be conducted by either concatenating the features from various sensors, then processing them, or by performing feature selection before combining and processing [28].

The fusion of data obtained by stationary and prototype online hyperspectral imaging systems was conducted by Mendoza et al. [29] to improve the prediction capability of firmness and soluble solid content (SSC) for Golden Delicious (GD), Jonagold (JD), and Red Delicious (RD) apple cultivars. Results showed a significant decrease of the standard error of prediction (SEP) values for firmness by

6.6, 16.1, and 13.7% for GD, JG, and RD, respectively. The values of SEP for SSC decreased for GD, JG, and RD by 11.2, 2.3%, and 3.0, respectively. Mendoza et al. [30] examined the fusion of visible and shortwave NIR spectroscopy (400–1100 nm), spectral scattering obtained from hyperspectral imaging (500–1000 nm), acoustic firmness, and bioyield firmness to assess the firmness and SSC of JG, GD, and RD apple cultivars. In such studies, fused data improved firmness prediction models by reducing SEP values by 14.6, 20.0, and 7.3% for JG, GD, and RD cultivars, respectively. In the case of SSC prediction models, the fusion of spectroscopic and hyperspectral imaging systems showed a decrease in SEP values by as much as 6.0%.

The data fusion approach has also been investigated with other agricultural products. Integrating electronic tongue (e-tongue) and UV-VIS-NIR spectroscopic data has been applied for determining the botanical origin of honey [31]. Ignat et al. [32] studied the fusion of VIS/NIR spectroscopic data with VIS hyperspectral imaging features, relaxation and ultrasonic data, and color measurements for predicting several maturity indices for bell peppers, including dry matter (DM), TSS, osmotic potential (OP), ascorbic acid (AA), total chlorophylls, carotenoids, the coefficient of elasticity for compression (CEc), and the coefficient of elasticity for rapture (CEr). Results illustrated the improvement of the determination coefficient (R^2) for fused data models. Values of R^2 increased from 0.93 to 0.95 for DM, 0.93 to 0.96 for TSS, 0.79 to 0.83 for AA, 0.87 to 0.90 for OP, 0.60 to 0.77 for total chlorophylls, 0.92 to 0.96 for carotenoids, 0.55 to 0.63 for CEc, and 0.52 to 0.54 for CEr. Several studies were also conducted to boost the evaluation of various quality attributes for fruits and vegetables using data fusion. Such commodities included bell peppers [33,34], tomatoes [35], apples [36–39], eggplants [40], peaches [41,42], and oranges [43].

The main objective of this study was to investigate the potential of combining data obtained from hyperspectral imaging and spectroscopic systems for building calibration and prediction models of leaf primordia of potato tubers during storage.

2. Materials and Methods

2.1. Raw Materials, Sampling, and Measurement of Primordial Leaf Count

Electronic measurements were conducted on Frito Lay 1879 (FL1879) and Russet Norkotah (R.Norkotah) potato cultivars used for chipping and baking, respectively. Samples were obtained from a commercial farm in Southwest Michigan, United States. After discarding defected and deteriorated tubers, samples were cleaned and stored at 7 °C for four weeks for periderm maturation [44]. Sampling was first examined on 20 tubers per cultivar. Tubers were then stored at 7, 10, and 15 °C, and sampled at 20, 80, and 130 days of storage with 60 tubers per cultivar. A total of 200 tubers tested form FL1879 or R.Norkotah were tested. The reason for choosing such storage temperatures was to create a broad distribution of leaf primordia, which increases the reliability of the prediction models. The measurements of primordial leaf count (LC) took place as stated in Rady et al. [45].

2.2. Electronic Measurements

Whole tubers were electronically scanned using a VIS/NIR spectroscopic system in the interactance mode and a VIS/NIR hyperspectral imaging in the reflectance mode. To obtain consistent measurements, each tuber was placed such that the light beam struck the middle area of the longitudinal axis. More detailed explanation of the scanning process for either system can be found in Rady et al. [45].

2.2.1. VIS/NIR Interactance System

The VIS/NIR spectroscopic system in the interactance mode was used to acquire spectral information of the whole tubers. The system, as shown in Figure 1, contained an Ocean optic spectrometer (model No. USB 4000, Ocean Optics, Inc., Dunedin, FL, United States) connected by a 200 μm diameter fiber optic cable, and has a 3648-element linear silicon CCD (charge-coupled device) array with an optical resolution of 0.3 nm (full width half maximum, or FWHM) and a detection range of 200-1100 nm, as well as a radiometric power supply with a maximum power of 250 watts (model No.68931, Oriel Inst., Irvine, CA, United States) and a light source (model No. 66881, Oriel Inst., Irvine, CA, United States) that contained a quartz tungsten halogen lamp and lens transmittance range of 350–2500 nm. In the interactance mode, light photons illuminate the sample through a probe with a concentric outer illumination ring and an inner receptor. A foam-sealing ring was placed between both components for a separation between the light ring and the detector [45]. Thus, only the light passing through the sample was measured. Using such a configuration, the incident light represents a circle with a diameter of 24.7 mm. The interactance spectra for each sample was normalized using a Teflon disc (~25 mm diameter) as a reference material, and the relative interactance was calculated as follows:

$$Relative\ Interactance = \frac{I_s - I_d}{I_r - I_d}$$

where I_s is the intensity of the reflected light from the sample, I_r is the intensity of the reflected light from the reference material, and I_d is the intensity of the reflected light from the background.

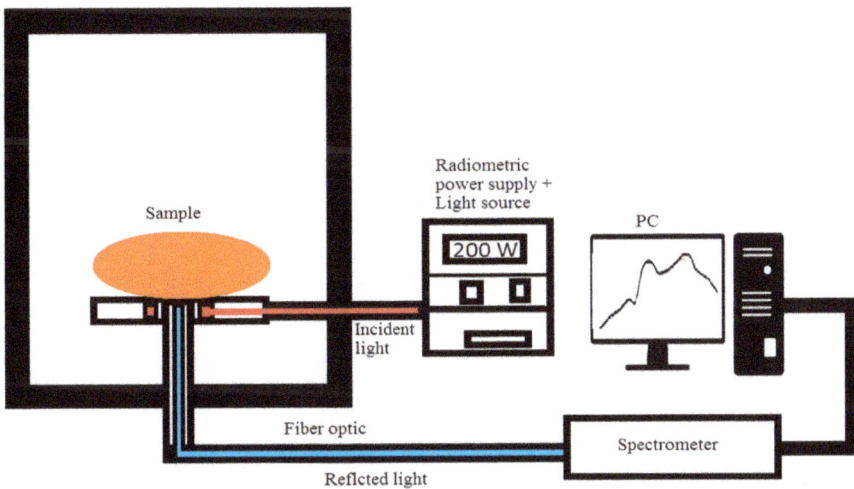

Figure 1. Schematic representation of the visible (VIS)/near-infrared (NIR) interactance testing of Frito Lay 1879 Frito Lay 1879 (FL1879) and Russet Norkotah (R.Burbank) potato cultivars.

2.2.2. VIS/NIR Hyperspectral Imaging System

The main target of using a hyperspectral imaging system (HSI) system in this study was to capture the diffuse scattered light in the range of 400–1000 nm under the reflection mode for whole tubers. The system, as shown in Figure 2, contained a Hamamatsu dual mode cooled CCD camera (model No. C4880, Hamamatsu Photonics, Hamamatsu, Japan), an imaging spectrograph directly attached to the CCD camera (ImSpector V10, Spectral Imaging Ltd., Oulu, Finland), a power supply control (model No. 69931, Oriel Instruments Irvine, CA, United States), a digital exposure controller (model No. 68945, Oriel Instruments, Irvine, CA, United States), and a light source (model No. 66881, Oriel Instruments, Irvine, CA, United States) holding a 250 W Quartz Tungsten Halogen lamp and having a lens material transmittance range of 350–2500 nm. A fiber optic cable coupled with a lens focusing assembly was used to deliver a broadband light beam of 1.5 mm diameter, making a 15° angle away from the vertical axis and 1.6 mm apart from the scanning line. The sample holder movement was controlled using a step motor, and each sample was scanned 10 times with a distance of 1 mm between two successive scans, which totally covered the 9 mm longitudinal distance along the sample. The acquisition time was adjusted for each sample at 200 ms, so the total scanning time for each scanning was 2 s. At each scanning line, the spectrograph acquired the spectral information represented by a 256 × 256 pixel image, with spatial and spectral resolutions of 0.2 mm/pixel and 2.35 nm, respectively.

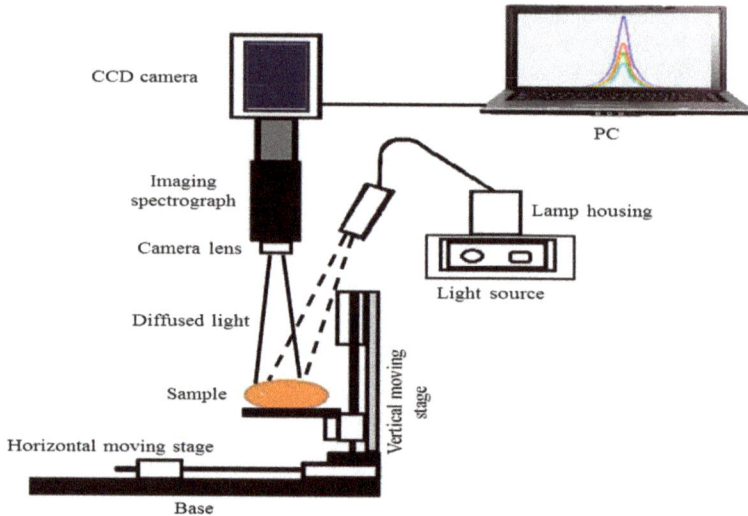

Figure 2. Schematic representation of the VIS/NIR hyperspectral reflectance system used to test whole FL1879 and R.Burbank potato cultivars.

2.3. Data Analysis and Fusion

2.3.1. Calculation of the Mean Reflectance Spectra and Wavelength Selection

The average reflectance spectra were calculated for the hyperspectral imaging data, using 256 wavelengths in the range of 400–1000 nm. For each image, the spectra were first averaged over the spatial coordinates. The relative reflectance (RR) spectrum was then calculated as follows:

$$RR = \frac{AS_s - AS_b}{AS_r - AS_b}$$

where AS_s, AS_b, and AS_r are the average spectra for the sample, background, and reference (Teflon cube), respectively.

Wavelength selection was conducted to reduce the number of variables involved in multivariate regression, to overcome the possibility of the overfitting problem related to relatively high dimensional data, such as spectroscopic data [46]. Therefore, using wavelength selection techniques improves the robustness of the calibration models and reduces the computational time [47].

Interval partial least squares (IPLS) was adopted as a variable selection technique on the data obtained from spectroscopic and hyperspectral imaging, following the results obtained by Rady and Guyer [48]. The configuration of the applied IPLS included the forward mode, window width (W) of one and two variables, and using 20 latent variables (LV).

2.3.2. Data Fusion

After obtaining the most influencing wavelengths, data from the spectroscopic and hyperspectral imaging systems were normalized at each wavelength (column) by dividing all values at such a wavelength by the maximum value at the same wavelength. For each sample (row), data obtained from both systems was then concatenated to form the fused data matrix.

2.3.3. Partial Least Squares Regression and the Preprocessing of Fused Data

Partial least squares regression (PLSR) was applied on the fused data to build calibration and prediction models. PLSR is a linear regression technique known for handling high dimensional data and overcoming the colinearity problem associated with such types of data [46].

According to Rinnan et al. [49], spectral data contains noisy signals resulting from various electronic sources, and consequently data preprocessing is necessary to reduce such undesirable electronic effects and increase the signal-to-noise ratio. Preprocessing was conducted in two stages. The first stage included, in addition to non-processing, smoothing using a first derivative, smoothing using a second derivative, normalization, a standard normal variate (SNV), multiplicative scattering correction (MSC), and the median center. The second stage included the mean center, multiplicative scattering correction, and orthogonal signal correction. Numerical transformation was also carried out on the reference data (leaf primordia count) to obtain uniform distribution. Logarithmic (base 10) and second degree power transformations were applied, in addition to the non-transformed reference values. The regression analysis was carried out on calibration (80% or 160 tubers) and prediction (20% or 40 tubers) sets of data. To reduce the possibility of overfitting and increase the robustness of calibration models, a four-fold cross-validation technique was implemented on the calibration data set, and the best calibration model was chosen as the one with the minimum root mean square error of calibration for cross validation (RMSEC$_{cv}$). Prediction models were then obtained by applying the optimal calibration models on the separate prediction data sets. A complete layout of the data analysis operations is shown in Figure 3. The best prediction model was chosen based on the values of the correlation coefficient (r), the root mean square error of prediction (RMSEP), and the ratio of the standard deviation to the root mean square error of prediction (RPD).

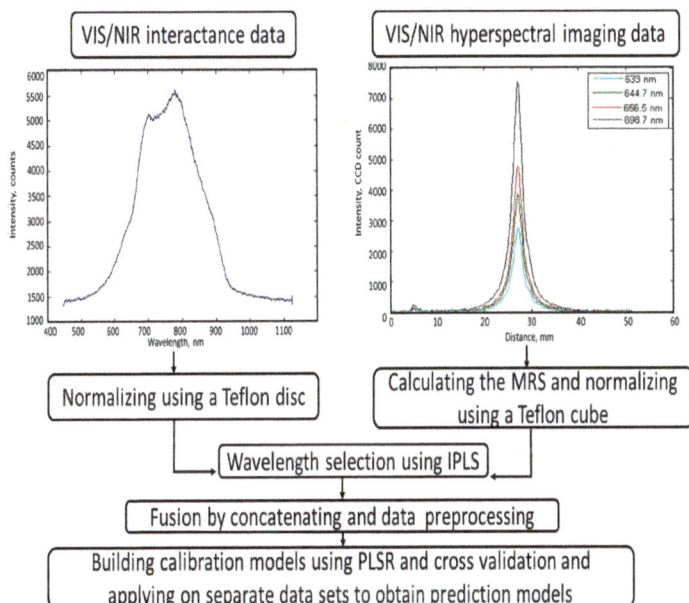

Figure 3. Flow chart of acquiring data from VIS/NIR spectroscopic and VIS/NIR hyperspectral imaging systems, wavelength selection, preprocessing, and building regression models of leaf primordia count for FL1879 and R.Norkotah potato cultivars.

3. Results

3.1. Constituent Distribution and Wavelength Selection Results

The minimum, maximum, mean, and standard deviation values of primordial leaf count (LC) were calculated for FL1879 and R.Norkotah cultivars as shown in Table 1. Both cultivars showed close minimum and mean values. Maximum and standard deviation, however, showed higher values in the case of FL1879, which possibly shows more sprouting. The average LC values were 13.47 and 12.96 for FL1879 and R.Norkotah, respectively. Whereas the standard deviation values were 13.62 for FL1879 and 8.61 for R.Norkotah. The high standard deviation values were intentionally conducted using relatively higher storage temperatures to obtain a broad LC range, which helps develop more comprehensive prediction models for LC.

Table 1. Statistical summary of primordial leaf count (LC) measured for Frito Lay 1879 (FL.1879) and Russet Norkotah (R.Norkotah) potato cultivars.

	Minimum	Maximum	Mean	Standard Deviation
FL1879	4.33	57.66	13.47	13.62
R.Norkotah	4.33	45.67	12.96	8.61

Results of wavelength selection shown in Table 2 indicated that the FL1879 spectral data yielded from the interactance system generally illustrated the highest number of selected wavelengths among all spectral data. In contrast, the number of selected wavelengths obtained from the hyperspectral imaging for R.Norkotah was higher than those obtained from the interactance system, except for W = 2, at which a similar number of wavelengths was selected for both electronic systems. Moreover, the number of selected wavelengths for the hyperspectral imaging was generally higher in the visible spectrum than in the NIR range for both cultivars. In the case of the interactance system, results showed a higher number of selected wavelengths in the NIR range, especially for the R.Norkotah.

Table 2. Number of selected wavelengths using the interval partial least squares (IPLS) technique for primordial leaf count, using data obtained from VIS/NIR interactance and hyperspectral imaging systems for Frito Lay 1879 (FL1879) and Russet Norkotah (R.Norkotah) potato cultivars. Shaded cells show optimal models.

		No. of Selected Wavelengths			No. of Wavelengths in the Visible Range			No. of Wavelengths in the NIR Range		
		W = 1	W = 2	W = 3	W = 1	W = 2	W = 3	W = 1	W = 2	W = 3
VIS/NIR	FL1879	94	106	93	49	40	45	45	66	48
interactance	R.Norkotah	26	34	33	1	10	18	25	24	15
VIS/NIR	FL1879	29	36	63	22	20	39	7	16	24
hyperspectral	R.Norkotah	59	34	60	47	21	45	12	13	15

3.2. Partial Least Squares Regression Results

To make a comparison between the performance of prediction models, based on data obtained from individual or fused sensors, we first illustrate the PLSR results using individual systems data for whole Frito Lay 1879 (FL1879) and Russet Norkotah (R.Norkotah) potato cultivars in Table 3.

On the other side, the best PLSR calibration and prediction models of primordial leaf count for FL1879 and R.Norkotah cultivars are shown in Table 4. The optimal models are shown in the shaded cells. In the case of FL1879, the values of r(RPD) of prediction models were 0.95(3.01), 0.91(2.27), and 0.91(2.49) for W = 1, 2, and 3, respectively. Whereas, in the case of R.Norkotah, the r(RPD) values were 0.96(3.55), 0.95(3.24), and 0.94(2.93), for W = 1, 2, and 3, respectively. The spectral preprocessing methods for the optimal models were first derivative and MSC for FL1879, and second derivative and mean center for R.Burbank. However, the preprocessing of the LC values for the same models was power transformation. The relationship between the measured and predicted LC values for FL1879 and R.Norkotah deduced from the optimal prediction models for W = 1 is shown in Figure 4a,b.

Table 3. Partial least squares regression (PLSR) results of the primordial leaf count, using data obtained from either VIS/NIR interactance or hyperspectral imaging systems for Frito Lay 1879 (FL1879) and Russet Norkotah (R.Norkotah) potato cultivars.

Optical System	Cultivar	Calibration *			Prediction **		
		R_{cal}	$RMSE_{CV}$	LVs	R_{pred}	$RMSEP_{pred}$	RPD_{pred}
VIS/NIR interactance system	FL1879	0.99	0.3055	18	0.89	0.3285	2.22
	R.Norkotah	0.91	0.4183	18	0.77	0.3560	1.5
VIS/NIR hyperspectral imaging	FL1879	0.49	13.124	7	0.47	11.7014	1.14
	R.Norkotah	0.78	9.5766	5	0.43	7.8047	1.10

* R_{cal}: correlation coefficient for the calibration model; $RMSE_{cv}$: root mean square error of calibration, using cross validation for the calibration model; LVs: number of latent variables. ** R_{pred}: correlation coefficient for the prediction model; $RMSE_{pred}$: root mean square error of calibration, using cross validation for the prediction model; RPD_{pred}: ratio between standard deviation and the $RMSEP_{pred}$.

Table 4. PLSR results for predicting primordial leaf count using data fused from VIS/NIR interactance and VIS/NIR hyperspectral imaging systems for whole tubers for Frito Lay 1879 (FL1879) and Russet Norkotah cultivars. Optimal results are shaded.

Interval Width (W)	Cultivar	Preprocessing Method [a]	Calibration			Prediction		
			R_{cal}	$RMSE_{CV}$	LVs	R_{pred}	RMSEP	RPD_{val}
W = 1	FL1879	$A_5, B_2; C_2$	0.99	0.1299	20	0.95	0.1662	3.01
	R.Norkotah	$A_6, B_1; C_2$	0.98	0.1401	12	0.96	0.1411	3.55
W = 2	FL1879	$A_3, B_1; C_2$	0.98	0.1815	13	0.91	0.2206	2.27
	R.Norkotah	$A_5, B_2; C_2$	0.96	0.1775	11	0.95	0.1547	3.24
W = 3	FL1879	$A_1, B_3; C_2$	0.98	0.1933	20	0.91	0.2012	2.49
	R.Norkotah	$A_5, B_3; C_2$	0.98	0.1504	17	0.94	0.1709	2.93

[a] A_x: First stage spectra preprocessing. A_0: No preprocessing. A_1: First derivative. A_2: Second derivative. A_3: Normalization. A_4: Standard normal variate (SNV). A_5: Multiplicative signal correction (MSC). A_6: Median center. B_x: Second stage spectra preprocessing. B_1: Mean center. B_2: Multiplicative scattering correction. B_3: Orthogonal signal correction. C_x: Reference data preprocessing. C_0: No reference transformation. C_1: Log reference transformation. C_2: Power reference transformation.

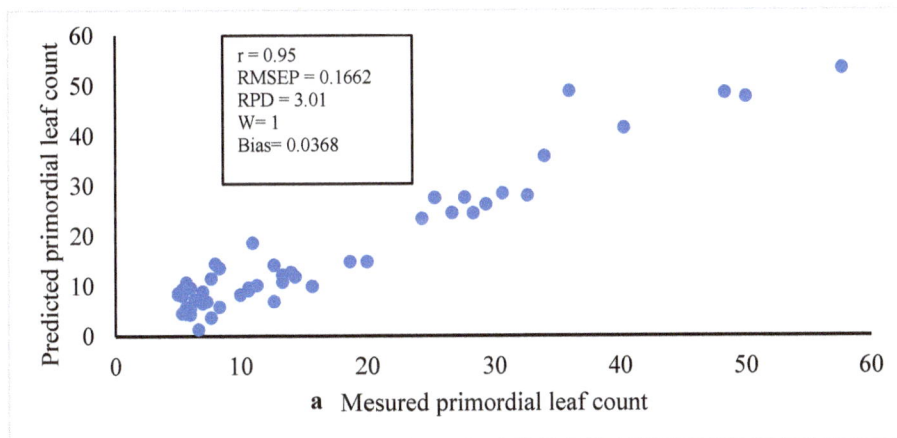

Figure 4. *Cont.*

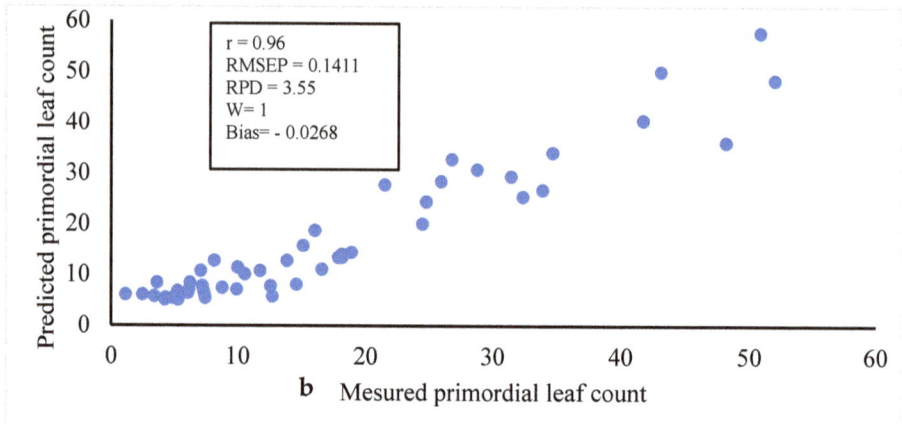

Figure 4. Relationship between measured and predicted primordial leaf count using combined VIS/NIR interactance spectroscopy and VIS/NIR hyperspectral imaging for (a) Frito Lay1879 and (b) Russet Norkotah.

4. Discussion

The number of wavelengths selected using the IPLS technique was generally proportional to the window size, especially for R.Norkotah cultivars in the case of data yielded from the two electronic systems; this is expected, as the larger the window size is, the higher the number of selected variables [50]. It was also noted that the interactance data for FL1879 required a higher number of selected wavelengths to explain the variation of LC in comparison to R.Norkotah, except when W = 1 for the hyperspectral imaging. Furthermore, selected wavelengths based on the window width of one variable (W = 1) that were almost the least compared to those obtained using W = 2 or W = 3 yielded the optimal prediction models. Such results illustrate that the small window width could eliminate redundant variables that might be included during the IPLS search algorithm. Zhao et al. [51] developed a modified IPLS method for variable selection, and their study showed a general conclusion that with the low window width, the number of selected variables decreased, and the root mean square error of prediction (RMSEP) improved. Moreover, Deng et al. [52] compared the number of variables selected using different methods, including synergy interval PLS (siPLS), moving window PLS (MWPLS), and genetic algorithm PLS (GA-PLS). Generally, it was obvious that the smaller the window width, the greater the performance of the prediction models.

Using fused data from the two systems, the prediction of LC significantly improved for both cultivars. In a previous study by Rady et al. [25], as shown in Table 1, the optimal prediction models using the VIS/NIR interactance system showed r(RPD) values of 0.89(2.22) and 0.77(1.50) for FL1879 and R.Norkotah, respectively. Whereas, the r(RPD) values obtained from the VIS/NIR hyperspectral imaging systems were 0.47(1.14) for FL1879 and 0.43(1.10) for R.Norkotah. Additionally, prediction results obtained from the fused data in this study are comparable to the work conducted by Jeong et al. [27] for estimating potato sprouting using NIR diffuse reflectance data. The latter study had r(RPD) values of 0.94(2.0) for the calibration models using cross validation. In our study, data fusion led to significant improvement of the prediction performance, which was mainly based on a separate set of data in which the boosted prediction models yielded r(RPD) values of 0.95(3.01) and 0.96(3.55) for FL1879 and R.Norkotah, respectively. The fusion of the data, along with wavelength selection, has not been investigated before for the sprouting prediction of potatoes.

The above results indicate that there is a possibility of obtaining a robust prediction of sprouting activity of potato tubers during the storage period, using fused from VIS/NIR spectroscopic and hyperspectral imaging systems. One of the main restrictions of applying hyperspectral imaging

J. Imaging **2019**, *5*, 10

systems in on-line sorting and quality inspection processes for food and agricultural products is the relatively long acquisition time. The prediction models obtained in this study, however, were based on selected wavelengths. Thus, decreasing the computation time is accomplished by using fewer wavelengths to build a multispectral imaging system.

5. Conclusions

The main objective of this research study was to investigate the potential of utilizing fused data from VIS/NIR spectroscopic and VIS/NIR hyperspectral imaging systems on predicting primordial leaf count of potatoes. Leaf count is an important factor assessing the sprouting capability of tubers; thus, continuous observation of such activity during storage is crucial to maintain the appropriate physiological status of tubers, especially for processing or seeds. Electronic measurements were performed on whole tubers of FL1879 and R.Norkotah potatoes stored at different temperatures, to stimulate the real storage conditions and obtain wide ranges of LC. After obtaining the most influential wavelengths from both electronic systems using IPLS, data from both systems were fused. Results obtained from PLSR indicated a feasible application of the fusion method to considerably improve LC prediction. Compared to the optimal results obtained from individual systems, values of r(RPD) have been boosted by 6.7%(35.6%) and 24.7%(136.7%) for FL1879 and R.Norkotah, respectively, which stands as a unique enhancement and application of data fusion for potato sprouting. Results deduced from this study initiate the possibility of developing an electronic system, either portable or stationary, that is composed from multispectral imaging along with an interactance sensors to obtain rapid and accurate prediction of sprouting activity of stored potatoes. However, future steps are still needed to reduce the number of selected wavelengths using different versions of IPLS, such as moving average IPLS, synergy IPLS, backward/forward IPLS, and a genetic algorithm. More cultivars should also be tested, and experiments should be conducted over several growing seasons to improve the robustness and reproducibility of the prediction models.

Author Contributions: Conceptualization: D.G., A.R. and I.R.D.-G.; methodology: D.G., W.K., I.R.D.-G., and A.R.; software: A.R., I.R.D.-G.; validation: A.R.; Formal Analysis, A.R.; investigation: D.G., A.R. and I.R.D.-G.; writing (original draft preparation): A.R.; writing (review and editing): D.G., I.R.D.-G., and W.K.; supervision: D.G.; funding acquisition: D.G.; project administration: D.G.

Funding: This research was partially funded by the United States Department of Agriculture (USDA)-(Agricultural Research Service)ARS-State Partnership Potato Program.

Acknowledgments: The authors wish to acknowledge the USDA-ARS postharvest lab of Renfu Lu; the Michigan State University (MSU) Plant Pathology lab of Dennis Fulbright, with special appreciation to Sara Stadt; with special appreciation to Rob Shafer and Walther Farms (Three Rivers, MI, United States); and the MSU AgBioResearch.

Conflicts of Interest: The authors declare no conflict of interest. The funding sponsors had no role in the design of the study; in the collection, analyses, or interpretations of data; in the writing of the manuscript; and in the decision to publish the results.

References

1. Pihlanto, A.; Mäkinen, S.; Mattila, P. Potential health-promoting properties of potato-derived proteins, peptides and phenolic compounds. In *Agriculture Issue and Policies: Production, Consumption and Health Benefits*; Claudio, C., Ed.; Nova Science Publishers, Inc.: Hauppauge, NY, USA, 2012; pp. 173–194.
2. United States Department of Agriculture (USDA); Economic Research Services (RRS). Potatoes: U.S. per capita availability 1970–2017. In *Vegetables and Pulses Yearbook*; USDA: Washington, DC, USA, 2018.
3. Copp, L.J.; Blenkinsop, R.W.; Yada, R.Y.; Marangoni, A.G. The relationship between respiration and chip color during long-term storage of potato tubers. *Am. J. Potato Res.* **2000**, *77*, 279–287. [CrossRef]
4. Stark, J.C.; Love, S.L. Tuber Quality, in Potato Production Systems; Stark, J.C., Love, S.L., Eds.; University of Idaho: Moscow, ID, USA, 2003; pp. 329–344.
5. Wohleb, C.H.; Knowles, N.R.; Pavek, M.J. Plant Growth and Development. In *The Potato, Botany, Production and Uses*; Navarre, R., Pavek, M., Eds.; CABI: Boston, MA, USA, 2014; pp. 64–82.

6. Cutter, E.G. Structure and development of the potato plant. In *The Potato Crop The Scientific Basis for Improvement*, 2nd ed.; Capman & Hall: London, UK, 1992; pp. 65–161.
7. Suttle, J.C. Dormancy and sprouting. In *Potato Biology and Biotechnology Advances and Perspectives*; Vreugdenhil, D., Bradshaw, J., Gebhardt, C., Governs, F., Mackerron, D.K.L., Taylor, M.A., Ross, H.A., Eds.; Elsevier: Oxford, UK, 2007; pp. 287–310.
8. Pringle, B.; Bishop, C.; Clayton, R. *Potatoes Postharvest*; CABI International: Oxfordshire, UK, 2009.
9. Spychalla, J.P.; Desborough, S.L. Fatty acids, membrane permeability, and sugars of stored potato tubers. *Plant Physiol.* **1990**, *94*, 1207–1213. [CrossRef]
10. Kirk, W.W.; Davis, H.V.; Marshall, B. The effect of temperature on the initiation of leaf primordia in developing potato sprouts. *J. Exp. Bot.* **1985**, *36*, 1634–1643. [CrossRef]
11. Afek, U.; Orenstein, J.; Nuriel, E. Using HPP (hydrogen peroxide plus) to inhibit potato sprouting during storage. *Am. J. Potato Res.* **2000**, *77*, 63–65. [CrossRef]
12. Jedhav, S.J.; Mazza, G.; Desai, U.T. Postharvest handling and storage. In *Potato: Production, Processing, and Products*; Salunkhe, D.K., Kadam, S.S., Jadhav, S.J., Eds.; CRC Press: Boca Raton, FL, USA, 1991; p. 69.
13. Pinhero, R.G.; Coffin, R.; Yada, Y.R. Post-harvest storage of potatoes. In *Advances in Potato Chemistry and Technology*; Singh, J., Kaur, L., Eds.; Academic Press: Cambridge, MA, USA, 2009; pp. 339–370.
14. Daniels-Lake, B.; Olsen, N.; Delgado, H.L.; Zink, R. *Potato Sprout Control Products to Minimize Sprout Production, NAPPO Science and Technology Documents*; North American Plant Protection Organization: Ottawa, ON, Canada, 2013.
15. McClure, W.F. *Near-Infrared Spectroscopy in Food Science and Technology*; Ozaki, Y., McClure, W.F., Christy, A.A., Eds.; John Willey & Sons, Inc.: Hoboken, NJ, USA, 2007; pp. 1–10.
16. Fassio, A.; Fernández, E.G.; Restaino, E.A.; La Manna, A.; Cozzolino, D. Predicting the nutritive value of high moisture grain corn by near infrared reflectance spectroscopy. *Comput. Electron. Agric.* **2009**, *67*, 59–63. [CrossRef]
17. Pearson, T. Hardware-based image processing for high-speed inspection of grains. *Comput. Electron. Agric.* **2009**, *69*, 12–18. [CrossRef]
18. Suphamitmongkol, W.; Nie, G.; Liu, R.; Kasemsumran, S.; Shi, Y. An alternative approach for the classification of orange varieties based on near infrared spectroscopy. *Comput. Electron. Agric.* **2013**, *91*, 87–93. [CrossRef]
19. Kumar, S.; McGlone, A.; Whitworth, C.; Volz, R. Postharvest performance of apple phenotypes predicted by near-infrared (NIR) spectral analysis. *Postharvest Biol. Technol.* **2015**, *100*, 16–22. [CrossRef]
20. Sánchez, M.; Garrido-Varo, A.; Pérez-Marín, D. NIRS technology for fast authentication of green asparagus grown under organic and conventional production systems. *Postharvest Biol. Technol.* **2013**, *85*, 116–123. [CrossRef]
21. Chen, J.Y.; Zhang, H.; Miao, Y.; Matsunaga, R. NIR measurement of specific gravity of potato. *Food Sci. Technol. Res.* **2005**, *11*, 26–31. [CrossRef]
22. Scanlon, M.G.; Pritchard, M.K.; Adam, L.R. Quality evaluations of processing potatoes by near infrared reflectance. *J. Sci. Food Agric.* **1999**, *79*, 763–771. [CrossRef]
23. Subedi, P.P.; Walsh, K.B. Assessment of potato dry matter concentration using short-wave near-infrared spectroscopy. *Potato Res.* **2009**, *52*, 67–77. [CrossRef]
24. Haase, N.U. Prediction of potato processing quality by near infrared reflectance spectroscopy of ground raw tubers. *J. Near Infrared Spectrosc.* **2011**, *19*, 37. [CrossRef]
25. Rady, A.M.; Guyer, D.E.; Kirk, W.; Donis-Gonzalez, I.R. The potential use of visible/near infrared spectroscopy and hyperspectral imaging to predict processing-related constituents of potatoes. *J. Food Eng.* **2014**, *135*, 11–25. [CrossRef]
26. Yaptenco, K.F.; Kawakamis, S.; Takano, K. Nondestructive determination of sugar content in 'Danshaku' potato (solanum tuberosum l.) by near infrared spectroscopy. *J. Agric. Sci. Tokyo Nogyo Daigaku* **2000**, *44*, 284–294.
27. Jeong, J.-C.; Ok, H.-C.; Hur, O.-S.; Kim, C.G. Prediction of sprouting capacity using near-infrared spectroscopy in potato tubers. *Am. J. Potato Res.* **2008**, *85*, 309–314. [CrossRef]
28. Manso, J.Y. Sensor Fusion of IR, NIR, and Raman Spectroscopic Data for Polymorph Quantitation of an Agrochemical Compound. Master's Thesis, Faculty of the University of Delaware, Newark, DE, USA, 2008.

29. Mendoza, F.; Lu, R.; Ariana, D.; Cen, H.; Bailey, B. Integrated spectral and image analysis of hyperspectral scattering data for prediction of apple fruit firmness and soluble solids content. *Postharvest Biol. Technol.* **2011**, *62*, 149–160. [CrossRef]

30. Mendoza, F.; Lu, R.; Cen, H. Comparison and fusion of four nondestructive sensors for predicting apple fruit firmness and soluble solids content. *Postharvest Biol. Technol.* **2012**, *73*, 89–98. [CrossRef]

31. Ulloa, P.A.; Guerra, R.; Cavaco, A.M.; Rosa da Costa, A.M.; Fifueira, A.C.; Brigas, A.F. Determination of the botanical origin of honey by sensor fusion of impedance e-tongue and optical spectroscopy. *Comput. Electron. Agric.* **2013**, *94*, 1–11. [CrossRef]

32. Ignat, T.; Alchanatis, V.; Schmilovitch, Z. Maturity prediction of intact bell peppers by sensor fusion. *Comput. Electron. Agric.* **2014**, *104*, 9–17. [CrossRef]

33. Mohebbi, M.; Amiryousefi, M.R.; Hasanpour, N.; Ansarifar, E. Employing an intelligence model and sensitivity analysis to investigate some physicochemical properties of coated bell pepper during storage. *Int. J. Food Sci. Technol.* **2011**, *47*, 299–305. [CrossRef]

34. Ignat, T.; Mizrach, A.; Schmilovitch, Z.; Fefoldi, J.; Egozi, H.; Hoffman, A. Bell pepper maturity determination by ultrasonic technique. *Prog. Agric. Eng. Sci.* **2010**, *6*, 17–34. [CrossRef]

35. Baltazar, A.; Aranda, J.I.; Gonzalez-Aguilar, G. Bayesian classification of ripening stages of tomato fruit using acoustic impact and colorimeter sensor data. *Comput. Electron. Agric.* **2008**, *60*, 113–121. [CrossRef]

36. Li, C.; Heinemann, P.; Sherry, R. Neural network and Bayesian network fusion models to fuse electronic nose and surface acoustic wave sensor data for apple defect detection. *Sens. Actuators B Chem.* **2007**, *125*, 301–310. [CrossRef]

37. Kavdir, I.; Guyer, D.E. Apple grading using fuzzy logic. *Turk. J. Agric.* **2003**, *27*, 375–382.

38. Zude, M.; Herold, B.; Roger, J.M.; Bellon-Maurel, V.; Landahl, S. Nondestructive tests on the prediction of apple fruit flesh firmness and soluble solids content on tree and in shelf life. *J. Food Eng.* **2006**, *77*, 254–260. [CrossRef]

39. Steinmetz, V.; Roger, J.M.; Molto, E.; Blasco, J. On-line fusion of colour camera and spectrophotometer for sugar content prediction of apples. *J. Agric. Eng. Res.* **1999**, *73*, 207–216. [CrossRef]

40. Saito, Y.; Hatanaka, T.; Uosaki, K.; Shigeto, K. Eggplant classification using artificial neural network. In Proceedings of the International Joint Conference on Neural Networks, Portland, OR, USA, 20–24 July 2003; Volume 2, pp. 1013–1018.

41. Natale, C.D.; Zude-Sasse, M.; Macagnano, A.; Paolesse, R.; Herold, B.; D'amico, A. Outer product analysis of electronic nose and visible spectra: Application to the measurement of peach fruit characteristics. *Anal. Chim. Acta* **2002**, *459*, 107–117. [CrossRef]

42. Ortiz, C.; Barreiro, P.; Correa, E.; Riquelme, F.; Ruiz-Altisent, M. Nondestructive identification of woolly peaches using impact response and nearinfrared spectroscopy. *J. Agric. Eng. Res.* **2001**, *78*, 281–289. [CrossRef]

43. Steinmetz, V.; Biavati, E.; Molto, E.; Pons, R.; Fornes, I. Predicting the maturity of oranges with non-destructive sensors. *Actae Hortic.* **1997**, *421*, 271–278. [CrossRef]

44. Knowles, N.R.; Plissey, E.S. Maintaing tuber health during harvest, storage, and post-storage handling. In *Potato Health Management*; Johnson, D.A., Ed.; American Phytopathological Society Press: St. Paul, MN, USA, 2007.

45. Rady, A.; Guyer, D.E.; Lu, R. Evaluation of sugar content of potatoes using hyperspectral imaging. *J. Food Bioprocess Technol.* **2015**, *8*, 995–1010. [CrossRef]

46. Varmuza, K.; Filmoser, P. *Introduction to Multivariate Statistical Analysis in Chemometrics*; CRC Press: Boca Raton, FL, USA, 2009.

47. Mark, H. Data Analysis: Multilinear regression and principal component analysis. In *Handbook of Near-Infrared Analysis*; Burns, D.A., Ciurczak, E.W., Eds.; Marcel Dekker, Inc.: New York, NY, USA, 2001.

48. Rady, A.; Guyer, D.E. Utilization of visible/near-infrared spectroscopic and wavelength selection methods in sugar prediction and potatoes classification. *J. Food Meas. Charact.* **2015**, *9*, 20–34. [CrossRef]

49. Rinnan, Å.; Berg, F.; Engelsen, S.B. Review of the most common pre-processing techniques for near-infrared spectra. *Trends Anal. Chem.* **2009**, *28*, 1201–1222. [CrossRef]

50. Wise, B.M.; Gallagher, N.B.; Bro, R.; Shaver, J.M.; Windig, W.; Kock, R.S. *PLS_Toolbox 4.0 for Use with Matlab*; Eigenvector Research, Inc.: Manson, WA, USA, 2006; pp. 137–192.

51. Zhao, Y.; Wang, S.; Li, Z.; Pei, Z.; Cao, F. An improved changeable size moving window partial least square applied for molecular spectroscopy. *Chemom. Intell. Lab. Syst.* **2016**, *152*, 118–124. [CrossRef]
52. Deng, B.-C.; Yun, Y.-H.; Ma, P.; Lin, C.-C.; Ren, D.-B.; Liang, Y.-Z. A new method for wavelength interval selection that intelligently optimizes the locations, widths and combinations of the intervals. *Analyst* **2015**, *140*, 1876–1885. [CrossRef] [PubMed]

Journal of
Imaging

MDPI

Article

Hyperspectral Imaging as Powerful Technique for Investigating the Stability of Painting Samples

Giuseppe Bonifazi [1], Giuseppe Capobianco [1], Claudia Pelosi [2,*] and Silvia Serranti [1]

[1] Department of Chemical Engineering Materials & Environment, Sapienza, Rome University,
    Via Eudossiana 18, 00184 Rome, Italy; giuseppe.bonifazi@uniroma1.it (G.B.);
    giuseppe.capobianco@uniroma1.it (G.C.); silvia.serranti@uniroma1.it (S.S.)
[2] Department of Economics, Engineering, Society and Business Organization, Laboratory of Diagnostics and
    Materials Science, University of Tuscia, Largo dell'Università, 01100 Viterbo, Italy
*   Correspondence: pelosi@unitus.it; Tel.: +39-0761-357673

Received: 26 October 2018; Accepted: 26 December 2018; Published: 3 January 2019

Abstract: The aim of this work is to present the utilization of Hyperspectral Imaging for studying the stability of painting samples to simulated solar radiation, in order to evaluate their use in the restoration field. In particular, ready-to-use commercial watercolours and powder pigments were tested, with these last ones being prepared for the experimental by gum Arabic in order to propose a possible substitute for traditional reintegration materials. Samples were investigated through Hyperspectral Imaging in the short wave infrared range before and after artificial ageing procedure performed in Solar Box chamber under controlled conditions. Data were treated and elaborated in order to evaluate the sensitivity of the Hyperspectral Imaging technique to identify the variations on paint layers, induced by photo-degradation, before they could be detected by eye. Furthermore, a supervised classification method for monitoring the painted surface changes, adopting a multivariate approach was successfully applied.

Keywords: Hyperspectral imaging; painting samples; retouching pigments; watercolours; multivariate analysis

1. Introduction

Hyperspectral imaging (HSI) is a diagnostic tool deserving great interest in the field of cultural heritage due to its non-invasive character and to the possibility of obtaining a lot of information with a single technique [1–3]. If coupled with chemometric techniques, it allows for gathering qualitative and/or quantitative information on the nature and physical-chemical characteristics of the investigated materials, and to combine imaging with spectroscopy for evaluating the distribution of materials on the surfaces [4–9]. By using classification methods, already applied in other research fields, it is possible to create a predictive model that is able to identify little variations of the painting layers due to the degradation phenomena of the constituent materials [10–15]. In conservation of cultural heritage, these classification methods could have great relevance because they allow to monitor in real time the surface changes by observing the spectra variation in respect to the calibration dataset. For these reasons, in the present work, HSI was applied with the aims to evaluate the sensitivity of the technique in order to identify the variations on paint layers, induced by photo-degradation, before they could be observed by eye and to use, following a multivariate based approach, the supervised classification methods for monitoring the painted surface changes [8,9]. As paint samples, a set of commercial watercolours was chosen together with various powder pigments, mainly iron oxide based materials, which were mixed with gum Arabic, without any additive, in order to verify their possible use in painting retouching. Iron oxide based pigments were chosen, as they are stable and widely used for millennia thanks to their durability [16–18]. However, when combined with gum Arabic and additives in commercial watercolours, the stability of paintings seems to be not the same [19,20]. The choice of testing watercolours derived from their wide use as materials for painting retouching, together with other products more or less recently introduced in the conservation field [21,22]. Watercolours are frequently used for retouching, especially by Italian conservators that, in particular, commonly choose Winsor&Newton as the preferred brand [20]. Watercolours are produced by the combination of a pigment with gum Arabic and other substances not specified by the manufacturer to safeguard the industrial patent [23–28]. The necessity to investigate the stability of retouching products is linked to the unknown and unpredictable behaviour of the commercial mixtures whose composition is not declared by suppliers [19,29–32]. Though watercolours are widely used in conservation, their stability in the long run has not been sufficiently studied [33–38] or it is limited to the investigation of pigment modification without examining the binder behaviour [39]. In general, even if retouching is a consolidated praxis in restoration, the monitoring of behaviour of retouched artworks is not widely applied, especially due to the high costs or lack of maintenance programs. However, several cases of chromatic alteration in areas retouched through watercolours were found, especially in red and brown painting zones where iron based pigments were used [20].

The evaluation of stability of commercial products, used in conservation, can be performed through different analytic and diagnostic techniques, requiring the preparation of a lot of micro-samples to perform the analyses [40]. The use of sampling based techniques is not always possible in conservation and monitoring, especially due to the difficulty or impossibility to repeat the measurements in the same points during the time. For this reason, non-invasive no-contact methods were chosen to study and monitor the photo degradation processes in watercolours and pigment powders. Specifically, HSI in the short ware infrared region (SWIR) was used with the aim to early detect and monitor the degradation of the investigated painting materials. HSI techniques were widely applied for the identification and characterization of paint layers but rarely for monitoring of degradation patterns [41–47]. Infrared reflectance spectroscopy is a well-known technique to obtain materials characterization and to set up a correct diagnostic plan based on non-destructive and non-invasive approach [4,48–55]. In particular, the SWIR range provides information about vibrational transitions, which are mostly overtones and combination bands whose fundamental transitions occur in the mid-IR. These features are often related to functional groups, like hydroxyl ($-OH$), carbonate ($-CO_3$), and sulphate ($-SO_4$) [56,57]. Absorption features from organic materials, like the paint binders, can also be observed and used to map their spatial distribution [44,58,59].

Based on a previously published paper, the present work extends the HSI results to the entire set of painting samples (totally 58) in order to make a comparison on a larger number of pigments [60].

In Section 2 (Materials and Methods), the experimental procedure will be reported. It is organized as follows: Section 2.1 sample preparation and ageing; Section 2.2 hyperspectral imaging (HSI), describing equipment and acquisition modalities; and, Section 2.3 spectral analysis, which reports in detail the data elaboration and the definition of prediction model used in the work. The Section 3 concerns results and shows spectra, PCA score plots, and prediction models. Section 4 reports discussion of the results shown in the Section 3. Last, Section 5 is devoted to the conclusions of the paper and to further possible research lines to develop.

2. Materials and Methods

2.1. Sample Preparation and Ageing

As painting materials, commercial watercolours, professional series supplied by Winsor&Newton both in form of tubes and pans, were selected in order to compare their stability, to light and UV radiation, with that of iron oxide pigments [61] supplied by Chroma with the specification of the country of origin, and two blue pigments in powder, all applied by gum Arabic (GA, by W&N) as binder in order to have the same binder of commercial watercolours (see Tables 1 and 2 for pigment abbreviation and description). Only cobalt blue and ultramarine blue pigments were supplied by Zecchi (Florence).

Table 1. W&N samples in pan and tube, abbreviation and description.

Abbreviation	Visible Colour	Description
GB1	Black	Ivory black in pan
TB1	Black	Ivory black in tube
GBr1	Dark brown	Burnt umber in pan
TBr1	Dark brown	Burnt umber in tube
GBr2	Light brown	Natural umber in pan
TBr2	Light brown	Natural umber in tube
GBr3	Reddish brown	Burnt Sienna in pan
GBr4	Yellow-orange	Natural Sienna in pan
TBr3	Reddish brown	Burnt Sienna in tube
TBr4	Yellow-orange	Natural Sienna in tube
GR1	Dark red	Indian red in pan
TR1	Dark red	Indian red in tube
GR2	Light red	Venetian red in pan
TR2	Light red	Venetian red in tube
GR3	Light red	Cadmium red in pan
TR3	Light red	Cadmium red in tube
GY1	Light Yellow	Yellow ochre in pan
TY1	Light yellow	Yellow ochre in tube
GG1	Green	Bladder green in pan
TG1	Green	Chrome green in tube
GG2	Green	Viridian in pan
TG2	Green	Viridian in tube
GC1	Blue	Cobalt blue in pan
TC1	Blue	Cobalt blue in tube
GU1	Blue	Ultramarine blue in pan
TU1	Blue	Ultramarine blue in pan

Table 2. Powder samples mixed with GA, abbreviation and description.

Abbreviation	Visible Colour	Description
Br1	Dark brown	Burnt umber in powder + GA
Br2	Dark brown	Burnt umber in powder + GA
Br3	Dark brown	Natural umber in powder + GA
Br4	Dark brown	Burnt umber in powder + GA
Br5	Dark brown	Burnt umber in powder + GA
Br6	Dark brown	Natural umber in powder + GA
Br7	Dark brown	Natural umber in powder + GA
Br8	Dark brown	Natural umber in powder + GA
Br9	Dark brown	Burnt umber in powder + GA
Br10	Dark brown	Natural umber in powder + GA
Br11	Dark brown	Natural umber in powder + GA
R1	Light red	Red ochre in powder + GA
R2	Dark red	Red ochre in powder + GA
R3	Dark red	Red ochre in powder + GA
R4	Dark red	Red ochre powder + GA
R5	Dark red	Red ochre powder + GA
R6	Light red	Red ochre in powder + GA
R7	Light red	Red ochre in powder + GA
Y1	Dark yellow	Yellow ochre in powder + GA
Y2	Light yellow	Yellow ochre in powder + GA
Y3	Light yellow	Yellow ochre in powder + GA
Y4	Dark yellow	Yellow ochre in powder + GA
Y5	Dark yellow	Yellow ochre in powder + GA
Y6	Light yellow	Yellow ochre in powder + GA
Y7	Dark yellow	Yellow ochre in powder + GA
Y8	Light yellow	Yellow ochre in powder + GA
Y9	Dark yellow	Yellow ochre in powder + GA
Y10	Dark yellow	Yellow ochre in powder + GA
Y11	Dark yellow	Yellow ochre in powder + GA
Y12	Dark yellow	Yellow ochre in powder + GA
CB1	Blue	Cobalt blue in powder + GA
UB1	Blue	Ultramarine blue in powder + GA

In the case of natural iron oxide pigments, more than one colour typology was found, for this reason multiple samples are available for each kind of materials, i.e., five powders for burnt umber, nine powders for dark yellow ochre, etc. (Table 2). Commercial watercolours were chosen in order to have, for each colour and when available, two typologies: tube and pan, which were chosen due to their wide use in retouching (Table 1) [20]. The choice of samples for stability testing was also made on the base of previous data reported in experimental theses [18,20].

According to these data, the traditional watercolours that were used by conservators were classified between the less stable mixtures. For this reason, it was chosen to test the possibility of substituting these watercolours with materials having the same or similar colour appearance but prepared with natural pigments in powder and gum Arabic, without any additive.

The above described painting materials were homogeneously applied by brush on traditional gypsum/glue ground in order to create colour check tables with the different chosen pigments. Gypsum ground was chosen in order to simulate a painting repair to be covered by retouching, as commonly occurs in the practice of restoration of painting lacuna (Figure 1) [20].

Figure 1. Painting samples with the corresponding abbreviations as explained in Tables 1 and 2. GA and GG indicate gum Arabic and GG ground layer, respectively [60].

Artificial ageing was performed by a model 1500E Solar Box chamber (Erichsen Instruments GmbH&Co, Hemer, Germany) simulating sunlight irradiation (visible and ultraviolet). The system is equipped with a 2.5 kW xenon-arc lamp and UV filter that cuts off the spectrum at 280 nm [62,63]. The samples were exposed in the Solar Box chamber from 1 to 504 h at 550 W/m^2, 55 °C, and the UV filter at 280 nm. In these conditions, ageing was performed, evaluating the effects of light and UV radiation without considering other environmental agents, such as relative humidity. Inside the Solar Box chamber, relative humidity was constant (50%) and determined by the irradiation conditions. Relative humidity was monitored by a data logger positioned inside the Solar Box.

Hyperspectral imaging data were acquired at the following ageing times: 0 h, 168 h, 336 h, and 504 h, corresponding to 0 J/m^2, 3.3 × 10^8 J/m^2, 6.7 × 10^8 J/m^2, and 1.0 × 10^9 J/m^2, respectively, of the total energy on the irradiated surfaces at the different times.

2.2. Hyperspectral Imaging (HSI)

Hyperspectral analyses were carried out on sample table at 0, 168, 336, and 504 h of exposure in the wavelength interval 1000–2500 nm (SWIR). The acquisitions were performed utilizing the SISUChema

XL^{TM} (Specim, Oulu, Finland) device, equipped with a 31 mm lens allowing the acquisition of the paint layer with a resolution of 300 micron/pixel.

The spectral resolution was 6.3 nm. Illumination was obtained by SPECIM's diffuse line illumination unit. Images were acquired through scanning each investigated sample line by line. Instrument is delivered with spectral calibration. Image data is automatically calibrated to reflectance by measuring an internal standard reference target before each sample scan.

The image correction was thus performed, adopting the following equation:

$$I = \frac{I_0 - B}{W - B} \times 100,$$ (1)

where I is the corrected hyperspectral image, I_0 is the original hyperspectral image, B is the black reference image (~0% reflectance), and W is the white reference image (~99.9% reflectance).

2.3. Spectral Analysis

HSI derived spectral data were analyzed by adopting standard chemometric methods [64,65], with the PLS_Toolbox (Version 8.2 Eigenvector Research, Inc., Manson, WA, USA) running inside Matlab (Version 8.4, The Mathworks, Inc., Natick, WA, USA). In more details, the spectra preprocessing was performed as follows: raw spectra were preliminary cut, at the beginning and at the end of the investigated wavelength range, in order to eliminate unwanted effects due to lighting/background noise. Preprocessing was adopted in order to reduce the noise and emphasize the spectral signal [66–69]. The following preprocessing algorithms were applied: standard normal variate (SNV) to reduce the effect of light scattering; 1st derivative to emphasize the spectral absorption of the investigated paint layers. Finally, Mean Center (MC) was adopted for centering the data before applying principal component analysis.

Principal Component Analysis (PCA) was applied as a powerful and versatile method that is capable of providing an overview of complex multivariate data. PCA can be used for revealing relations existing between variables and samples (e.g., clustering), detecting outliers, finding and quantifying patterns, generating new hypotheses, as well as many other things [70]. In this work, PCA was used to decompose the "processed" spectral data into several principal components (PCs), embedding the spectral variations of each collected spectral data set. The first few PCs, resulting from PCA, are generally utilized to analyze the common features among samples and their grouping: in fact, samples that are characterized by similar spectral signatures tend to aggregate in the score plot of the first two or three components.

k-Nearest Neighbor (k-NN) is one of the most fundamental and simple "non parametric" algorithm that is used in classification methods [71]. This algorithm has been used for creating the prediction model that is used in the present work to establish the variations of painting surfaces over time and at the different solar irradiation dose. Specifically, the proposed classification model is based on the identification of a pictorial layer before ageing, i.e., at time 0 h.

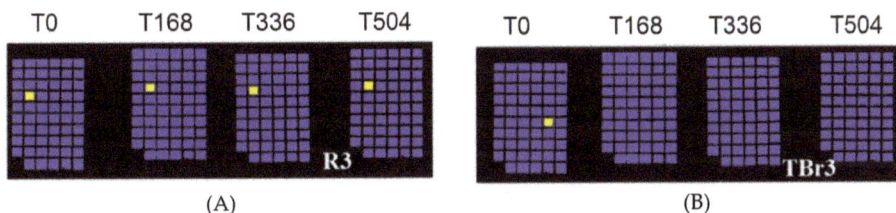

Figure 2. Example of paint layer (R3) which does not exhibit variations during the whole ageing cycle (**A**); example of paint layer (TBr3) showing significant spectral change after 168 h (**B**).

If no spectral variations occur during irradiation in Solar Box (at times: 168 h, 336 h, and 504 h), then the prediction model will identify the unchanged painting sample in all ageing times (Figure 2A). Otherwise, if Solar Box irradiation causes chemical changes in a certain sample, the prediction model will be able to identify the spectral variations in respect to time 0 and will highlight them (Figure 2B).

k-NN assumes that the data are in a feature space. More exactly, the data points are in a metric space. The data can be scalars or possibly even multidimensional vectors. Since the points are in feature space, they have a notion of distance. Each of the training data consists of a set of vectors and class label associated with each vector [72]. Given a query vector x0 and a set of N labeled instances {xi, yi} N1, the task of the classifier is to predict the class label of x0 on the predefined P classes [73]. The k-NN classification algorithm tries to find the k nearest neighbors of x0 and it uses a majority vote to determine the class label of x0. Without prior knowledge, the k-NN classifier usually applies Euclidean distances as the distance metric [74]. The performance of a k-NN classifier is primarily determined by the choice of k as well as the distance metric applied [75]. This number decides how many neighbors (where neighbors are defined by a distance metric) influence the classification. However, it has been shown that, when the points are not uniformly distributed, predetermining the value of k becomes difficult. Generally, larger values of k are more immune to the noise presented and make boundaries smoother between classes. The k-NN classification approach has been widely used in various types of classification tasks. This classification approach has gained popularity based on low implementation cost and high degree of classification effectiveness. However, its sample similarity computing is very large, which limits its applications in some cases that have high dimensional spaces or very large training sets [5,76]. In order to reduce the computation time and memory requirement without sacrificing classification capability, we apply the k-NN algorithm to the score matrix T computed with the PCA model.

3. Results

Preliminary results about the application of traditional techniques and HSI showed that, between the four colours examined (burnt umber, raw Sienna, Venetian red, and yellow ochre), the pigment powders mixed with gum Arabic are the most stable in regard to the artificial ageing [60]. This result encouraged to apply the same elaboration of HSI data on the entire set of samples in order to verify if effectively the pigment powders mixed with gum Arabic can be considered, in general, to be more stable in respect to the commercial products of similar colours, but of different composition. In fact, commercial mixtures contain additives and un-specified substances, added by supplier for improving the characteristics of the products, which are not present in the mixtures prepared in our laboratory only with powder and pure GA dissolved in water.

In Figure 3, the image of the sample set is shown, together with the false colour image representing the acquired raw hypercube and the selected regions of interest (ROI). The false colours in Figure 3B depend on the spectral information contained in each sample; as the data are not elaborated, it is not possible, through these false colours, to see the little differences between the pigments that in many cases are very similar in composition.

An example of ROI selection is also detailed in Figure 4. Different regions of interest were selected in order to define, for each studied pigment, a specific area to investigate.

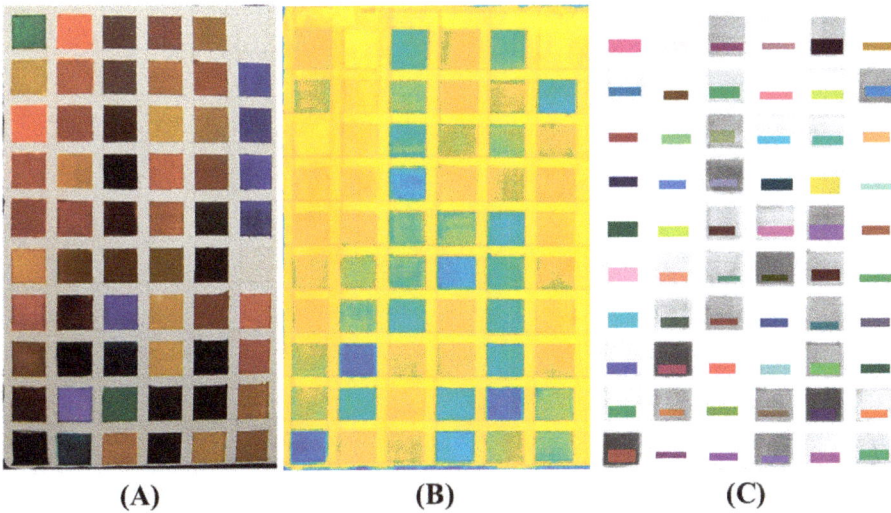

Figure 3. Visible image of the sample set (**A**); false colour image representing the average values of the raw hypercube after having imported it into Matlab (**B**) and the selected regions of interest (ROI) (**C**).

Figure 4. An example of ROI selection.

Subsequently, SWIR acquisition was performed, and the results have been shown as average and pre-processed spectra. Pre-processing was adopted in order to better highlight the spectral differences between the paintings (Figure 5A,B). The results of PCA are displayed in Figure 5C,D.

The full colour RGB (red, green and blue) images of painting samples, at the different ageing times, are shown in Figure 6. They display the changes visible by eyes in painting samples, but these images are not able to show little colour differences that could be associated to spectral variations. The prediction/classification model, as described in the Section 2.3 named Spectral analysis, was obtained in respect to time 0 h and displayed at 168 h, 336 h, and 504 h in order to highlight the similarities or differences for each sample at the chosen ageing times. The calibration dataset of the PCA-KNN classification model has been set by considering the spectra at time 0. PCA-KNN model, applied in this modality, allows for seeing the variation of paint layers for each ageing time in respect to time 0, based on the proximity of unknowns to the different groups in the training set [60,77–81].

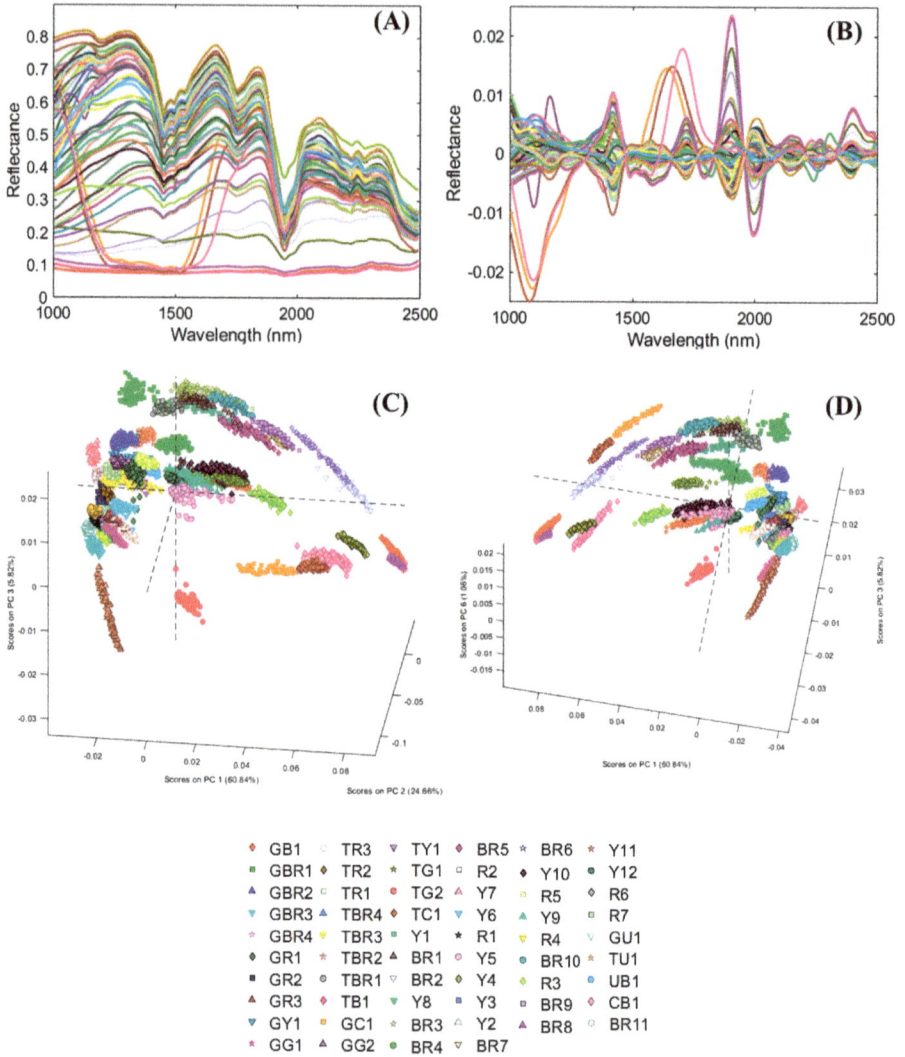

Figure 5. Average spectra of all painting samples (**A**); pre-processed spectra (**B**) and Principal Component Analysis (PCA) score plots (**C,D**).

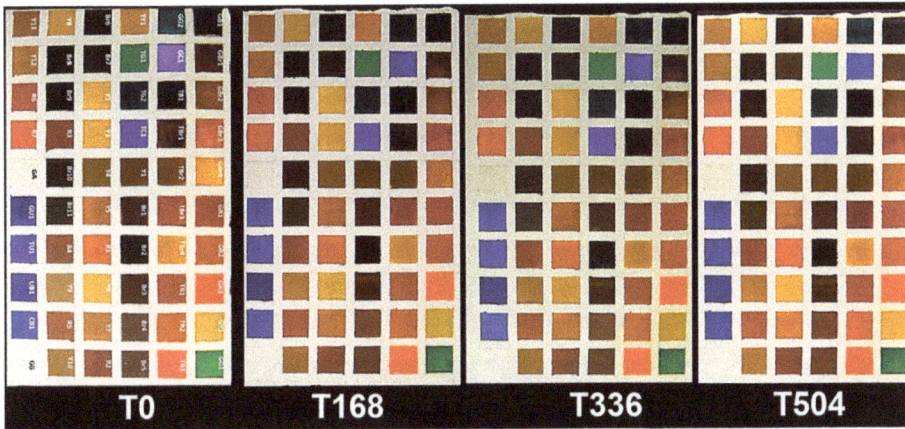

Figure 6. Full colour RGB (red, green and blue) images of the sample set at the different ageing times.

The prediction for each painting sample is obtained and reported in Figures 7–11. Samples were grouped, in each figure, according to their stability, as observed by applying the prediction/ classification model. The prediction map of each painting sample shows a logical (true/false) class assignment to each specific class based on strict multiple-class assignment rules. The yellow colour in Figures 7–11 identifies a specific painting sample at time zero and the painting samples with the same spectral fingerprint. The blue colour is assigned to painting samples with different composition and/or painting samples that degrade during ageing. i.e., that change the spectral profile.

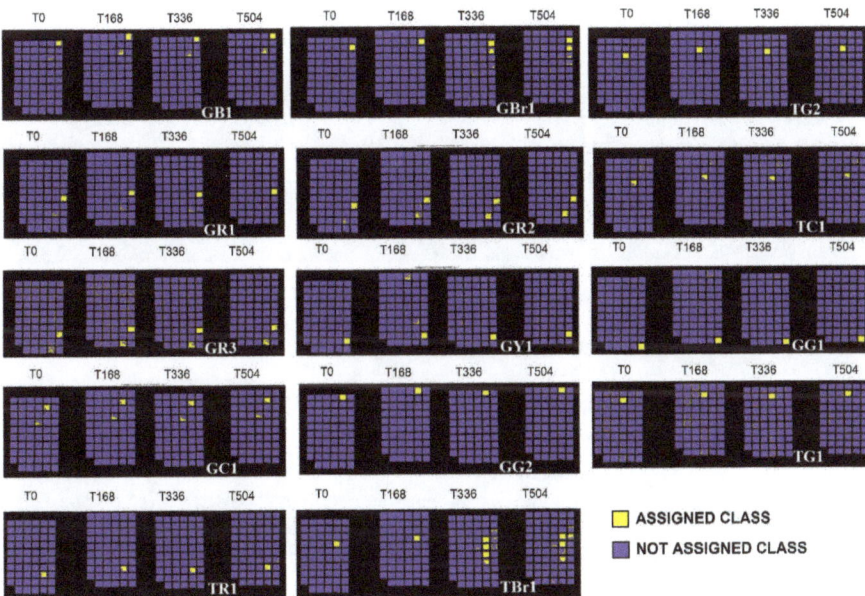

Figure 7. The prediction model results for commercial watercolours exhibiting no variations with ageing.

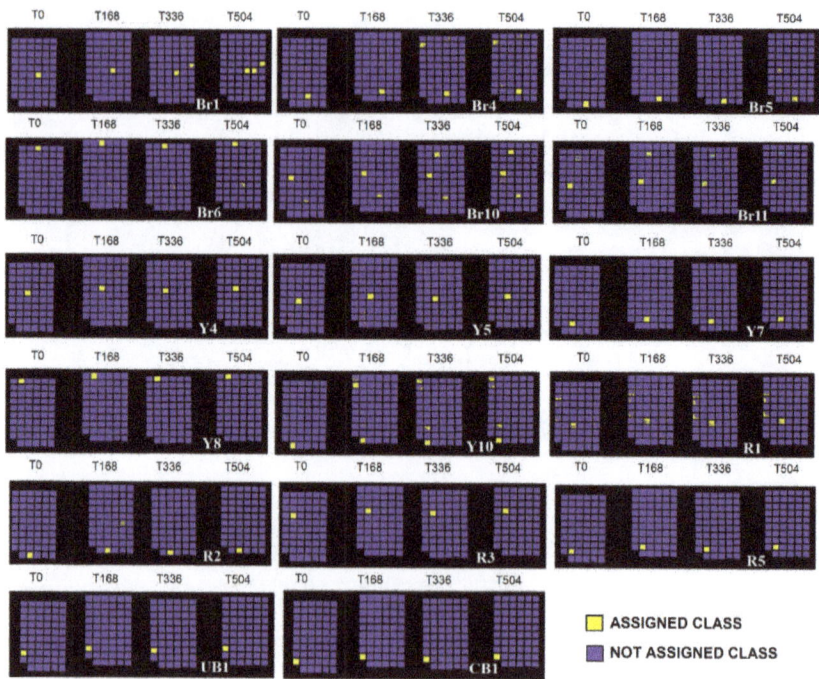

Figure 8. Prediction model results for pigment powders, mixed with GA, exhibiting no variations with ageing.

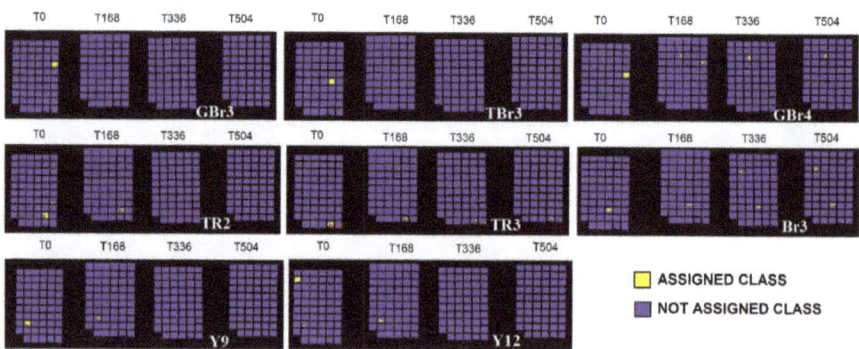

Figure 9. Prediction model results for painting samples exhibiting variations after 168 h of ageing.

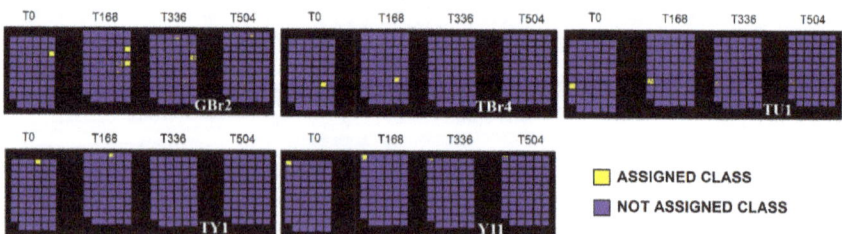

Figure 10. Prediction model results for painting samples exhibiting variations after 336 h of ageing.

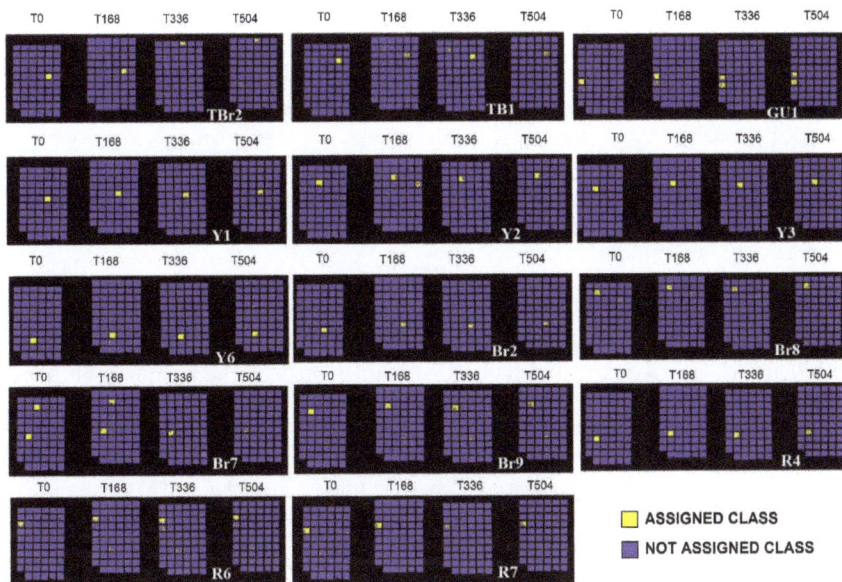

Figure 11. Prediction model results for painting samples exhibiting gradual variations until 504 h of ageing.

4. Discussion

The average spectra in Figure 5A show the contribution of gypsum in all samples with the SWIR absorption near 1450, 1490, and 1535 nm, and the OH/H_2O features around 1750 and 1950 nm [44,77]. The presence, in the region 2000–2350 nm, of some absorption can be attributed to other inorganic fractions. In this region, in fact, there are the absorption of calcium carbonate (around 2230, 2341, 2373 nm) and silicates (around 2200 and 2250 nm) [78]. In some cases, $CaCO_3$ is contained in the sample, as it was revealed by Fourier transform infrared (FT-IR) spectroscopy [60].

Results of PCA are displayed in Figure 5C,D. They show a variance of 91.32% with the first three principal components. The samples are grouped in different areas of the PCA score plot according to the pigment colours and typologies. In some cases, pigments that are similar in colour and declared composition, locate themselves in different regions of the score plot, such as, for example, samples Br1 and Br2 (both being burnt umber + GA). This behaviour can also be observed for samples in pan and tube but with the same compositions, as also previously highlighted [60]. This is the case, for example, of samples GY1 and TY1 (yellow ochre in pan and tube), samples GBr2 and TBr2 (natural umber in pan and tube), GR2 and TR2 (Venetian red in pan and tube) and better in GC1 and TC1 (cobalt blue in pan and tube) and in GG2/TG2 (Viridian in pan and tube). The difference that was observed between pan and tube samples, of the same colour and typology has been attributed to the differences in additives influencing the behaviour concerning ageing [60].

Differences can be observed also between natural iron oxide pigments, similar in colour and composition, such as, for example, between the Y series (Y1-Y11). Samples Y4 and Y12 are well grouped and differentiated in the PCA score plots (Figure 5C,D, respectively) as well as samples Y5 and Y7 (Figure 5C). These results demonstrated the great potentiality of PCA in separating and grouping materials having very similar characteristics. GA and GG behaviour was widely discussed in the previous paper and are not considered here [60].

Concerning the prediction model applied on the samples set, the results give double information. Firstly, the prediction identifies the painting layer, starting from a calibration dataset composed of 58 pigments, with a low error related to pigments having similar fingerprint (i.e., similar composition).

In the second step, the model evaluates the variation of the fingerprint in each sample in respect to time 0, highlighting how the spectral changes of the pigments occur during time. This double information is particularly relevant in the field of restoration, because it allows for differentiating the retouched areas in respect to the original painting, and also for monitoring the restored surfaces in order to evaluate the possible degradation during the time. Moreover, another advantage of the prediction model is the possibility for the restorer to select the retouching material with better performance in regard to ageing.

To deepen the behaviour of each single pigment, the test was applied on the different painting areas at the chosen ageing time, as shown in Figures 7–11.

By observing the sample behaviour in Figures 7 and 8, it can be derived that the highest stability against the artificial ageing can be seen in the following samples: GR1 (ivory black in pan), GBr1 (burnt umber 1 in pan), GR1 (Indian red 1 in pan), GR2 (Venetian red 2 in pan), GBr3 (Burnt Sienna 3 in pan), GY1 (yellow ochre 1 in pan), GG1 (bladder green 1 in pan), TR1 (Indian red 1 in tube), TBr1 (burnt umber 1 in tube), GC1 (cobalt blue 1 in pan), GG2 (Viridian 2 in pan), TG1 (chrome green 1 in tube), TG2 (Viridian 2 in tube), TC1 (cobalt blue 1 in tube), Br1 (burnt umber 1), Br4 (burnt umber 4), Br5 (burnt umber 5), Y4 (yellow ochre 4), Y5 (yellow ochre 5), Y7 (yellow ochre 7), Y8 (yellow ochre 8), Y10 (yellow ochre), R1 (red ochre 1), R2 (red ochre 2), R3 (natural umber 3), R5 (red ochre 5), Br6 (natural umber 6), Br10 (natural umber 10), and UB1 (ultramarine blue 1). These samples, in fact, exhibit no variations during irradiation times, resulting in being stable also at high total energy dose, i.e., 1.0×10^9 J/m^2 reached at 504 h of ageing.

Samples having a similar composition seem to reduce the variability with the increasing of ageing times. For example, the spectral signature of sample GBr2 (Figure 10) tends to overlap with that of GBr1 (both burnt umber based pigments) after 336 h of ageing, corresponding to a total dose of energy equal to 6.7×10^8 J/m^2. The spectral signature of sample TR2 (Figure 9) overlaps with that of sample GR2 (both Venetian red) after 168 h of ageing (total energy applied 3.3×10^8 J/m^2). Sample TR3 (Figure 9) spectral signature overlaps with that of GR3 (both cadmium red based pigments) after 168 h of ageing. Similar behaviour can be observed in samples TBr2 and TBr1 (two natural umber watercolours in tube): the spectral signature of TBr2 (Figure 11) overlap to that of sample TBr1 after 336 h of ageing. Spectra of samples GBr4 (natural Sienna 4 in pan, Figure 9) and TBr3 (burnt Sienna 3 in tube, Figure 9) partially overlap with the spectral signature of sample Br1 (burnt umber 1) at 504 h of ageing. Another observation can be derived by observing the spectral signature of sample Y12 (yellow ochre 12, Figure 9), with the results being similar to that of Y7 (yellow ochre 7) at 336 h of irradiation in Solar box. The spectral signature of sample Y11 (yellow ochre 11, Figure 10) appears similar to that of R2 (red ochre 2) after 336 h of ageing. At last, the spectra of sample Br7 (natural umber 7, Figure 11) and of sample Br3 (natural umber 3, Figure 9) appear to be similar to that of sample Br10 (natural umber 10, Figure 8) after 336 h of ageing.

Samples GBr3 (burnt umber 3 in pan), TBr3 (burnt Sienna 3 in tube), Y9 (yellow ochre 9), and Y12 (yellow ochre 12) exhibit a definite variation after 168 h of ageing, whereas samples TR3 (cadmium red 3 in tube), Br3 (natural umber 3), and TR2 (Venetian red 2 in tube) show a partial degradation at 168 h that becomes definite at 336 h of irradiation (Figure 9). Such degradation has been associated to Arabic gum deterioration occurring between 168 h and 336 h of irradiation, combined with that of pigment components, as previously discussed [60].

Some samples undergo degradation at 336 h of irradiation in Solar box (Figure 10). Specifically, sample TBr4 (natural Sienna 4 in tube), TY1 (yellow ochre 1 in tube), GR2 (Venetian red 2 in pan), TU1 (ultramarine blue 1 in tube), and Y11 (yellow ochre 11) show stability until 168 h of ageing and then have a definite change at 336 h. Also, in this case, the changes can be associated to both gum Arabic degradation (between 168 h and 336 h, energy range 3.3×10^8–6.7×10^8 J/m^2) and to the components of the watercolour and pigment mixtures.

A series of samples exhibits gradual variation until 504 h of ageing (Figure 11). In particular, samples TB1 (ivory black 1 in tube), TBr2 (natural umber 2 in tube), Y1 (yellow ochre 1), Y2 (yellow ochre 2), Y3 (yellow ochre 3), Y6 (yellow ochre 6), GU1 (ultramarine blue 1 in pan), R4 (red ochre 4),

R6 (red ochre 6), R7 (red ochre 7), Br2 (burnt umber 2), Br7 (natural umber 7), Br8 (natural umber 8), and Br9 (burnt umber 9) exhibit partial degradation varying during the ageing times. It can be hypothesized that the slow variation of the spectral signature is due to the unique degradation of gum Arabic, whereas pigments seem stable and do not give clear changes at the measured time intervals.

For a better comprehension of the prediction test results, being a lot of samples, a final table is reported (Table 3).

A general assessment derived from the Table 3 is that pigment powders mixed with gum Arabic are stable to ageing apart from three samples: Br3, Y9, and Y12. This can be due to the presence of additives in the powders or to the predominance of degradation of gum Arabic that could become relevant in relation to the ratio pigment/binder.

The group of green watercolours exhibits high stability. In the case of the other watercolours, the stability varies as function of colour and of watercolour typology, i.e., pan and tube, as also previously found [60]. Sienna-based watercolours have in general low stability, both in pan and tube.

Table 3. Sample stability evaluation derived from the prediction model.

Commercial Watercolours		Pigment Powders+Gum Arabic	
Abbreviation	Stability	Abbreviation	Stability
GB1	High	Br1	High
TB1	Medium-high	Br2	Medium-high
GBr1	High	Br3	Medium-low
TBr1	High	Br4	High
GBr2	Medium	Br5	High
TBr2	Medium-high	Br6	High
GBr3	Low	Br7	Medium-high
GBr4	Low	Br8	Medium-high
TBr3	Low	Br9	Medium-high
TBr4	Medium	Br10	High
GR1	High	Br11	High
TR1	High	R1	High
GR2	High	R2	High
TR2	Medium-low	R3	High
GR3	High	R4	Medium-high
TR3	Medium-low	R5	High
GY1	High	R6	Medium-high
TY1	Medium	R7	Medium-high
GG1	High	Y1	Medium-high
TG1	High	Y2	Medium-high
GG2	High	Y3	Medium-high
TG2	High	Y4	High
GC1	High	Y5	High
TC1	High	Y6	Medium-high
GU1	Medium-high	Y7	High
TU1	Medium	Y8	High
		Y9	Low
		Y10	High
		Y11	Medium
		Y12	Low
		CB1	High
		UB1	High

5. Conclusions

Hyperspectral Imaging (HSI) in the short-wave infrared was utilized to evaluate the stability to light and UV ageing of a conspicuous number of painting materials, in particular powder pigments and commercial watercolours to be used in retouching.

The new methodological approach that was chosen for monitoring the ageing behaviour of watercolour samples and pigment powder applied by gum Arabic, produced interesting results that should be further discussed and investigated. Different degradation patterns have been observed for the different pigments and also between tube and pan of the same watercolour. As previously demonstrated, gum Arabic alone clearly showed degradation occurring between 168 and 336 h of irradiation. However, when we observe the behaviour of gum Arabic/pigments mixtures, we found that it depends both on the pigment itself but also probably on the combination pigment-gum Arabic and furthermore on the presence of organic additives.

In some cases, the variation of paintings with ageing times is due to gum Arabic, such as in Y1-3, R4, R6-7, Br7-9, Br2, TB1, TBr2, and GU1 and appear as a gradually occurring phenomenon from time 0 h to 504 h of irradiation in Solar box. In this case, the combination pigment-GA seems to create a mixture that degrades slowly during time.

For other samples, the degradation of gum Arabic is the main cause of the observed variation in the prediction model, specifically in samples TBr4, TY1, TU1, GBr2, and Y11. These samples, in fact, show variations between 168 h and 336 h in the prediction model, i.e., the same range of degradation of gum Arabic.

A group of painting samples exhibits a definite change at 168 h (GBr2, TBr3, Y9, and Y12) that can be associated in part to the degradation of gum Arabic but also to that of pigment components. This time corresponds to a total solar dose of 3.3×10^8 J/m^2. In this same group we included other three samples, TR3, TR2, and Br3, whose changes are observed at 168 h but become complete at 336 h of ageing.

A conspicuous group (totally 31) of painting samples demonstrated high stability to ageing, as shown in Figures 6 and 7 and Table 3, demonstrating. Fourteen of these samples are commercial watercolours and seventeen are pigment powders mixed with gum Arabic. In this case, the combination of gum Arabic, pigments, and additives (for commercial watercolours) creates stable mixtures that also prevent the degradation of gum Arabic. In fact, some authors, through surface investigations, suggested that a thin gum binder layer is present on the surface of watercolour paintings and that other components, such as pigments and additives, are located within the gum layer [82]. So, they concluded that the main changes should be attributed to gum Arabic binder. However, this result depends on pigment typology, on extender, such as calcium carbonate, and additives, which could influence the response of gum Arabic to ageing.

In general, it can be affirmed that the thirty-two investigated powder pigments mixed with gum Arabic have high or medium-high stability to ageing under simulated solar radiation, apart from three samples exhibiting low and medium-low stability and one having medium stability.

Tube and pan samples have different behaviour in relation to pigment. For example, green watercolours were demonstrated to be very stable to ageing, whereas Sienna-based mixtures have, in general, low stability. Some differences have been observed also between pan and tube of the same watercolour, such as the case of GR2 and TR2, GR3, and TR3. As previously discussed, these differences can be due to the different composition of pan and tube mixtures [60], in particular to the presence of additives in the tube watercolours necessary for obtaining the desired rheological characteristics.

The results point out the potentiality of powder pigments to be used for obtaining stable watercolours, without additives: these ones, in fact, as highlighted in other papers [19,20], are responsible for the variability and degradation in watercolours and they should be better known in order to evaluate the overall stability to ageing of these commercial materials [37,83], especially if they should be used in retouching of artworks, as commonly occurs, especially in the case of wall paintings.

As final conclusion, it can be affirmed that HSI coupled with chemometric approach allow for monitoring paint layers modifications during ageing time. Furthermore, the classification techniques based on PCA-KNN, utilizing the hyperspectral data collected by HSI, clearly outlined the potentiality of this approach for monitoring the changes occurring in the painting layers; this was possible thanks to the evaluation of little variations in the spectra during ageing times before changes can be seen by eyes.

J. Imaging **2019**, *5*, 8

We think that this result has great relevance in the cultural heritage field because it demonstrated the possibility of detecting damages before they become irreversible. This approach could be particularly useful in monitoring artworks and restoration interventions over times at relatively low cost in respect to other analytical methods.

As future research lines, we think to apply the developed approach to other restoration materials and reintegration products based on synthetic resins that have been introduced in the conservation applications, as also suggested by conservators. The same approach in classification and predication of material behaviour in regard to ageing can be applied on protective products for cultural heritage artifacts, with the aim of testing both traditional and innovative products.

Author Contributions: Conceptualization C.P. and G.C.; methodology, G.B., G.C., C.P. and S.S.; software, G.C.; validation, G.C.; formal analysis, G.C.; investigation, C.P. and G.C.; resources, G.B., S.S. and C.P.; data curation, C.P. and G.C.; writing—original draft preparation, C.P. and G.C.; writing—review and editing, G.B., G.C., C.P. and S.S.; visualization, C.P. and G.C.; supervision, G.B. and S.S.; project administration, G.B., S.S. and C.P.; funding acquisition, G.B., S.S. and C.P.

Funding: This research was partially founded by MIUR (Italian Ministry for Education, University and Research) with the special funding for the basic research activities of Claudia Pelosi (Law 232/2016), and by Lazio Region (Grant No. G06970, 30 May 2018) for the Project ADAMO.

Acknowledgments: The authors would like to thank the staff members of the Laboratory of Diagnostics and Materials Science of University of Tuscia for the technical support in preparing the samples used for the experimental tests.

Conflicts of Interest: The authors declare no conflict of interest.

References

1. Polak, A.; Kelman, T.; Murray, P.; Marshall, S.; Stothard, D.J.M.; Eastaugh, N.; Eastaugh, F. Hyperspectral imaging combined with data classification techniques as an aid for artwork authentication. *J. Cult. Herit.* **2017**, *26*, 1–11. [CrossRef]

2. Zucco, M.; Pisani, M.; Cavaleri, T. Fourier transform hyperspectral imaging for cultural heritage. In *Fourier Transforms–High-Tech Application and Current Trends*; INTECH: London, UK, 2017; pp. 215–234.

3. Bonifazi, G.; Serranti, S.; Capobianco, G.; Agresti, G.; Calienno, L.; Picchio, R.; Lo Monaco, A.; Santamaria, U.; Pelosi, C. Hyperspectral imaging as a technique for investigating the effect of consolidating materials on wood. *J. Electron. Imaging* **2016**, *26*, 011003. [CrossRef]

4. Capobianco, G.; Bonifazi, G.; Prestileo, F.; Serranti, S. Pigment identification in pictorial layers by Hyper-spectral Imaging. *Proc. SPIE* **2014**, *9106*, 91060B. [CrossRef]

5. Capobianco, G.; Bracciale, M.P.; Sali, D.; Sbardella, F.; Belloni, P.; Bonifazi, G.; Serranti, S.; Santarelli, M.L.; Cestelli Guidi, M. Chemometrics approach to FT-IR hyperspectral imaging analysis of degradation products in artwork cross-section. *Microchem. J.* **2017**, *132*, 69–76. [CrossRef]

6. Fischer, C.; Kakoulli, I. Multispectral and hyperspectral imaging technologies in conservation: Current research and potential applications. *Stud. Conserv.* **2006**, *51*, 3–16. [CrossRef]

7. Catelli, E.; Randeberg, L.L.; Alsberg, B.K.; Gebremariam, K.F.; Bracci, S. An explorative chemometric approach applied to hyperspectral images for the study of illuminated manuscripts. *Spectrochim. Acta A* **2017**, *177*, 69–78. [CrossRef]

8. Agresti, G.; Bonifazi, G.; Calienno, L.; Capobianco, G.; Lo Monaco, A.; Pelosi, C.; Picchio, R.; Serranti, S. Surface investigation of photo-degraded wood by colour monitoring, infrared spectroscopy and hyperspectral imaging. *J. Spectrosc.* **2013**, *1*, 380536. [CrossRef]

9. Bonifazi, G.; Calienno, L.; Capobianco, G.; Lo Monaco, A.; Pelosi, C.; Picchio, R.; Serranti, S. Modelling color and chemical changes on normal and red heart beech wood by reflectance spectrophotometry, Fourier Transform infrared spectroscopy and hyperspectral imaging. *Polym. Degrad. Stab.* **2015**, *113*, 10–21. [CrossRef]

10. Westad, F.; Marini, F. Validation of chemometric models—A tutorial. *Anal. Chim. Acta* **2015**, *893*, 14–24. [CrossRef]

11. De la Ossa, M.Á.; García-Ruiz, C.; Amigo, J.M. Near infrared spectral imaging for the analysis of dynamite residues on human handprints. *Talanta* **2014**, *130*, 315–321. [CrossRef]

12. Bonifazi, G.; Capobianco, G.; Serranti, S. Asbestos containing materials detection and classification by the use of hyperspectral imaging. *J. Hazard. Mater.* **2017**, *344*, 981–993. [CrossRef] [PubMed]
13. Burger, J.; Geladi, P. Hyperspectral NIR image regression part II: Dataset preprocessing diagnostics. *J. Chemometr.* **2006**, *20*, 106–119. [CrossRef]
14. Van Ruth, S.M.; Villegas, B.; Akkermans, W.; Rozijn, M.; van der Kamp, H.; Koot, A. Prediction of the identity of fats and oils by their fatty acid, triacylglycerol and volatile compositions using PLS-DA. *Food Chem.* **2010**, *118*, 948–955. [CrossRef]
15. Basri, K.N.; Hussain, M.N.; Bakar, J.; Sharif, Z.; Khir, M.F.A.; Zoolfakar, A.S. Classification and quantification of palm oil adulteration via portable NIR spectroscopy. *Spectrochim. Acta A* **2017**, *173*, 335–342. [CrossRef] [PubMed]
16. Clarke, J. Two aboriginal rock art pigments from western Australia: Their properties, use, and durability. *Stud. Conserv.* **1976**, *21*, 134–142. [CrossRef]
17. Levison, H.W. *Artists' Pigments: Lightfastness Tests and Ratings: The Permanency of Artists' Colors and an Evaluation of Modern Pigments*; Colorlab: Hallandale, FL, USA, 1976; p. 107.
18. Rossi, S. Studio Della Stabilità di Acquerelli Commerciali e Pigmenti Naturali per la Reintegrazione dei Beni Culturali. Bachelor's Thesis, University of Tuscia, Viterbo, Italy, 2016.
19. Lo Monaco, A.; Marabelli, M.; Pelosi, C.; Picchio, R. Colour measurements of surfaces to evaluate the restoration materials. *Proc. SPIE* **2011**, *8084*, 1–14.
20. Di Marcello, S.; Notarstefano, C. La verifica della durabilità dei colori ad acquerello impiegati nella reitegrazione pittorica dei dipinti murali. In *A Scuola di Restauro*; Bonelli, M., D'Agostino, L., Mercalli, M., Eds.; Gangemi Editore: Roma, Italy, 2011; pp. 71–81, ISBN 978-88-492-2111-4.
21. Stoner, J.H.; Rischfield, R. *Conservation of Easel Paintings (Routledge Series in Conservation and Museology)*, 1st ed.; Taylor & Francis: London, UK, 2012; ISBN 9780750681995.
22. Rebecca, E.; Smithen, P.; Turnbull, R. (Eds.) *Mixing and Matching. Approaches to Retouching Paintings*; Archetype Publications in Association with the Icon Paintings Group and the British Association of Paintings Conservators-Restorers (BAPCR): London, UK, 2010; ISBN 9781904982500.
23. Größl, M.; Harrison, S.; Kaml, I.; Kenndler, E. Characterisation of natural polysaccharides (plant gums) used as binding media for artistic and historic works by capillary zone electrophoresis. *J. Chromatogr. A* **2005**, *1077*, 80–89. [CrossRef] [PubMed]
24. Ormsby, B.A.; Townsend, J.H.; Singer, B.W.; Dean, J.R. British watercolour cakes from the eighteenth to the early twentieth century. *Stud. Conserv.* **2005**, *50*, 45–66. [CrossRef]
25. Caruso, S. Caratterizzazione e Invecchiamento di Leganti Pittorici a Base di Gomme Vegetali. Ph.D. Thesis, University of Torino, Torino, Italy, 2006.
26. Bonaduce, I.; Brecoulaki, H.; Colombini, M.P.; Lluveras, A.; Restivo, V.; Ribechini, E. Gas chromatographic-mass spectrometric characterisation of plant gums in samples from painted works of art. *J. Chromatogr. A* **2007**, *1175*, 275–282. [CrossRef]
27. Kokla, V.; Psarrou, A.; Konstantinou, V. Watercolour identification based on machine vision analysis. *e-Preserv. Sci.* **2010**, *7*, 22–28.
28. Riedo, C.; Scalarone, D.; Chiantore, O. Advances in identification of plant gums in cultural heritage by thermally assisted hydrolysis and methylation. *Anal. Bioanal. Chem.* **2010**, *396*, 1159–1569. [CrossRef]
29. Russell, W.; de Abney, W. *Report on the Action of Light on Watercolours to the Science and Art Department of the Committee of Council on Education*; HMSO: London, UK, 1888.
30. Brommelle, N.S. The Russell and Abney report on the action of light on watercolours. *Stud. Conserv.* **1964**, *9*. [CrossRef]
31. Whitmore, P.M.; Bailie, C. Studies on the photochemical stability of synthetic resin-based retouching paints: The effect of white pigments and extenders. *Stud. Conserv.* **1990**, *35*, 144–149. [CrossRef]
32. Digney-Peer, S.; Thomas, K.; Perry, R.; Townsend, J.; Gritt, S. The imitative retouching of easel paintings. In *Conservation of Easel Paintings (Routledge Series in Conservation and Museology)*, 1st ed.; Routledge: Abingdon-on-Thames, UK, 2012; pp. 607–634, ISBN 9780750681995.
33. Lerwill, A.; Townsend, J.H.; Thomas, J.; Hackney, S.; Caspers, C.; Liang, H. Photochemical colour change for traditional watercolour pigments in low oxygen levels. *Stud. Conserv.* **2015**, *60*, 15–32. [CrossRef]
34. Lewill, A. Micro-Fading Spectrometry: An Investigation into the Display of Traditional Watercolour Pigments in Anoxia. Ph.D. Thesis, Nottingham Trent University, Nottingham, UK, 2011.

35. Callede, B. *Stabilité des Couleurs Utilisées en Restauration, Pigments Bleus*; Comité pour la Conservation de l'ICOM 4éme Reunion Triennale, ICOM: Venise, Italy, 1975.

36. De La Rie, E.R.; Lomax, S.Q.; Palmer, M.; Deming Glinsman, L.; Maines, C.A. An investigation of the photochemical stability of urea-aldehyde resin retouching paints: Removability tests and colour spectroscopy. *Stud. Conserv.* **2000**, *45*, 51–59. [CrossRef]

37. Korenberg, C. The photo-ageing behaviour of selected watercolour paints under anoxic conditions. *Br. Mus. Tech. Res. Bull.* **2008**, *2*, 49–57.

38. Dellaportas, P.; Papageorgiou, E.; Panagiaris, G. Museum factors affecting the ageing process of organic materials: Review on experimental designs and the INVENVORG project as a pilot study. *Herit. Sci.* **2014**, *2*, 1–11. [CrossRef]

39. Bailão San Andrés, A.; Calvo, A. Colorimetric analysis of two watercolours used in retouching. *Int. J. Conserv. Sci.* **2014**, *5*, 329–342.

40. Ropret, P.; Zoubek, R.; Sever Škapin, A.; Bukovec, P. Effects of ageing on different binders for retouching and on some binder–pigment combinations used for restoration of wall paintings. *Mater. Charact.* **2007**, *58*, 1148–1159. [CrossRef]

41. Pelosi, C.; Marabelli, M.; Patrizi, F.; Ortenzi, F.; Giurlanda, F.; Falcucci, C. *Valutazione della Stabilità degli Acquerelli nel Restauro Attraverso Misure di Colore*; Atti della V Conferenza Nazionale del Gruppo Colore, StarryLink Editrice: Brescia, Italy, 2009; pp. 141–149.

42. Kubik, M. Hyperspectral imaging: A new technique for the non-invasive study of artworks. *Phys. Tech. Study Art Archaeol. Cult. Herit.* **2007**, *2*, 199–259. [CrossRef]

43. Aceto, M.; Agostino, A.; Fenoglio, G.; Idone, A.; Gulmini, M.; Picollo, M.; Ricciardi, P.; Delaney, J.K. Characterisation of colourants on illuminated manuscripts by portable fibre optic UV-visible-NIR reflectance spectrophotometry. *Anal. Methods* **2014**, *6*, 1488–1500. [CrossRef]

44. Delaney, J.K.; Zeibel, J.G.; Thoury, M.; Littleton, R.; Palmer, M.; Morales, K.M.; De la Rie, E.R.; Hoenigswald, A. Visible and infrared imaging spectroscopy of Picasso's Harlequin musician: Mapping and identification of artist materials in situ. *Appl. Spectrosc.* **2010**, *64*, 584–594. [CrossRef]

45. Rosi, F.; Grazia, C.; Gabrieli, F.; Romani, A.; Paolantoni, M.; Vivani, R.; Brunetti, B.G.; Colomban, P.; Miliani, C. UV–Vis-NIR and micro Raman spectroscopies for the non destructive identification of Cd1-xZnxS solid solutions in cadmium yellow pigments. *Microchem. J.* **2016**, *124*, 856–867. [CrossRef]

46. Miliani, C.; Rosi, F.; Brunetti, B.G.; Sgamellotti, A. In situ noninvasive study of artworks: The MOLAB multitechnique approach. *Acc. Chem. Res.* **2010**, *43*, 728–738. [CrossRef] [PubMed]

47. Balas, C.; Papadakis, V.; Papadakis, N.; Papadakis, A.; Vazgiouraki, E.; Themelis, G. A novel hyper-spectral imaging apparatus for the non-destructive analysis of objects of artistic and historic value. *J. Cult. Herit.* **2003**, *4*, 330–337. [CrossRef]

48. Rusu, R.D.; Simionescu, B.; Oancea, A.V.; Geba, M.; Stratulat, L.; Salajan, D.; Ursu, L.E.; Popescu, M.C.; Dobromir, M.; Murariu, M.; et al. Analysis and structural characterization of pigments and materials used in Nicolae Grigorescu heritage paintings. *Spectrochim. Acta A* **2016**, *168*, 218–229. [CrossRef]

49. Arrizabalaga, I.; Gómez-Laserna, O.; Aramendia, J.; Arana, G.; Madariaga, J.M. Applicability of a diffuse reflectance infrared Fourier transform handheld spectrometer to perform in situ analyses on cultural heritage materials. *Spectrochim. Acta A* **2014**, *129*, 259–267. [CrossRef]

50. Arrizabalaga, I.; Gómez-Laserna, O.; Aramendia, J.; Arana, G.; Madariaga, J.M. Determination of the pigments present in a wallpaper of the middle nineteenth century: The combination of mid-diffuse reflectance and far infrared spectroscopies. *Spectrochim. Acta A* **2014**, *124*, 308–314. [CrossRef]

51. Von Aderkas, E.L.; Barsan, M.M.; Gilson, D.F.; Butler, I.S. Application of photoacoustic infrared spectroscopy in the forensic analysis of artists' inorganic pigments. *Spectrochim. Acta A* **2010**, *77*, 954–959. [CrossRef]

52. Maynez-Rojas, M.A.; Casanova-González, E.; Ruvalcaba-Sil, J.L. Identification of natural red and purple dyes on textiles by fiber-optics reflectance spectroscopy. *Spectrochim. Acta A* **2017**, *178*, 239–250. [CrossRef]

53. Trzcińska, B.; Kowalski, R.; Zięba-Palus, J. Comparison of pigment content of paint samples using spectrometric methods. *Spectrochim. Acta A* **2014**, *130*, 534–538. [CrossRef] [PubMed]

54. Rampazzi, L.; Brunello, V.; Corti, C.; Lissoni, E. Non-invasive techniques for revealing the palette of the Romantic painter Francesco Hayez. *Spectrochim. Acta A* **2017**, *176*, 142–154. [CrossRef] [PubMed]

55. Carlesi, S.; Bartolozzi, G.; Cucci, C.; Marchiafava, V.; Picollo, M.; La Nasa, J.; Di Girolamo, F.; Dilillo, M.; Modugno, F.; Degano, I.; et al. Discovering "The Italian Flag" by Fernando Melani (1907–1985). *Spectrochim. Acta A* **2016**, *168*, 52–59. [CrossRef] [PubMed]

56. Delaney, J.K.; Thoury, M.; Zeibel, J.G.; Ricciardi, P.; Morales, K.M.; Dooley, K.A. Visible and infrared imaging spectroscopy of paintings and improved reflectography. *Herit. Sci.* **2016**, *4*, 1–10. [CrossRef]

57. Clark, R.N. Spectroscopy of rocks and minerals, and principles of spectroscopy. In *Manual of Remote Sensing, Remote Sensing for the Earth Sciences*; Wiley: New York, NY, USA, 1999; Volume 3, pp. 3–58, ISBN 9781601196620.

58. Ricciardi, P.; Delaney, J.K.; Facini, M.; Zeibel, J.G.; Picollo, M.; Lomax, S.; Loew, M. Near infrared reflectance imaging spectroscopy to map paint binders in situ on illuminated manuscripts. *Angew. Chem. Int. Edit.* **2012**, *51*, 5607–5610. [CrossRef]

59. Dooley, K.A.; Lomax, S.; Zeibel, J.G.; Miliani, C.; Ricciardi, P.; Hoenigswald, A.; Loew, M.; Delaney, J.K. Mapping of egg yolk and animal skin glue paint binders in Early Renaissance paintings using near infrared reflectance imaging spectroscopy. *Analyst* **2013**, *138*, 4838–4848. [CrossRef]

60. Pelosi, C.; Capobianco, G.; Agresti, G.; Bonifazi, G.; Morresi, F.; Rossi, S.; Santamaria, U.; Serranti, S. A methodological approach to study the stability of selected watercolours for painting reintegration, through reflectance spectrophotometry, Fourier transform infrared spectroscopy and hyperspectral imaging. *Spectroch. Acta A* **2018**, *198*, 92–106. [CrossRef]

61. Helwig, K. Iron oxide pigments. In *Artistis' Pigments. A Handbook of Their History and Characteristics*, 1st ed.; Barrie, B.H., Ed.; National Gallery of Art: Washington, DC, USA; Archetype Publications: London, UK, 2007; Volume 4, pp. 39–109, ISBN 9781904982234.

62. *Paint and Varnishes: Methods of Exposures to Laboratory Light Sources-Part 1: General Guidance*; ISO 16474-1; ISO: Geneva, Switzerland, 2013.

63. *Paint and Varnishes—Methods of Exposure to Laboratory Light Sources-Part 1: Xenon-Arc Lamp*; ISO16474-2; ISO: Geneva, Switzerland, 2013.

64. Grahn, H.; Geladi, P. (Eds.) *Techniques and Applications of Hyperspectral Image Analysis*; John Wiley & Sons: West Sussex, UK, 2007; pp. 1–15, ISBN 9780470010860.

65. Otto, M. *Chemometrics, Statistics and Computer Application in Analytical Chemistry*; Wiley-VCH: New York, NY, USA, 1999.

66. Vidal, M.; Amigo, J.M. Pre-processing of hyperspectral images. Essential steps before image analysis. *Chemometr. Intell. Lab.* **2012**, *117*, 138–148. [CrossRef]

67. Rinnan, Å.; van den Berg, F.; Engelsen, S.B. Review of the most common pre-processing techniques for near-infrared spectra. *TrAC-Trend Anal. Chem.* **2009**, *28*, 1201–1222. [CrossRef]

68. Amigo, J.M.; Babamoradi, H.; Elcoroaristizabal, S. Hyperspectral image analysis. A tutorial. *Anal. Chim. Acta* **2015**, *896*, 34–51. [CrossRef]

69. Rinnan, Å; Nørgaard, L.; van den Berg, F.; Thygesen, J.; Bro, R.; Engelsen, S.B. Data pre-processing. In *Infrared Spectroscopy for Food Quality Analysis and Control*; Academic Press: New York, NY, USA, 2009; Chapter 2; pp. 29–50.

70. Bro, R.; Smilde, A.K. Principal component analysis. *Anal. Methods* **2014**, *6*, 2812–2831. [CrossRef]

71. Parvin, H.; Alizadeh, H.; Minaei-Bidgoli, B. MKNN: Modified k-nearest neighbour. In *Proceedings of the World Congress on Engineering and Computer Science 2008 (WCECS 2008)*; Newswood Limited: San Francisco, CA, USA, 2008; pp. 831–834.

72. Chitari, A.; Bag, V.V. Detection of brain tumor using classification algorithm. *Int. J. Invent. Comput. Sci. Eng.* **2014**, *1*, 2348–3539.

73. Duda, R.O.; Hart, P.E.; Stork, D.G. *Pattern Classification*; John Wiley & Sons: Hoboken, NJ, USA, 2012.

74. Siuly, S.; Kabir, E.; Wang, H.; Zhang, Y.; Siuly, S.; Kabir, E.; Wang, H.; Zhang, Y. Exploring sampling in the detection of multicategory EEG signals. *Comput. Math. Methods Med.* **2015**, *2015*, 576437. [CrossRef] [PubMed]

75. Imandoust, S.B.; Bolandraftar, M. Application of K-Nearest Neighbor (KNN) approach for predicting economic events: Theoretical background. *Int. J. Eng. Res. Appl.* **2013**, *3*, 605–610.

76. Du, M.; Chen, X. Accelerated k-nearest neighbours algorithm based on principal component analysis for text categorization. *J. Zhejiang Univ. Sci. C* **2013**, *14*, 407–416. [CrossRef]

77. Velasco, F.; Alvaro, A.; Suarez, S.; Herrero, J.-M.; Yusta, I. Mapping Fe-bearing hydrated sulphate minerals with short wave infrared (SWIR) spectral analysis at San Miguel mine environment, Iberian Pyrite Belt (SW Spain). *J. Geochem. Explor.* **2005**, *87*, 45–72. [CrossRef]

78. Fremout, W.; Kuckova, S.; Crhova, M.; Sanyova, J.; Saverwyns, S.; Hynek, R.; Kodicek, M.; Vandenabeele, P.; Moens, L. Classification of protein binders in artist's paints by matrix-assisted laser desorption/ionisation time-of-flight mass spectrometry: And evaluation of principal component analysis (PCA) and soft independent modelling of class analogy (SIMCA). *Rapid Commun. Mass Spectrom.* **2011**, *25*, 1631–1640. [CrossRef]

79. Duchêne, S.; Detalle, V.; Bruder, R.; Sirven, J.B. Chemometrics and Laser Induced Breakdown Spectroscopy (LIBS) analyses for identification of wall paintings pigments. *Curr. Anal. Chem.* **2010**, *6*, 60–65. [CrossRef]

80. Checa-Moreno, R.; Manzano, E.; Mirón, G.; Capitan-Vallvey, L.F. Comparison between traditional strategies and classification technique (SIMCA) in the identification of old proteinaceous binders. *Talanta* **2008**, *75*, 697–704. [CrossRef]

81. Hoehse, H.; Paul, A.; Gornushkin, I.; Panne, U. Multivariate classification of pigments and inks using combined Raman spectroscopy and LIBS. *Anal. Bioanal. Chem.* **2012**, *402*, 1443–1450. [CrossRef]

82. Clark, R.N.; King, T.V.V.; Klejwa, M.; Swayze, G.A.; Vergo, N. High spectral resolution reflectance spectroscopy of minerals. *J. Geophys. Res.* **1990**, *95*, 12653–12680. [CrossRef]

83. Kirby, J.; Saunders, D. Fading and colour change of Prussian blue: Methods of manufacture and the influence of extenders. *Natl. Gallery Tech. Bull.* **2004**, *25*, 73–99.

Journal of
Imaging

MDPI

Article

Investigation of the Performance of Hyperspectral Imaging by Principal Component Analysis in the Prediction of Healing of Diabetic Foot Ulcers

Qian Yang [1], Shen Sun [2], William J. Jeffcoate [3], Daniel J. Clark [3], Alison Musgove [3], Fran L. Game [4] and Stephen P. Morgan [1,*]

[1] Optics and Photonics Research Group, Faculty of Engineering, University of Nottingham, Nottingham NG7 2RD, UK; eexqy3@icloud.com
[2] Biomedical Information Processing Lab, College of Life Science and Bioengineering, Beijing University of Technology, Beijing 100124, China; sunshen@bjut.edu.cn
[3] Nottingham University Hospitals NHS Trust, Nottingham NG5 1PB, UK; william.jeffcoate@gmail.com (W.J.J.); daniel.clark@nuh.nhs.uk (D.J.C.); awlmusgrove@hotmail.co.uk (A.M.)
[4] University Hospitals of Derby and Burton NHS Foundation Trust, Derby DE22 3NE, UK; frances.game@nhs.net
* Correspondence: steve.morgan@nottingham.ac.uk; Tel.: +44-115-951-5570

Received: 22 September 2018; Accepted: 4 December 2018; Published: 7 December 2018

Abstract: Diabetic foot ulcers are a major complication of diabetes and present a considerable burden for both patients and health care providers. As healing often takes many months, a method of determining which ulcers would be most likely to heal would be of great value in identifying patients who require further intervention at an early stage. Hyperspectral imaging (HSI) is a tool that has the potential to meet this clinical need. Due to the different absorption spectra of oxy- and deoxyhemoglobin, in biomedical HSI the majority of research has utilized reflectance spectra to estimate oxygen saturation (SpO_2) values from peripheral tissue. In an earlier study, HSI of 43 patients with diabetic foot ulcers at the time of presentation revealed that ulcer healing by 12 weeks could be predicted by the assessment of SpO_2 calculated from these images. Principal component analysis (PCA) is an alternative approach to analyzing HSI data. Although frequently applied in other fields, mapping of SpO_2 is more common in biomedical HSI. It is therefore valuable to compare the performance of PCA with SpO_2 measurement in the prediction of wound healing. Data from the same study group have now been used to examine the relationship between ulcer healing by 12 weeks when the results of the original HSI are analyzed using PCA. At the optimum thresholds, the sensitivity of prediction of healing by 12 weeks using PCA (87.5%) was greater than that of SpO_2 (50.0%), with both approaches showing equal specificity (88.2%). The positive predictive value of PCA and oxygen saturation analysis was 0.91 and 0.86, respectively, and a comparison by receiver operating characteristic curve analysis revealed an area under the curve of 0.88 for PCA compared with 0.66 using SpO_2 analysis. It is concluded that HSI may be a better predictor of healing when analyzed by PCA than by SpO_2.

Keywords: hyperspectral imaging; principal component analysis; oxygen saturation; wound healing; diabetic foot ulcer

1. Introduction

Diabetic foot ulcers are thought to affect 15–25% of people with diabetes during their lifetime [1] and are a major source of suffering and cost. The principal pathological conditions contributing to foot ulceration are peripheral neuropathy, peripheral artery disease, pre-existing deformity, and trauma, but the contributions of each vary considerably. While some ulcers heal relatively quickly, others fail to heal and deteriorate. An accurate prediction of those ulcers least likely to heal quickly can therefore be useful, because it would enable consideration of more intensive intervention at an earlier stage and thereby improve overall outcome.

Peripheral artery disease can both cause ulceration and delay its healing by reducing the delivery of oxygen to peripheral tissues. While there are a number of approaches to assessing the extent of disease in larger arteries (including pulse palpation, pressure measurements, and angiography), none are routinely used to investigate associated dysfunction of smaller arteries, arterioles, and capillaries. One option that has been explored is hyperspectral imaging (HSI).

HSI is a noninvasive technique by which images are formed at different wavelengths to produce a hypercube (x, y, λ). Due to the different absorption spectra of oxy- and deoxyhemoglobin, in biomedical HSI the majority of research has utilized the reflectance hypercube to estimate oxygen saturation (SpO$_2$) values from peripheral tissue [2–9].

Greenman and colleagues used HSI to investigate whether oxygen delivery and muscle metabolism were factors in diabetic foot disease. That study included 108 individuals without ulceration, comparing three groups: volunteers without diabetes, diabetic patients without neuropathy, and patients with both diabetes and neuropathy [5]. SpO$_2$ was reduced in people with diabetes, and especially in those with neuropathy. Yudovsky et al. [6] also used HSI in the visible spectrum (450–700 nm) to predict tissue breakdown. A two-layer (epidermis, dermis) skin model was used to fit to the measured data and obtain an index of SpO$_2$. The algorithm was able to predict tissue at risk of ulceration with a sensitivity and specificity of 95% and 80% respectively, 58 days before breakdown is visible to the naked eye [6].

A formula to derive an indication of SpO$_2$ from HSI was also used by Khaodhiar et al. [7] to estimate oxyhemoglobin and deoxyhemoglobin of 10 patients with type 1 diabetes with foot ulcers, 13 without ulcers, and 14 subjects without diabetes. A spectrum for each pixel was compared with standard tissue to determine measures of oxyhemoglobin and deoxyhemoglobin. Using this approach, the sensitivity and specificity of HSI in predicting ulcer healing were 93% and 86%, while the positive and negative predictive values for ulceration were calculated as 93% and 86%. Nouvong et al. [8] used a similar approach to estimate relative values of tissue oxyhemoglobin and deoxyhemoglobin in 66 people with diabetic foot ulcers and reported that the sensitivity of HSI to predict healing within 6 months was 80% and the specificity was 74%. As discussed in our previous work [9], both of these papers had weaknesses, which helps to explain the differences between results of [7–9]. The first study [7] was very small and acknowledged to be simply a pilot, and both studies based their analysis on outcome per ulcer rather than outcome per person. The population included in the second study [8] was also highly selected, with a mean age of participants of just over 50 years, much younger than a representative population with diabetic foot ulcers.

Principal component analysis (PCA) is an alternative approach to analyzing HSI data. Although frequently applied in other fields, such as remote sensing and the food industry, mapping of SpO$_2$ is more common in biomedical HSI. It is therefore valuable to compare the performance of PCA with SpO$_2$ measurement in the prediction of wound healing. PCA is a process that converts a number of possibly correlated variables into a set of linearly uncorrelated variables called principal components. PCA has been demonstrated to be an effective and efficient preprocessing method, as retaining only the first few principal components significantly reduces data [10]. In the food industry, PCA and HSI have been applied to tea classification [11], detection of bruise damage on mushrooms [12], and estimation of the quality of pork [13]. Some examples of the application of PCA in in vivo biomedical HSI are provided in useful reviews [14–16], with a focus on laparoscopic imaging [11,12]. PCA has also been used as a dimension-reduction algorithm for wavelet-based segmentation of hyperspectral colon tissue imagery [17]. For tissue measurement, the contiguous bands of a hypercube are highly correlated, as they are dominated by the oxy- and deoxyhemoglobin spectra. This has the benefit of being a data-reduction method for the hypercubes obtained from the tissues of feet affected by ulcers.

This study performed a novel investigation by comparing the performance of PCA with more widely used SpO$_2$ measurements in predicting whether a wound will heal within 12 weeks of presentation. More accurate prediction of wound healing will support earlier intervention and better treatment.

2. Method

2.1. Experimental Setup

The HSI setup is shown in Figure 1. Illumination of the foot was via 16 × 1 W white light-emitting diodes (LEDs) (LXHL-MWEC, Lumileds^{TM} Lighting, San Jose, CA, USA) with 8 units placed on either side of the camera. Light scattered from the foot was passed through an aperture, which controlled the amount of collected light and was focused onto a detector by a C-mount lens (f = 15 mm, f# = 2.2; Schneider).

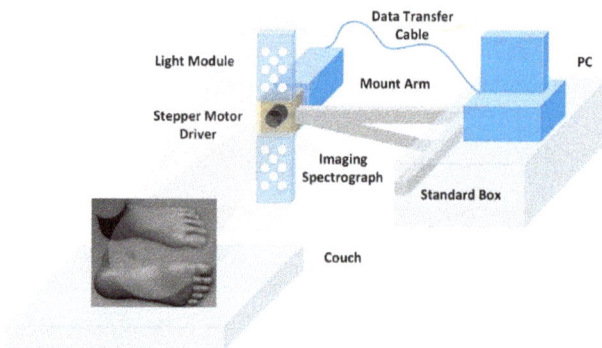

Figure 1. Hyperspectral imaging setup for imaging the foot (foot-to-lens distance typically 25–30 cm).

The HSI camera is a "push-broom" type that images a line from the scene onto a diffraction grating, which splits the light into a range of colors across the photosensor array. The camera comprises a Peltier cooled charge-coupled device (CCD) (Sensicam QE, PCO imaging, Cooke Corporation, Auburn Hills, MI, USA) coupled to an imaging spectrograph (ImSpector V10E, Specim Ltd., Oulu, Finland), which contains an input slit and a prism–grating–prism system. The input slit defines the field of view for the spatial scan, while the prism–grating–prism diffracts the light from the aperture into its spectrum. Scanning the line allows a 3D data cube that is transmitted to a PC via a peripheral component interface (PCI) for storage and future analysis. For the measurements taken in this study, each 3D data cube contained 2D spatial images (120 × 170 pixels) over a wavelength range from 430 nm to 750 nm (272 values). The sweep of the system moves from heel to toe and takes ~30 s to obtain an image, with an exposure time of 100 ms per row. Calibration images of white (99% reflectance Spectralon; Labsphere, Inc., North Sutton, NH, USA) and black (lens cap on camera) targets are recorded to take into account the effects of the nonuniform spectrum of the light source and dark noise, respectively. In order to reduce noise but not lose significant features in the reflected light spectra, a 9 point moving average filter is applied to the spectra.

For a certain position in the image plane (x, y) at a wavelength λ, the calibrated attenuation value is calculated as:

$$A(x, y, \lambda) = -\log \frac{I_{sample}(x, y, \lambda) - I_{dark}(x, y, \lambda)}{I_{white}(x, y, \lambda) - I_{dark}(x, y, \lambda)} \tag{1}$$

where I_{sample} is the intensity measured from the raw image, and I_{white} and I_{dark} are the intensities of the white and dark references, respectively.

2.2. Clinical Protocol

The data used to compare prediction of wound healing via SpO$_2$ and PCA were obtained in a clinical study described previously [9]. The published study received research ethics approval and all participants provided written informed consent. Recruitment was of a consecutive cohort, and the only major prespecified exclusions were those with a unilateral major amputation and those who withheld or were unable to give informed consent [9]. There was therefore no control of gender balance, as one would expect a predominance of male patients in all studies of foot ulcers. There was also no control of diabetes type, as this is not a recognized significant factor associated with the outcome of diabetic foot ulcers.

Participants attending an imaging session were required to avoid drinking tea or coffee and smoking tobacco for at least 2 hours prior to the visit, as these stimulants could lead to a change in blood flow. Capillary glucose was determined on arrival, and patients who were hypoglycemic were excluded from the study. All the studies were undertaken in a temperature-controlled clinic room at 22 °C and the test time was between 09:00 and 12:00. Prior to any assessments, participants lay on an examination couch for at least 15 minutes after removing their shoes and socks. Intensity hypercubes of the ulcer site were obtained for each participant, and the data were processed using SpO$_2$ algorithms and PCA, as described in Sections 2.3 and 2.4, respectively.

2.3. SpO$_2$ Data Processing

Oxygen saturation is defined as:

$$SpO_2 = \frac{HbO_2}{HbO_2 + Hb} \times 100\% \tag{2}$$

where HbO_2 is the concentration of oxyhemoglobin (mole L^{-1}) and Hb is the concentration of deoxyhemoglobin (mole L^{-1}).

Due to the different absorption spectra of the dominant absorbers, oxy- and deoxyhemoglobin, it is possible to extract information about the oxygen saturation of tissues based on optical measurements such as HSI.

The absorption coefficient ($\mu_a(\lambda)$) and attenuation ($A(\lambda)$) can be expressed as [18]:

$$\mu_a(\lambda) = \alpha(\lambda)[HbO_2] + \beta(\lambda)[Hb] \tag{3}$$

$$A(\lambda) = (\alpha(\lambda)[HbO_2] + \beta(\lambda)[Hb])d \tag{4}$$

where $\alpha(\lambda)$ is the specific absorption of oxyhemoglobin (cm^{-1} mole^{-1} L), $\beta(\lambda)$ is the specific absorption of deoxyhemoglobin (cm^{-1} mole^{-1} L), and d is the path length of the light (cm).

If $\mu_a(\lambda)$ is known at 2 wavelengths, then it is straightforward to calculate SpO$_2$ from Equations (3) and (2), as $\alpha(\lambda)$ and $\beta(\lambda)$ are known from literature values. A challenge is to relate measurements of $A(\lambda)$ and $\mu_a(\lambda)$. In the absence of light scattering, the path length is the geometric path length through the sample and the relationship is the Lambert–Beer law. In practice, the relationship between attenuation and absorption is nonlinear due to light scattering. An approximation is therefore needed to relate $A(\lambda)$ and $\mu_a(\lambda)$ in the presence of light scattering. The most commonly applied is the modified Lambert–Beer law [19,20]:

$$A(\lambda) = \mu_a(\lambda)d + G \tag{5}$$

where an offset G is used to take into account attenuation due to scattering. Alternative relationships include a parabolic model [21]:

$$A(\lambda) = -a(\mu_a(\lambda)d)^2 + b\mu_a(\lambda)d + G \tag{6}$$

where a and b are fitting parameters, as well as a power law model derived from photon diffusion theory [22]:

$$A(\lambda) = a\mu_a^{\frac{1}{2}}(\lambda) + G \qquad (7)$$

Here, a model is applied based on the power law approximation [18]:

$$A(\lambda) = a\mu_a^b(\lambda) + G \qquad (8)$$

where a and b are fitting parameters. It should be noted that when $b = 1$, this equation is same as the modified Lambert–Beer Law (Equation (5)) and when $b = 0.5$ the equation becomes the power law model described in Equation (7). Using Equation (8) to fit to the measured data and applying Equations (3) and (2) enabled images of oxygen saturation to be obtained. The fit is performed with a nonlinear search algorithm that uses the simplex search method [18,23]. This is a direct search method that does not use numerical or analytic gradients. For n unknown parameters of the fitting equation, the simplex in n-dimensional space is characterized by the $n + 1$ distinct vectors that are its vertices. At each step of the search, a new point in or near the current simplex is generated. The function's value at the new point is compared with its values at the vertices of the simplex, and usually one of the vertices is replaced by the new point, giving a new simplex. This step is repeated until the diameter of the simplex is less than the specified tolerance. The fitting algorithm and model have been previously validated using Monte Carlo data that simulate light propagation in tissue [18]. In the presence of noise, this method was found to be robust and was subsequently applied to tissue measurement.

Similar to [9], tissue oxygenation was assessed by HSI at a site measuring 1 cm^2 in an area of intact skin adjacent (typically 1–5 mm) to the edge of the ulcer and unaffected by callus.

2.4. Principal Component Analysis

The method applied in the wound study follows a similar approach to that described in the literature for applications in the food industry [11–13]. A cropped region of interest of 50 × 50 pixels was selected, as this was found to be sufficient to extract the wound and surrounding tissue for all the images obtained. Where background pixels remained in the image (e.g., when a wound was close to or on a toe), these were removed by thresholding, as the attenuation of the background was lower than that of tissue.

The process of converting the three-dimensional data cube into images of each of the principal components is shown in Figure 2. The data cube is first unfolded into a two-dimensional matrix, where each column represents all pixels contained in one spectral band of the original image cube and each row represents each pixel's spectrum (Figure 2). Mathematically this is expressed as [24]:

$$A_i = [A(\lambda_i)_1, A(\lambda_i)_2, \ldots, A(\lambda_i)_N]_i^T \qquad (9)$$

where N is the total number of pixels in the image, $A(\lambda_i)$ is the attenuation at each pixel, i represents the wavelength bin number of the spectrum, and T denotes the transpose.

To calculate the principal components, it is necessary to calculate the eigenvectors and eigenvalues of the 2D matrix (Figure 2c). The mean vector is given by:

$$m = \frac{1}{N} \sum_{i=1}^{N} A_i \qquad (10)$$

The covariance matrix of A_i is expressed as:

$$Cov(A) = \frac{1}{N} \sum_{i=1}^{N} (A_i - m)(A_i - m)^T \qquad (11)$$

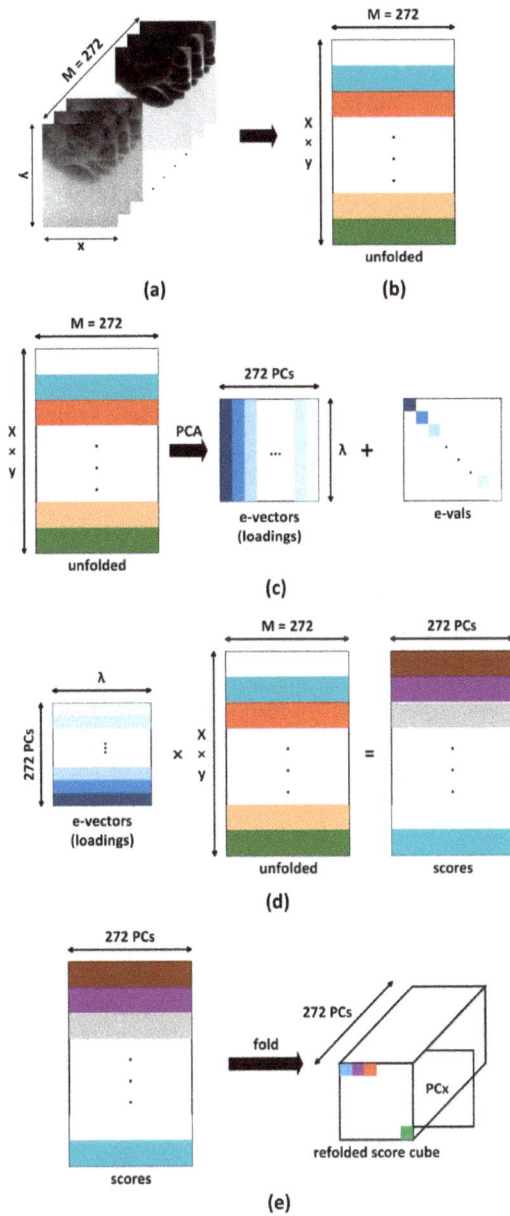

Figure 2. Principal component analysis (PCA) applied to a hypercube, where e-vector means eigenvector and e-value means eigenvalue: unfold (**a**) 3-D datacube into (**b**) 2-D matrix; (**c**) obtain eigenvectors and eigenvalues from covariance matrix; (**d**) multiply the 2-D matrix by the eigenvectors to obtain a score matrix; (**e**) refold the score matrix to form images at each principal component.

PCA is dependent on the eigenvalue decomposition of the covariance matrix, and *Cov*(x) can be denoted in another form as:

$$Cov(A) = UDU^T \tag{12}$$

$$D = diag\ (P_1,\ P_2,\ \ldots,\ P_N) \tag{13}$$

$$U = (u_1, u_2, \ldots u_N) \tag{14}$$

where D is the diagonal matrix, P is the eigenvalues of the covariance matrix, and U is the orthonormal matrix that contains the eigenvectors of the covariance matrix.

Multiplying the 2-D matrix A_i by the eigenvector matrix provides a score matrix v_i (Figure 2d), which can then be refolded to form a data cube that represents images of principal components.

$$v_i = U^T A_i\ (i = 1, 2, \ldots, M) \tag{15}$$

Arranging these images according to the magnitude of the eigenvector ($P_1 \gg P_2 \gg \ldots \gg P_N$) enables data reduction, as usually only information is contained in the first few principal components. In this case, the oxy- and deoxyhemoglobin spectra are correlated and only the first two principal components (PC1 and PC2) are used for image classification.

In order to compare the classification performance of using SpO$_2$ values or PCA, receiver operating characteristic (ROC) curves are used to express the performance of a binary classifier system due to a varying discrimination threshold. An ROC curve is obtained by plotting true positive rate (TPR) against false positive rate (FPR). TPR is the fraction of true positives out of the total actual positives. FPR is the fraction of false positives out of the total actual negatives.

3. Results

A total of 43 volunteers participated in the clinical study, as previously reported [9]. There were 12 women and 31 men; mean age was 62.7 years. Six of the 43 patients had type 1 diabetes and 37 had type 2 diabetes; 9 were smokers and 39 patients were judged to have neuropathy. Median (range) ankle brachial pressure index (ABPI) was 1.06 (0.15–1.63). Median (range) estimated duration of ulcers prior to assessment was 4.97 (1–26) weeks. 24 healed by 12 weeks and a further 7 healed between 12 and 24 weeks. Ten ulcers did not heal within 24 weeks of follow-up.

3.1. Oxygen Saturation Analysis

As previously reported [9], the SpO$_2$ results from baseline were significantly different between ulcers that did and did not heal within 12 weeks, but not between those that did and did not by 24 weeks. Figure 3 shows measured SpO$_2$ at a point adjacent to the wound site against healing time (healed by 12 weeks represented by blue diamonds, unhealed at 12 weeks represented by red triangles). The dashed line shows the optimum threshold using Youden's index [25] obtained from the ROC curve shown in the next section. An R^2 value of 0.4 was obtained when applying a linear fit to the data obtained for healing within the first 12 weeks.

For the SpO$_2$ classifier with the threshold set to 59.5%, the black dashed line (shown in Figure 3) can be used as the decision line where patients with SpO$_2$ values adjacent to the wound site lower than the threshold are classified as healing by 12 weeks. Only two of the unhealed ulcers were grouped incorrectly. The TPR was 50.0% (12 of 24) and the FPR was 11.8% (2 of 17). When using SpO$_2$ values to predict the healing of diabetic ulcers in 12 weeks, the sensitivity was 50% (12 of 24), the specificity was 88.2% (15 of 17), and the positive predictive value was 85.7% (12 of 14).

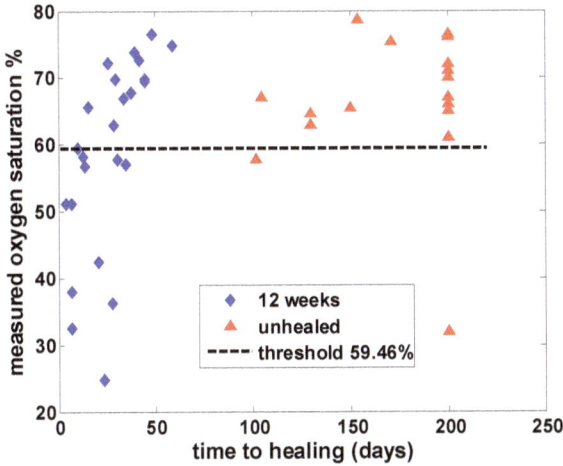

Figure 3. Relationship between time to healing (days) and oxygen saturation in a region adjacent to the wound. In order to plot unhealed ulcers, their healing days were set at 200 (higher than all the healing days of the healed ulcers). The dashed line represents the optimum threshold calculated from Youden's index.

3.2. Principal Component Analysis

PC1 and PCs greater than 2 did not provide any indication of wound healing, so it was not possible to identify a threshold for classification of wound healing in these cases. There was, however, clustering of data corresponding to healing within 12 weeks in the second principal component (PC2) scores and thresholds could be selected. Figure 4 shows PC2 against time to healing for all patients (healed by 12 weeks represented by blue diamonds, unhealed at 12 weeks represented by red triangles). Using Youden's index sets the upper and lower PC2 thresholds to +0.62 and −0.62. In this case, the TPR was 87.5% and the FPR was 11.8%. The sensitivity was 87.5% (21 of 24), the specificity was 88.2% (15 of 17), and the positive predictive value was 91.3% (21 of 23).

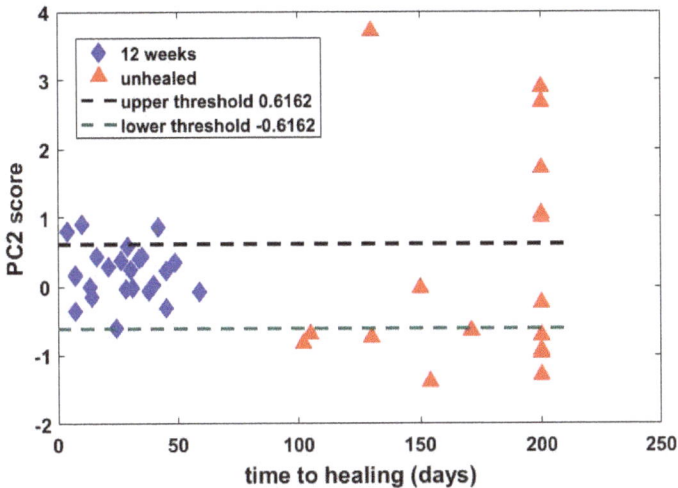

Figure 4. Relationship between time to healing (days) and principal component 2 (PC2). In order to plot unhealed ulcers, healing days was set at 200 (higher than all the healing days of the healed ulcers).

In order to further compare the performance of SpO_2 and PC2 classifiers, the ROC curve in Figure 5 shows the PCA classifier (blue line) much closer to the ideal right-angled case than the SpO_2 classifier (red line). A common method to compare classifiers in a single scalar value is to calculate the area under the ROC curve (AUC) [26]. The AUC under the PCA classifier is 0.88, which is 33% more than the AUC for the SpO_2 classifier (0.66).

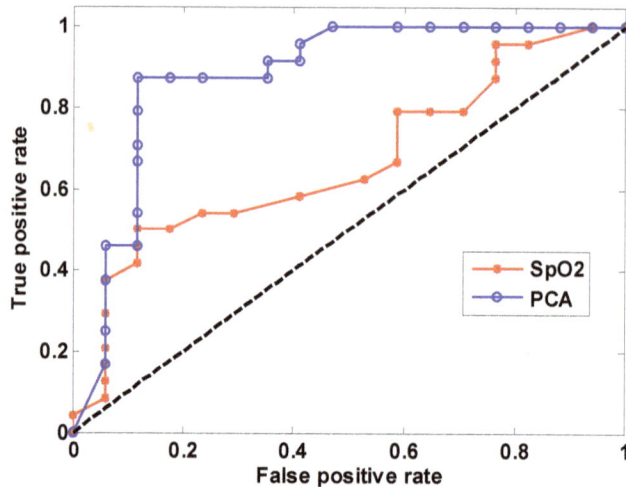

Figure 5. Receiver operating characteristic (ROC) analysis: red line indicates classification based on SpO_2 values adjacent to the wound site and blue line represents classification based on the absolute value of the PC2 score. Black dashed line is the worst case.

4. Discussion and Conclusions

Hyperspectral imaging is a tool that has the potential to predict healing of diabetic foot ulcers. Such a tool would be highly beneficial, as foot ulcers represent a major complication of diabetes and are a considerable burden for both patients and health care providers. Healing often takes many months and accurate prediction of those ulcers least likely to heal quickly can therefore be useful, because it would enable more intensive intervention at an earlier stage, which could improve overall outcome. Due to the different absorption spectra of oxy- and deoxyhemoglobin, biomedical HSI has previously predicted wound healing based on SpO_2 values. Principal component analysis is an alternative approach that has not been investigated in the prediction of wound healing. It is therefore of value to investigate whether PCA improves the prediction of wound healing and to compare this with the performance of SpO_2 mapping.

Hyperspectral images from a previous study of 43 patients with wounds were analyzed by using both SpO_2 values and PCA, and the principal finding was that classification of time to healing by 12 weeks based on PCA (sensitivity = 87.5%, specificity = 88.2%) outperformed that using SpO_2 (sensitivity = 50%, specificity = 88.2%). Comparison by receiver operating characteristic (ROC) analysis revealed an area under the curve of 0.88 for PCA, compared with 0.66 using oxygen saturation analysis. Thus, PCA based on the second principal component appeared superior to analysis using SpO_2 values in predicting healing of wounds by 12 weeks based on hyperspectral images taken at baseline.

Although one cannot uniquely map a physical property onto a principal component, it is interesting to consider how physical properties influence PCs. The absorption spectra over the range of interest (430–750 nm) are dominated by oxy- and deoxyhemoglobin. These have broadly similar features, i.e., high absorption in the blue/green region, reducing into the red/near-infrared range. We believe that these features are captured by PC1. Differences in the oxy- and deoxyhemoglobin spectra are then characterized by PC2, which provides discrimination of wound healing with superior performance to that achieved by more widely applied SpO_2 measurement approaches. Due to the dominance of oxy- and deoxyhemoglobin, PCs greater than 2 provide no discriminatory value.

The classification performance obtained in this study is slightly better than that of our earlier publication [9] and is comparable to that obtained by another [7], which reported estimates of sensitivity and specificity of 93% and 86%, respectively, in a rather smaller group of patients. SpO_2 values may still be useful in cases where a hyperspectral camera is not available (as, for example, when making single point measurements using a lower-cost spectrometer-based method or when making measurements with a wound dressing with a fiber optic probe placed adjacent to the wound site). Furthermore, the previous demonstration that SpO_2 values on the top of the foot are well correlated with those on the underside means that precisely locating the fiber optic probe may not be necessary.

Author Contributions: Data curation, Q.Y.; Formal analysis, Q.Y., S.S., W.J.J., and F.L.G.; Funding acquisition, W.J.J.; Investigation, Q.Y., D.J.C., A.M., and S.P.M.; Methodology, W.J.J., D.J.C., F.L.G., and S.P.M.; Project administration, S.P.M.; Resources, S.P.M.; Software, Q.Y. and S.S.; Supervision, S.P.M.; Validation, Q.Y.; Visualization, S.S.; Writing–original draft, Q.Y.; Writing–review and editing, S.S., W.J.J., A.M., and S.P.M.

Funding: Q.Y. was funded by a studentship from the University of Nottingham. The original clinical study was funded by the UK National Institute for Health Research (NEAT/i4i) programme grant FSG027, and Nottingham University Hospitals Trust acted as sponsor.

Acknowledgments: We thank Nada Savic for her contribution to the original analysis and helpful discussion.

Conflicts of Interest: The authors declare no conflict of interest.

References

1. Boulton, A.J.; Vileikyte, L.; Ragnarsson-Tennvall, G.; Apelqvist, J. The global burden of diabetic foot disease. *Lancet* **2005**, *366*, 1719–1724. [CrossRef]
2. Johnson, W.R.; Humayun, M.; Bearman, G.; Wilson, D.W.; Fink, W. Snapshot hyperspectral imaging in ophthalmology. *J. Biomed. Opt.* **2007**, *12*, 014036. [CrossRef]
3. Gao, L.; Smith, R.T.; Tkaczyk, T.S. Snapshot hyperspectral retinal camera with the Image Mapping Spectrometer (IMS). *Biomed. Opt. Express* **2012**, *3*, 48–54. [CrossRef]
4. Mordant, D.J.; Al-Abboud, I.; Muyo, G.; Gorman, A.; Sallam, A.; Rodmell, P.; Crowe, J.; Morgan, S.; Ritchie, P.; Harvey, A.R.; et al. Validation of human whole blood oximetry, using a hyperspectral fundus camera with a model eye. *Investig. Ophthalmol. Vis. Sci.* **2011**, *52*, 2851–2859. [CrossRef] [PubMed]
5. Greenman, R.L.; Panasyuk, S.; Wang, X.; Lyons, T.E.; Dinh, T.; Longoria, L.; Giurini, J.M.; Freeman, J.; Khaodhiar, L.; Veves, A. Early changes in the skin microcirculation and muscle metabolism of the diabetic foot. *Lancet* **2005**, *366*, 1711–1717. [CrossRef]
6. Yudovsky, D.; Nouvong, A.; Schomacker, K.; Pilon, L. Assessing diabetic foot ulcer development risk with hyperspectral tissue oximetry. *J. Biomed. Opt.* **2011**, *16*, 026009. [CrossRef] [PubMed]
7. Khaodhiar, L.; Dinh, T.; Schomacker, K.T.; Panasyuk, S.V.; Freeman, J.E.; Lew, R.; Vo, T.; Panasyuk, A.A.; Lima, C.; Giurini, J.M.; et al. The use of medical hyperspectral technology to evaluate microcirculatory changes in diabetic foot ulcers and to predict clinical outcomes. *Diabetes Care* **2007**, *30*, 903–910. [CrossRef]
8. Nouvong, A.; Hoogwerf, B.; Mohler, E.; Davis, B.; Tajaddini, A.; Medenilla, E. Evaluation of diabetic foot ulcer healing with hyperspectral imaging of oxyhaemoglobin and deoxyhaemoglobin. *Diabetes Care* **2009**, *32*, 2056–2061. [CrossRef]
9. Jeffcoate, W.J.; Clark, D.J.; Savic, N.; Rodmell, P.I.; Hinchliffe, R.J.; Musgrove, A.; Game, F.L. Use of HSI to measure oxygen saturation in the lower limb and its correlation with healing of foot ulcers in diabetes. *Diabet. Med.* **2015**, *32*, 798–802. [CrossRef]
10. Koonsanit, K.; Jaruskulchai, C.; Eiumnoh, A. Band Selection for Hyperspectral Imagery with PCA-MIG. In *Web-Age Information Management*; Springer: Berlin/Heidelberg, Germany, 2012; pp. 119–127.
11. Kelman, T.; Ren, J.; Marshall, S. Effective classification of Chinese tea samples in hyperspectral imaging. *Artif. Intell. Res.* **2013**, *2*, 87. [CrossRef]

12. Gowen, A.A.; O'Donnell, C.P.; Taghizadeh, M.; Cullen, P.J.; Frias, J.M.; Downey, G. Hyperspectral imaging combined with principal component analysis for bruise damage detection on white mushrooms (Agaricus bisporus). *J. Chemom.* **2008**, *22*, 259–267. [CrossRef]
13. Liu, L.; Ngadi, M.O. Computerized pork quality evaluation system. In *Medical Biometrics*; Springer: Berlin/Heidelberg, Germany, 2010; pp. 145–152.
14. Lu, G.; Fei, B. Medical hyperspectral imaging: A review. *J. Biomed. Opt.* **2014**, *19*, 10901. [CrossRef] [PubMed]
15. Zuzak, K.J.; Naik, S.C.; Alexandrakis, G.; Hawkins, D.; Behbehani, K.; Livingston, E.H. Characterization of a Near-Infrared Laparoscopic Hyperspectral Imaging System for Minimally Invasive Surgery. *Anal. Chem.* **2007**, *79*, 4709–4715. [CrossRef] [PubMed]
16. Gerstner, A.O.H.; Laffers, W.; Bootz, F.; Farkas, D.L.; Martin, R.; Bendix, J.; Thies, B. Hyperspectral imaging of mucosal surfaces in patients. *J. Biophotonics* **2012**, *5*, 255–262. [CrossRef] [PubMed]
17. Rajpoot, K.M.; Rajpoot, N.M. Wavelet based segmentation of hyperspectral colon tissue imagery. In Proceedings of the 7th International Multi Topic Conference, Islamabad, Pakistan, 8–9 December 2003; pp. 38–43.
18. Rodmell, P.I. A Novel Oximeter. Ph.D. Thesis, University of Nottingham, Nottingham, UK, 2005.
19. Lübbers, D.W.; Wodick, R. The examination of multicomponent systems in biological materials by means of a rapid scanning photometer. *Appl. Opt.* **1969**, *8*, 1055–1062. [CrossRef] [PubMed]
20. Stockford, I.M.; Lu, B.; Crowe, J.A.; Morgan, S.P.; Morris, D.E. Reduction of Error in spectrophotometry of scattering media using polarization techniques. *Appl. Spectrosc.* **2007**, *61*, 1379–1389. [CrossRef] [PubMed]
21. Loyalka, S.; Riggs, C. Inverse problem in diffuse reflectance spectroscopy: Accuracy of the Kubelka-Munk equations. *Appl. Spectrosc.* **1995**, *49*, 1107–1110. [CrossRef]
22. Arridge, S.R.; Cope, M.; Delpy, D. The theoretical basis for the determination of optical pathlengths in tissue: Temporal and frequency analysis. *Phys. Med. Biol.* **1992**, *37*, 1531. [CrossRef]
23. Lagarias, J.C.; Wright, J.A.R.M.H.; Wright, P.E. Convergence properties of the nelder-mead simplex method in low dimensions. *SIAM J. Optim.* **1998**, *9*, 112–147. [CrossRef]
24. Gonzales, R.C.; Woods, R.E. *Digital Image Processing*, 2nd ed.; Prentice Hall: Upper Saddle River, NJ, USA, 2002.
25. Fluss, R.; Farragi, D.; Resiser, B. Estimation of the Youden Index and its associated cutoff point. *Biom. J.* **2005**, *47*, 458–472. [CrossRef]
26. Fawcett, T. An introduction to ROC analysis. *Pattern Recognit. Lett.* **2006**, *27*, 861–874. [CrossRef]

Journal of
Imaging

MDPI

Article

Age Determination of Blood-Stained Fingerprints Using Visible Wavelength Reflectance Hyperspectral Imaging

Samuel Cadd, Bo Li, Peter Beveridge, William T. O'Hare and Meez Islam *

School of Science Engineering and Design, Teesside University, Borough Road, Middlesbrough TS1 3BA, UK;
cadd.sam@gmail.com (S.C.); bo.li@emr.ac.uk (B.L.); p.beveridge@tees.ac.uk (P.B.);
w.t.o.hare@gmail.com (W.T.O.)
* Correspondence: m.islam@tees.ac.uk; Tel.: +44-16-4234-2410

Received: 21 October 2018; Accepted: 27 November 2018; Published: 29 November 2018

Abstract: The ability to establish the exact time a crime was committed is one of the fundamental aims of forensic science. The analysis of recovered evidence can provide information to assist in age determination, such as blood, which is one of the most commonly encountered types of biological evidence and the most common fingerprint contaminant. There are currently no accepted methods to establish the age of a blood-stained fingerprint, so progress in this area would be of considerable benefit for forensic investigations. A novel application of visible wavelength reflectance, hyperspectral imaging (HSI), is used for the detection and age determination of blood-stained fingerprints on white ceramic tiles. Both identification and age determination are based on the unique visible absorption spectrum of haemoglobin between 400 and 680 nm and the presence of the Soret peak at 415 nm. In this study, blood-stained fingerprints were aged over 30 days and analysed using HSI. False colour aging scales were produced from a 30-day scale and a 24 h scale, allowing for a clear visual method for age estimations for deposited blood-stained fingerprints. Nine blood-stained fingerprints of varying ages deposited on one white ceramic tile were easily distinguishable using the 30-day false colour scale.

Keywords: fingerprints; blood detection; age determination; hyperspectral imaging

1. Introduction

The reliable and accurate determination of when a crime was committed is one of the fundamental aims of forensic research. The analysis of recovered evidence from a crime scene can provide information to assist in this determination. At violent crime scenes, blood is one of the most commonly encountered types of biological evidence [1] and is the most common fingerprint contaminant [2]. The initial objective when dealing with suspected blood evidence is to conclusively establish that the substance is actually blood [3]. Dark substrates can pose considerable problems, due to the low contrast between the substrate and the fingerprint, due to the high amount of incident light absorbed by the surface [3]. Other colours or patterns that are particularly similar to the stain can also cause issues for identification through visual examination alone. Presumptive tests are therefore used as part of the current forensic workflow to indicate the presence of blood [2]. Despite a high sensitivity to blood, these wet chemical tests are not specific to blood and can generate false positives [2]. Wet chemical testing can also contaminate the stain, potentially having a detrimental effect on subsequent DNA analysis [4]. Previous and current research has therefore focused on the development of alternate methods for the non-destructive identification of blood [1,5–13].

The ability to establish the age of a fingerprint is a highly relevant factor in criminal investigations [14]. Convictions can largely depend on the ability to show whether a fingerprint was deposited at the time a crime was committed or from a previous legitimate visit, as is often claimed by the defence team [15,16]. There are currently no accepted analytical methods for reliably establishing a time frame when a fingerprint was deposited, and speculation around age is subject to considerable error [17]. This is primarily due to the unreliability of previously proposed methods [14]. Successful identification of blood and estimation of the age of a blood-stained fingerprint could provide the first indication to investigators as to when a crime was committed [18]. This could be especially beneficial if a blood-stained fingerprint is the only evidence available.

Early research made small steps forward in scientific knowledge, including exploring the solubility of blood in water over time, as the solubility rapidly drops over time before decreasing more slowly [19]. Other techniques have also furthered the understanding of the effect of time on blood stains, including the use of oxygen electrodes to establish the changes that occur to the oxyhaemoglobin–haemoglobin ratio in blood stains [20]. Other methods have also been explored, including the analysis of RNA degradation to establish the age of blood stains [21–23]. Spectroscopic methods were explored over the past decade, including atomic force microscopy, which was used to explore the elasticity of blood stains on glass slides through coagulation over time [11]. A clear increase in the stiffness of blood samples over time was identified, although several limitations to the method were identified. Research exploring electron spin resonance spectroscopy established a relationship between the electron paramagnetic resonance (EPR) of ferric non-heme species and the number of days from bleeding [24]. An error range within 25% of the actual number of days was obtained under controlled conditions, but environmental factors, such as light exposure and temperature, had effects on the analysis of EPR-active compounds.

It has been established that the colour of a blood stain changes from red to brown over time [6]. This indicates that optical methods could be used to quantify the colour of blood stains. This was first explored using the reflectance spectra of blood stains, whereby the effects of environmental variables on the colour of the blood stain were recognised [25]. Further research quantified absorption bands independent of the amount of blood present as a possible approach for age determination [26] or for the use of a small spectral window [27].

Previous research has clearly established changes to the physical and chemical properties of blood over time [6]. The optimum technique will require a high selectivity to blood; a high level of sensitivity, even with diluted blood; and a high level of precision to determine the age of a blood stain or blood-stained fingerprint in practice. One method is the use of visible wavelength hyperspectral imaging. This was first reported for the detection and age determination of horse blood stains between 442 and 585 nm as proof of concept research [28]. The determination of age was obtained through linear discriminant analysis from data based on the progressive changes in the absorption spectra over time as the composition of the blood stain altered. This approach used training and test datasets from the same blood stain in order to determine the age with a high level of accuracy. With different blood stains, the accuracy was considerably lower, although this research demonstrated the potential of the method to establish an age estimation non-destructively. A similar method by a different research group also successfully demonstrated the identification of blood stains [29,30]. The proposed method allowed for rapid, non-destructive presumptive blood stain detection. Other research has explored forensic traces across a range of substrates [31]. Most recently, a new blood stain identification and age determination approach was proposed based on the Soret γ band absorption in haemoglobin [3,32] and indicated a higher sensitivity and specificity for the detection and identification of blood stains over previously proposed methods [3].

Previous research has identified the need for a non-contact and non-destructive method for the determination of the age of a blood-stained fingerprint. An ideal method should function across a practical range of ages of blood, have a high specificity to blood so as to prevent false positives, and have a clear and accurate method for determining the age of a blood-stained fingerprint, so as to allow for reliable age estimations.

The visible wavelength hyperspectral imaging method proposed in this paper meets all of these requirements. In this study, we present a novel application of visible wavelength hyperspectral imaging

(HSI) based on the absorption spectrum of haemoglobin between 400 and 680 nm for the non-contact, non-destructive detection, identification and age determination of blood-stained fingerprints on white tiles. False colour scales are presented based on the ratio of the 525 nm peak to the trough at 550 nm in the absorption spectrum. The 30-day colour scale is demonstrated with nine blood-stained fingerprints deposited on a single white tile to demonstrate the effectiveness of such a method for age determination.

This work follows on from [33,34] where hyperspectral imaging was used for the first time to detect, identify, and visualise ridge detail in blood-stained fingerprints across a wide range of substrates. The research presented in this paper again demonstrates the potential of HSI, through successful non-destructive detection, identification, and age determination of blood-stained fingerprints.

2. Materials and Methods

2.1. Production of Blood-Stained Fingerprints

Horse blood was used to produce blood-stained fingerprints in this study and was deposited into a Petri dish containing a small sponge. The right middle finger was pressed against the sponge to evenly coat the digit, and the blood-stained fingerprint was then deposited onto the white tile. The fingerprints were left to age under controlled conditions in an environmental chamber at 23–24 °C (Qualicool LR202, LTE Scientific, Oldham, UK) between analyses.

2.2. HSI System

The HSI system used in this study had the same setup as that detailed in [3,33,34], consisting of a liquid crystal tuneable filter (LCTF) coupled to a 2.3 megapixel Point Grey camera and a light source for scene illumination. The light source was comprised of two 40 W LEDs—one violet giving an output at 410 nm and one white giving an output between 450 and 700 nm. Control of the LCTF and image capture was performed using custom developed software written in C++ (Microsoft, Redmond, WA, USA). Images were captured between 400 and 680 nm with spectral sub-sampling at 5 nm intervals, resulting in an image cube at 56 wavelengths for each scan. Spectra from the image cube were subsequently analysed using custom routines developed in Visual Studio (Microsoft, USA) and Spyder (Python, Wilmington, DE, USA). The time required to acquire and process each image was approximately 30 s.

2.3. Hyperspectral Reflectance Image Acquisition and Pre-Processing

The hyperspectral reflectance measurements were made following the method detailed in [3] and [33,34]. A reference image (R_0) was obtained using a blank ceramic tile. This image was recorded in a 5 nm series of 56 discrete wavelengths between 400 and 680 nm. The sample image (R_s) was recorded at the same wavelengths under the same illumination conditions and integration time settings on the camera. The hyperspectral reflectance image (R) consisted of a data cube of 1280×1024 pixel values at 56 discrete wavelengths.

2.4. Criteria for the Identification of Blood Stains

The blood reflectance spectrum in the visible region is dominated by the presence of haemoglobin [28,32]. The spectrum contains a strong narrow absorption at 415 nm called the Soret or γ band with two weaker and broader absorptions between 500 and 600 nm known as the β and α bands [3]. The Soret band results in the distinctive red colour of blood, due to the absorption in the blue part of the visible spectrum. Other red substances also absorb between 400 and 680 nm in the blue region, although these absorption features tend to not be centred at 415 nm and are much broader. This is the basis of the methodology used to identify and discriminate blood stains from other similarly coloured substances. Further information is detailed in [3]. From the reflectance images obtained, the pixels which satisfied the criteria were marked in black, whilst all other pixels were marked in white.

This allowed regions of the image where the blood-stained fingerprint was present to be identified, as well as clear distinction of the ridge detail.

2.5. Age Determination Methodology for Blood-Stained Fingerprints

The age of the blood-stained fingerprints was explored through the effect of time on the composition of the fingerprint. This has already been established as the only potentially viable method for age determination, due to the numerous variables that affect fingerprint composition over time [14]. This research explored the compositional changes that occur within a blood-stained fingerprint using hyperspectral imaging. After deposition of a blood-stained fingerprint, specific chemical changes occur which result in a colour change from bright red to dark brown. This is attributed to the complete oxidation of haemoglobin (Hb) to oxy-haemoglobin (HbO_2), which then auto-oxidises to met-haemoglobin (met-Hb) and denatures to hemichrome (HC) [6,35], as shown in Figure 1. As this process occurs, the concentration of haemoglobin decreases, which can be observed in the visible spectrum through the decrease of the Soret band at 415 nm, as shown in Figure 2.

After the HSI analysis, the absorption spectra of the blood-stained fingerprints were analysed, and the ratio of the peak at 525 nm to the trough at 550 nm was determined. This ratio was used to produce false colour Red-Green-Blue aging scales, as shown with ten false coloured fingerprints in Figure 3, with the values determining the values assigned to red, green, and blue in the image.

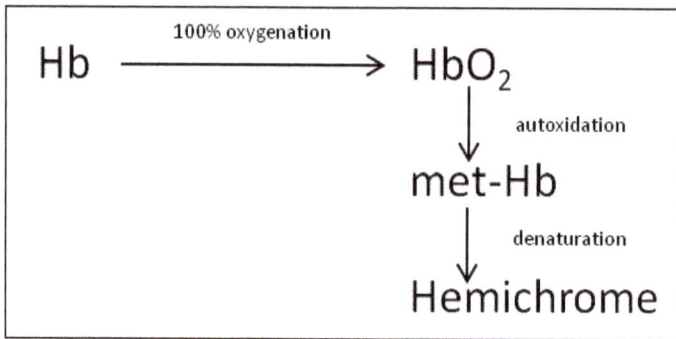

Figure 1. Reaction of haemoglobin in blood stains.

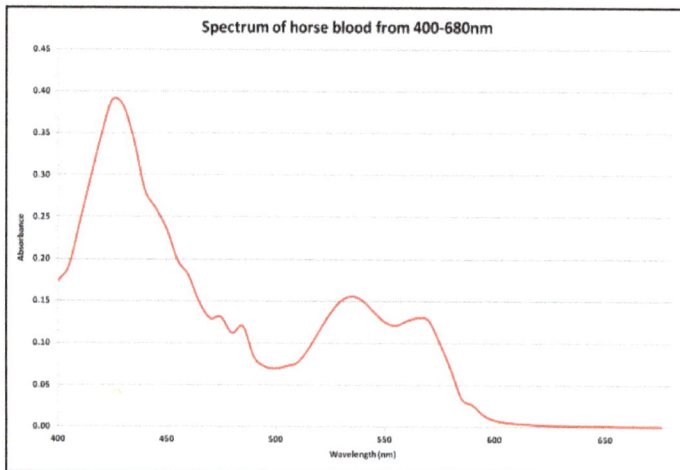

Figure 2. Spectrum of blood from 400–680 nm.

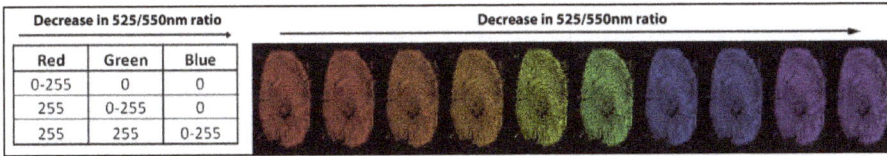

Figure 3. Use of Red-Green-Blue colours to produce false colour scales based on the 525/550 nm ratio.

2.6. Digital Single Lens Reflex (DSLR) Setup

The images used in this report were taken using a digital single lens reflex (DSLR) camera mounted on a Kaiser RS1 copy stand. The DSLR was a Canon EOS 700D which was fitted with a Canon Angle Finder C 90° viewfinder with a 1.25–2.5× optical magnification and a Canon TC-80N3 remote control external shutter release to avoid camera motion. Images were taken using two sizes of macro-lenses—a 50 mm lens for overview shots of the substrates and a 100 mm lens for high magnification macro-shots of individual fingerprints. The 50 mm lens was used as it is recognised as being generally equivalent to the view seen by the human eye [36]. The lenses used were a Canon EF 50 mm f2.5 macro lens and a Canon EF 100 mm f2.8L macro IS USM lens. Substrates were lit using oblique lighting from two Daylight Twist Portable Lamps with a white light output of 6500 K.

2.7. Age Estimation Intervals

2.7.1. Aging of Blood-Stained Fingerprints over 30 Days

Deposited blood fingerprints were aged under controlled conditions in an environmental chamber at 23–24 °C, as detailed in Section 2.1. Analyses were carried out daily for thirty days and false colour aging scales were produced after 7, 16, 24, and 30 days.

2.7.2. Aging of Blood-Stained Fingerprints over 24 h

Deposited blood fingerprints were also analysed over a 24 h period. Blood-stained fingerprints were deposited and analysed every hour for 12 h. A second set of fingerprints was then deposited and left overnight for 12 h. This set was then analysed from 12 h to 24 h. The results from both sets were combined to produce the 24 h aging scale.

2.7.3. Age Estimations of Nine Blood-Stained Fingerprints

Nine blood-stained fingerprints were deposited between 0 to 28 days onto one white tile, as shown in Figure 4. These were deposited in a random arrangement on the white tile. The fingerprints were aged under controlled conditions in an environmental chamber at 23–24 °C, as detailed in Section 2.1. Analyses were carried out using his, and the results were manually coloured using the 30-day false colour scale for a visual representation of the ages.

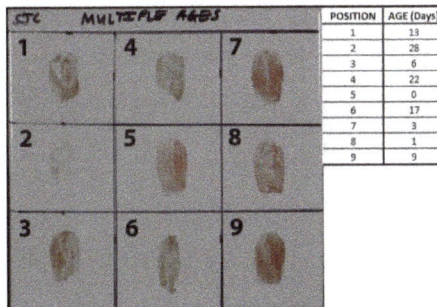

Figure 4. Nine blood-stained fingerprints from 0 to 28 days old.

3. Results

3.1. *Aging of Blood-Stained Fingerprints over 30 Days*

Blood-stained fingerprints were successfully detected and identified using hyperspectral imaging for the full 30 days explored. Clear ridge detail was identified for all scans, a selection of which are shown in seven day increments from 0 to 28 days in Figure 5. The level of clear ridge detail observable even after 30 days demonstrates the advantage of HSI over existing chemical methods, as not only can blood be conclusively identified, as opposed to an indication as occurs with presumptive tests, but ridge detail is preserved and photographed for potential comparison in one step.

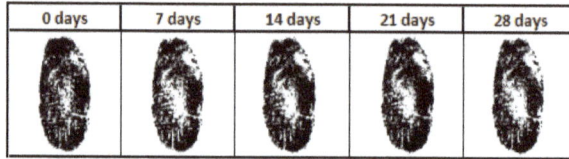

Figure 5. Visible ridge detail for blood-stained fingerprints with hyperspectral imaging (HSI) on day 0, 7, 14, 21, and 28.

The analysis of the absorption spectrum between 400 and 680 nm showed a clear decrease in the Soret band and the β and α bands between 500 and 600 nm over the aging period. This change forms the basis for the age estimation methodology. False colour scales were produced to represent the changes in the 525/550 nm ratio over 30 days, as shown in Figure 6.

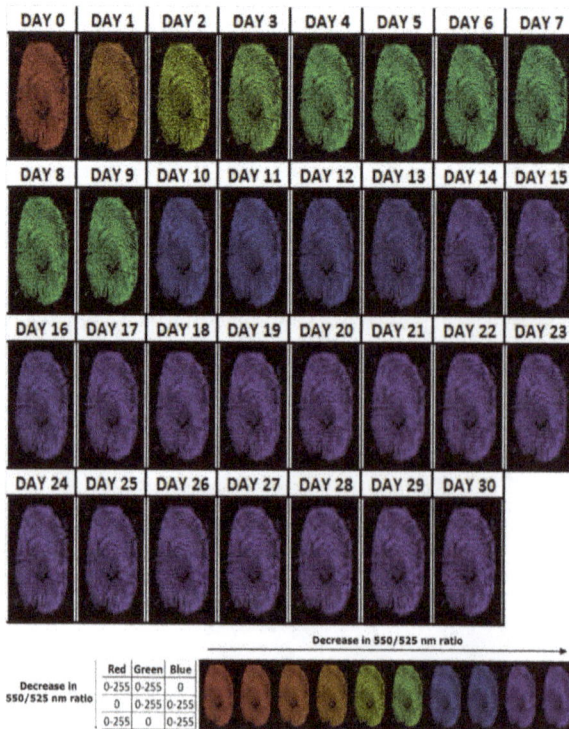

Figure 6. 30-day false colour aging scale.

3.2. Aging of Blood-Stained Fingerprints over 24 h

Blood-stained fingerprints were successfully detected using hyperspectral imaging for the full 24 h explored. Clear ridge detail was identified for all scans and the absorption spectrum was analysed to produce a false colour scale, as shown in Figure 7. This scale represents the changes that occurred in the absorption spectrum over the 24 h aging period, as shown in Figure 8. The logarithmic conversion also shown demonstrates the clear relationship between the 525/550 nm ratio and time.

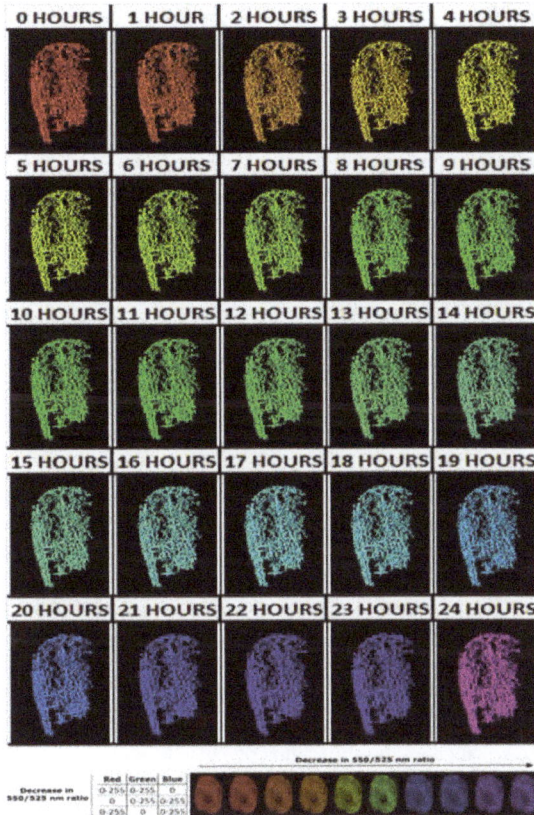

Figure 7. 24 h false colour aging scale.

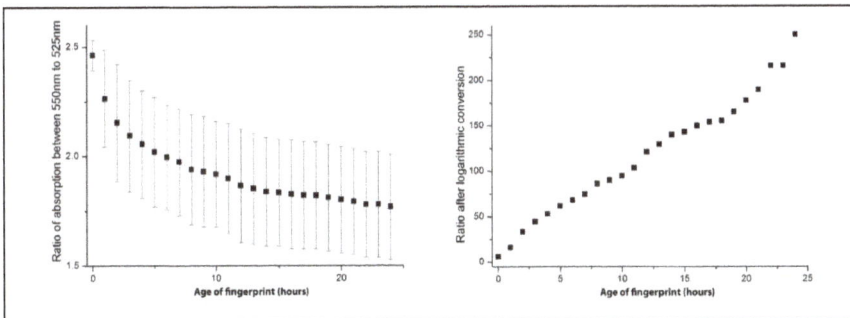

Figure 8. Effect of time on the absorption ratio of 550 and 525 nm (**left**) and the logarithmic conversion (**right**) over 24 h.

3.3. Age Estimations of Nine Blood-Stained Fingerprints

From a visual examination of the nine blood stains alone, it was very difficult to determine any significant differences that may allow for age estimations. The analysis carried out using HSI successfully detected and conclusively identified the blood ridge detail. Using the data obtained from the HSI analysis and the false colour scales produced, a clear visual representation of the different ages of the blood-stained fingerprints was produced, as shown in Figure 9. The DSLR images show minimal variation between the fingerprints. Using the false colour scales, all recently deposited fingerprints could be easily distinguished due to the significant differences in colour, such as fingerprints 5, 8, 7, and 3, which correspond to 0, 1, 3, and 6 days respectively. After 14 days, the variation in the composition of the blood-stained fingerprint was less, so the difference between the colours was smaller and harder to distinguish by eye. This was apparent for fingerprints 2, 4, and 6, which were all shades of purple, despite varying by eleven days. The use of this false colour method for age estimations is therefore most effective for blood-stained fingerprints deposited within 14 days, as the increased variation over the first seven days results in significant differences in the false colour images produced.

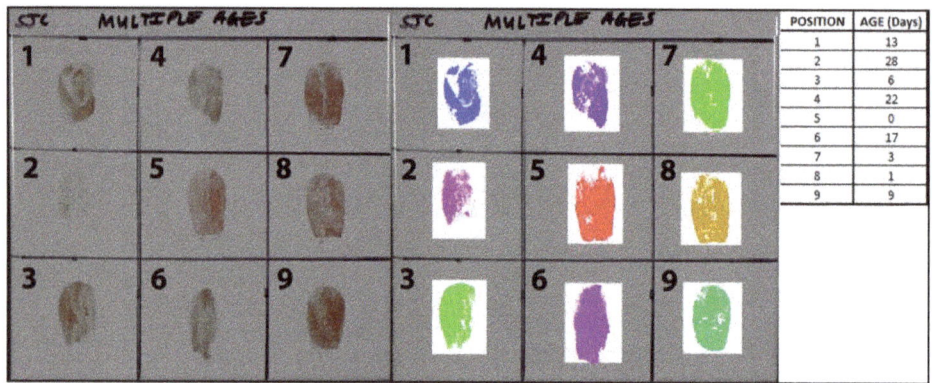

Figure 9. Digital single lens reflex (DSLR) (**left**) and false colour images manually coloured from HSI analysis (**right**) of blood-stained fingerprints based on a 30-day false colour scale.

4. Discussion

Results have been presented on the application of a visible wavelength reflectance hyperspectral imaging system for the detection of blood-stained fingerprints and the determination of the fingerprint age on white tiles. The method used for the detection and identification of blood as well as the age determination was based on use of the main narrow absorption peak in blood at approximately 415 nm (Soret band), and the ratio of the peak at 525 nm to the trough at 550 nm was also determined. This was used to produce a false colour aging scale over 30 days. A second false colour scale was also produced over 24 h, allowing for highly accurate estimations of the age of freshly deposited blood-stained fingerprints.

Nine blood stains of varying ages deposited on one tile were able to be distinguished using the 30-day false colour scale. With visual examination alone, there were minimal differences between deposited fingerprints in the DSLR image, making it very difficult to establish any significant differences for age estimations. The analysis carried out using HSI successfully detected and identified the ridge detail. The analysis using the 30-day false colour scale established a clear visual representation of the different ages of the blood-stained fingerprints. Recently deposited fingerprints were easier to distinguish due to the greater differences in colour. After 14 days, this variation reduced, due to the slower changes in the blood composition, and the colours were slightly more difficult to distinguish. Previous research has demonstrated the success of using hyperspectral imaging as a method for establishing the age of blood stains, although a clear visual approach using false colour scales has not been carried out. The approach presented in this study demonstrates the considerable potential of HSI to both conclusively identify a blood-stained fingerprint and to determine its age.

5. Conclusions

A novel application of visible wavelength reflectance hyperspectral imaging (HSI) was used for both the detection and age determination of blood-stained fingerprints on white ceramic tiles. Both the identification of blood and the age determination were based on the unique visible absorption spectrum of haemoglobin between 400 and 680 nm and the presence of the Soret peak at 415 nm. In the processed hyperspectral images, pixels where blood was identified were coloured black, whilst all other pixels were coloured white, thus enhancing the location of ridge detail in blood fingerprints.

Blood-stained fingerprints were aged over 30 days and analysed using HSI. From these results, a 30-day scale and a 24 h scale were produced, allowing for a clear visual method for age estimations of deposited blood-stained fingerprints. This was demonstrated with nine blood-stained fingerprints deposited on one white tile. From a visual examination using DSLR, no significant differences between the deposited fingerprints could be identified to allow for age estimations. The application of HSI demonstrated several advantages, as the analysis successfully detected and identified the blood ridge detail. Additionally, the application of the 30-day false colour scale allowed for the deposited blood-stained fingerprints to be coloured corresponding to their age. This identified clear distinctions between the nine blood-stained fingerprints in the false colour image produced, with the greatest differences in colour occurring among the most recently deposited fingerprints as the greatest differences in composition occur over the first 24 h after deposition [37,38].

Overall, HSI has significant benefits for both the detection and age determination of blood-stained fingerprints and blood stains. Preliminary work exploring both human and horse blood demonstrated minimal differences, indicating that the findings of this research are applicable to crime scenes involving human blood. Further work is required, however, to confirm that the age determination methodology demonstrated here is effective across both blood stains and blood-stained fingerprints with both human and horse blood. Large scale blind tests are also required to establish the effectiveness and reliability of the age determination method on white tiles, as well as other substrates, including other colours and porosities, and as a function of environmental variables, such as temperature and humidity. Previous work with this setup has already demonstrated the successful detection and identification of blood-stained fingerprints on a range of substrates [33,34]. A full comparison of the technique against existing chemical enhancement methods would be beneficial to allow a comparison of the sensitivity of the setups. Development of a more rugged instrument could allow for the production of a robust portable device for use at crime scenes, which would be particularly beneficial for criminal investigations. HSI could then be used for the detection and identification of both blood stains and blood-stained fingerprints, as well as for the reliable establishment of the age of a stain or fingerprint.

Author Contributions: S.C. and B.L. conducted the research presented in this study. M.I. and S.C. wrote the paper. M.I., W.T.O., P.B. and A.C. contributed to the development of the overall research design, provided guidance along the way and aided in the writing of the paper.

Funding: This research was partly funded by the Home Office through a SBRI phase 1 grant. S.C. would also like to acknowledge the Forensic Science Society, who kindly provided financial support through a Research Scholarship.

Acknowledgments: We would to thanks Steve Bleay at the Home Office, Centre for Applied Science and Technology (CAST).

Conflicts of Interest: The authors declare no conflict of interest.

References

1. Finnis, J.; Lewis, J.; Davidson, A. Comparison of methods for visualizing blood on dark surfaces. *Sci. Justice* **2013**, *53*, 178–186. [CrossRef] [PubMed]
2. Home Office CAST. *Fingerprint Sourcebook, Chapter 3, 3.1 Acid Dyes*, 1st ed.; Home Office: London, UK, 2013.

3. Li, B.; Beveridge, P.; O'Hare, W.T.; Islam, M. The application of visible wavelength reflectance hyperspectral imaging for the detection and identification of blood stains. *Sci. Justice* **2014**, *54*, 432–438. [CrossRef] [PubMed]

4. Passi, N.; Garg, R.K.; Yadav, M.; Singh, R.S.; Kharoshah, M.A. Effect of luminol and bleaching agent on the serological and DNA analysis from bloodstain. *Egypt. J. Forensic Sci.* **2012**, *2*, 54–61. [CrossRef]

5. Anderson, S.; Howard, B.; Hobbs, G.R.; Bishop, C.P. A method for determining the age of a bloodstain. *Forensic Sci. Int.* **2005**, *148*, 37–45. [CrossRef] [PubMed]

6. Bremmer, R.H.; Nadort, A.; van Leeuwen, T.G.; van Gemert, M.J.C.; Aalders, M.C.G. Age estimation of blood stains by hemoglobin derivative determination using reflectance spectroscopy. *Forensic Sci. Int.* **2011**, *206*, 166–171. [CrossRef] [PubMed]

7. de Wael, K.; Lepot, L.; Gason, F.; Gilbert, B. In search of blood—Detection of minute particles using spectroscopic methods. *Forensic Sci. Int.* **2008**, *180*, 37–42. [CrossRef] [PubMed]

8. Lin, A.C.; Hsieh, H.S.; Li, T.; Linacre, A.; Lee, J.C. Forensic Applications of Infrared Imaging for the Detection and Recording of Latent Evidence. *J. Forensic Sci.* **2007**, *52*, 1148–1150. [CrossRef] [PubMed]

9. McLaughlin, G.; Sikirzhytski, V.; Lednev, I.K. Circumventing substrate interference in the Raman spectroscopic identification of blood stains. *Forensic Sci. Int.* **2013**, *231*, 157–166. [CrossRef] [PubMed]

10. Stoilovic, M. Detection of semen and blood stains using polilight as a light source. *Forensic Sci. Int.* **1991**, *51*, 289–296. [CrossRef]

11. Strasser, S.; Zink, A.; Kada, G.; Hinterdorfer, P.; Peschel, O.; Heckl, W.M.; Nerlich, A.G.; Thalhammer, S. Age determination of blood spots in forensic medicine by force spectroscopy. *Forensic Sci. Int.* **2007**, *170*, 8–14. [CrossRef] [PubMed]

12. Turrina, S.; Filippini, G.; Atzei, R.; Zaglia, E.; de Leo, D. Validation studies of rapid stain identification-blood (RSID-blood) kit in forensic caseworks. *Forensic Sci. Int. Genet. Suppl. Ser.* **2008**, *1*, 74–75. [CrossRef]

13. Wawryk, J.; Odell, M. Fluorescent identification of biological and other stains on skin by the use of alternative light sources. *J. Clin. Forensic Med.* **2005**, *12*, 296–301. [CrossRef] [PubMed]

14. Cadd, S.; Islam, M.; Manson, P.; Bleay, S. Fingerprint composition and aging: A literature review. *Sci. Justice* **2015**. [CrossRef] [PubMed]

15. Gardner, T.; Anderson, T. *Criminal Evidence: Principles and Cases*, 7th ed.; Cengage Learning: Belmont, CA, USA, 2009.

16. Adebsi, S. Fingerprint Studies—The Recent Challenges And Advancements: A Literary View. *Internet J. Boil. Anthr.* **2008**, *2*, 1–9.

17. Midkiff, C. Lifetime of a Latent Print How Long? Can You Tell? *J. Forensic Identif.* **1993**, *43*, 386–396.

18. Bremmer, R.H.; de Bruin, K.G.; van Gemert, M.J.C.; van Leeuwen, T.G.; Aalders, M.C.G. Forensic quest for age determination of bloodstains. *Forensic Sci. Int.* **2012**, *216*, 1–11. [CrossRef] [PubMed]

19. Schwarzacher, D. Determination of the age of bloodstains. *Am. J. Police Sci.* **1930**. [CrossRef]

20. Matsuoka, T.; Taguchi, T.; Okuda, J. Estimation of bloodstain age by rapid determinations of oxyhemoglobin by use of oxygen electrode and total hemoglobin. *Boil. Pharm. Bull.* **1995**, *18*, 1031–1035. [CrossRef]

21. Bauer, M.; Polzin, S.; Patzelt, D. Quantification of RNA degradation by semi-quantitative duplex and competitive RT-PCR: A possible indicator of the age of bloodstains? *Forensic Sci. Int.* **2003**, *138*, 94–103. [CrossRef] [PubMed]

22. Bauer, M. RNA in forensic science. *Forensic Sci. Int. Genet.* **2007**, *1*, 69–74. [CrossRef] [PubMed]

23. Virkler, K.; Lednev, I.K. Analysis of body fluids for forensic purposes: From laboratory testing to non-destructive rapid confirmatory identification at a crime scene. *Forensic Sci. Int.* **2009**, *188*, 1–17. [CrossRef] [PubMed]

24. Fujita, Y.; Tsuchiya, K.; Abe, S.; Takiguchi, Y.; Kubo, S.; Sakurai, H. Estimation of the age of human bloodstains by electron paramagnetic resonance spectroscopy: Long-term controlled experiment on the effects of environmental factors. *Forensic Sci. Int.* **2005**, *152*, 39–43. [CrossRef] [PubMed]

25. Patterson, D. Use of reflectance measurements in assessing the colour changes of ageing bloodstains. *Nature* **1960**, *20*, 688–689. [CrossRef]

26. Kind, S.S.; Patterson, D.; Owen, G.W. Estimation of the age of dried blood stains by a spectrophotometric method. *Forensic Sci.* **1972**, *1*, 27–54. [CrossRef]

27. Blazek, V.; Lins, G. Spectroscopic age determination of blood stains: New technical aspects. *Acta Med. Leg. Et Soc.* **1982**, *32*, 613–616.

28. Li, B.; Beveridge, P.; O'Hare, W.T.; Islam, M. The estimation of the age of a blood stain using reflectance spectroscopy with a microspectrophotometer, spectral pre-processing and linear discriminant analysis. *Forensic Sci. Int.* **2011**, *212*, 198–204. [CrossRef] [PubMed]

29. Janchaysang, S.; Sumriddetchkajorn, S.; Buranasiri, P. Tunable filter-based multispectral imaging for detection of blood stains on construction material substrates Part 1: Developing blood stain discrimination criteria. *Appl. Opt.* **2012**, *51*, 6984–6996. [CrossRef] [PubMed]

30. Janchaysang, S.; Sumriddetchkajorn, S.; Buranasiri, P. Tunable filter-based multispectral imaging for detection of blood stains on construction material substrates Part 2: Realization of rapid blood stain detection. *Appl. Opt.* **2013**, *52*, 4898–4910. [CrossRef] [PubMed]

31. Edelman, G.J.; Gaston, E.; van Leeuwen, T.G.; Cullen, P.J.; Aalders, M.C.G. Hyperspectral imaging for non-contact analysis of forensic traces. *Forensic Sci. Int.* **2012**, *223*, 28–39. [CrossRef] [PubMed]

32. Li, B.; Beveridge, P.; O'Hare, W.T.; Islam, M. The age estimation of blood stains up to 30 days old using visible wavelength hyperspectral image analysis and linear discriminant analysis. *Sci. Justice* **2013**, *53*, 270–277. [CrossRef] [PubMed]

33. Cadd, S.; Li, B.; Beveridge, P.; O'Hare, W.T.; Campbell, A.; Islam, M. Non-contact detection and identification of blood stained fingerprints using visible wavelength reflectance hyperspectral imaging: Part 1. *Sci. Justice* **2016**, *56*, 181–190. [CrossRef] [PubMed]

34. Cadd, S.; Li, B.; Beveridge, P.; O'Hare, W.T.; Campbell, A.; Islam, M. Non-contact detection and identification of blood stained fingerprints using visible wavelength reflectance hyperspectral imaging: Part II effectiveness on a range of substrates. *Sci. Justice* **2016**, *56*, 191–200. [CrossRef] [PubMed]

35. Bremmer, R.H.; de Bruin, D.M.; de Joode, M.; jan Buma, W.; van Leeuwen, T.G.; Aalders, M.C.G. Biphasic oxidation of oxy-hemoglobin in bloodstains. *PLoS ONE* **2011**, *6*, e21845. [CrossRef] [PubMed]

36. Langford, M.; Fox, A.; Smith, R.S. *Langford's Basic Photography*, 8th ed.; Focal Press: Oxford, UK, 2009.

37. Cadd, S.; Bleay, S.; Sears, V. Evaluation of the solvent black 3 fingermark enhancement reagent: Part 2—Investigation of the optimum formulation and application parameters. *Sci. Justice* **2013**, *53*, 131–143. [CrossRef] [PubMed]

38. Archer, N.; Charles, Y.; Elliott, J.; Jickells, S. Changes in the lipid composition of latent fingerprint residue with time after deposition on a surface. *Forensic Sci. Int.* **2005**, *154*, 224–239. [CrossRef] [PubMed]

Journal of
Imaging

MDPI

Article

Hyperspectral Imaging Using Laser Excitation for Fast Raman and Fluorescence Hyperspectral Imaging for Sorting and Quality Control Applications

Florian Gruber [1,*], **Philipp Wollmann** [2], **Wulf Grählert** [1] and **Stefan Kaskel** [1]

[1] Department Chemical Surface and Reaction Technology, Fraunhofer Institute for Material and Beam Technology (IWS) Dresden, Winterbergstr. 28, 01277 Dresden, Germany; wulf.graehlert@iws.fraunhofer.de (W.G.); stefan.kaskel@iws.fraunhofer.de (S.K.)

[2] Chair for Electrochemistry, Technische Universitaet Dresden, Zellescher Weg 19, 01069 Dresden, Germany; philipp.wollmann@tu-dresden.de

* Correspondence: florian.gruber@iws.fraunhofer.de; Tel.: +49-351-83391-3721

Received: 24 August 2018; Accepted: 19 September 2018; Published: 21 September 2018

Abstract: A hyperspectral measurement system for the fast and large area measurement of Raman and fluorescence signals was developed, characterized and tested. This laser hyperspectral imaging system (Laser-HSI) can be used for sorting tasks and for continuous quality monitoring. The system uses a 532 nm Nd:YAG laser and a standard pushbroom HSI camera. Depending on the lens selected, it is possible to cover large areas (e.g., field of view (FOV) = 386 mm) or to achieve high spatial resolutions (e.g., 0.02 mm). The developed Laser-HSI was used for four exemplary experiments: (a) the measurement and classification of a mixture of sulphur and naphthalene; (b) the measurement of carotenoid distribution in a carrot slice; (c) the classification of black polymer particles; and, (d) the localization of impurities on a lead zirconate titanate (PZT) piezoelectric actuator. It could be shown that the measurement data obtained were in good agreement with reference measurements taken with a high-resolution Raman microscope. Furthermore, the suitability of the measurements for classification using machine learning algorithms was also demonstrated. The developed Laser-HSI could be used in the future for complex quality control or sorting tasks where conventional HSI systems fail.

Keywords: hyperspectral imaging; Raman; fluorescence; sorting; quality control; black polymers; PZT; classification; machine learning

1. Introduction

Hyperspectral imaging (HSI) technologies are increasingly being used in the fields of remote sensing [1], agriculture [2,3], pharmaceuticals [4,5], and medicine [6,7]. Example uses include analyzing nitrogen and water stress in wheat fields [8], measuring the coating thickness of tablets [9], and the in vivo detection of cancer [10].

For industrial applications, hyperspectral imaging methods are mainly used in the ultraviolet (UV), visible (VIS) and near-infrared (NIR) spectral ranges, because they are reasonably inexpensive, readily available, and ideally suited and proven for on-line or in-line quality control [11,12]. These methods allow a fast and extensive inspection of samples or continuous processes. Common HSI methods use broadband lighting for a sufficient excitation in the whole spectral range of interest, while other methods, such as laser-induced fluorescence (LIF) and Raman his, are not very common in the field of macroscopic HSI, although they are widely used for microscopic imaging [13–15].

LIF is a spectroscopic technology with high sensitivity, a wide dynamic range, and low detection limits [16]. It can be utilized, for example, for live cell microscopy of cells, using the high spectral resolution to increase the number of fluorophores that can be measured simultaneously [17].

Raman spectroscopy (RS) is a highly selective technology providing narrow and non-overlapping peaks of the measured sample. In contrast to near infrared spectroscopy, the signals are not interfered with by carbon or water absorption and nearly no sample pre-treatment is necessary. The main drawback of RS is the low fraction of scattered photons (typically < 0.0001%), which leads to low signal-to-noise ratios and long acquisition times. Some applications are, for example label-free imaging of cells [18] or high-resolution imaging of single-walled carbon nanotubes [19].

Fluorescence and Raman scattering are two competing effects and often occur together. Therefore, it is sometimes difficult to detect the weak Raman signals due to the high fluorescence background [20].

The aim of this paper is to use the simultaneously appearing fluorescence and Raman signals after laser excitation for chemical imaging of large surfaces in short periods of time. Because HSI for evaluation or classification tasks requires the highest possible variance in the underlying spectral data, the exact knowledge of the origin of the signals is of secondary importance. A distinction between fluorescence and Raman events is desirable but not absolutely necessary. The idea is that the technology, from here on called Laser-HSI, can be used for various applications in the field of process and food monitoring or sorting. Previous work has mainly focused on Raman imaging using NIR lasers to reduce the fluorescence background [21,22], while the focus of this work is the combination of Raman and fluorescence imaging. This is reflected mainly in the choice of the excitation laser and the imaging spectrometer used.

In this paper, the design of the Laser-HSI system is described, and the system is calibrated and characterized. Furthermore, some sample measurements are presented to illustrate possible applications in different areas.

2. Materials and Methods

2.1. System Design and Software

The developed Laser-HSI system is schematically illustrated in Figure 1. The illumination is accomplished using a frequency-doubled Nd:YAG laser with a wavelength of 5322 nm and a maximum output power of 300 mW (GLK 32XX TS, LASOS Lasertechnik, Jena, Germany). The spectral linewidth of the laser is $\leq$ 1 MHz, the beam quality is M2 < 1.2, and the power stability is stated by the manufacturer as $\leq\pm2$%. The laser beam is guided through a laser clean-up filter (LL01-532, Semrock, Rochester, NY, USA), which is used to remove spontaneous emission noise. The beam then impinges a scanning mirror (dynAxis S, Scanlab, Puchheim, Germany) with variable scanning length. The scanning rate can be adjusted to the measurement task. The scanning mirror reflects the beam again and spreads the laser to form a divergent line. After passing an achromatic collimation lens with a focal length of f = 300 mm (AC508-300-A-ML, Thorlabs, Newton, MA, USA), the line passes a dichroic mirror (LPD02-532RU, Semrock, USA) with a length of 6 cm, which reflects light with a wavelength below 537 nm and allows the transmittance of light coming back from the sample with a wavelength greater than 537 nm. The dichroitic mirror is mounted at 45° to the laser plane and projects the beam to the sample surface, creating an excitation line of variable length, ranging from 0.5 cm up to 10 cm.

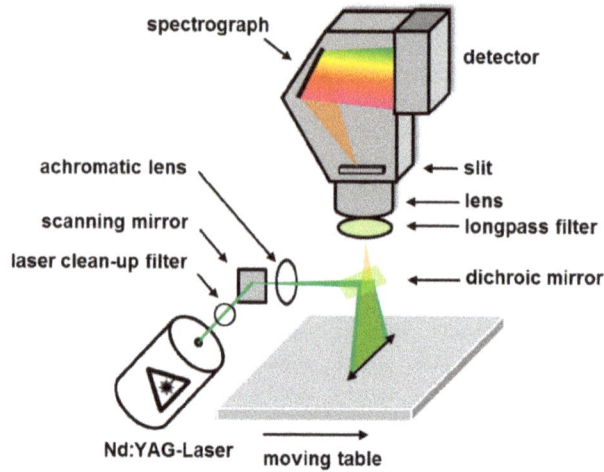

Figure 1. Schematic illustration of the developed laser hyperspectral imaging system (Laser-HSI) system.

The hyperspectral imaging system is mounted directly above the excitation line. It consists of a holographic imaging spectrometer (Hyperspec VNIR, Headwall Photonics, Bolton, MA, USA) and a high-performance EMCCD (electron multiplying charge-coupled device) camera (Andor Luca R, Andor Technology, Belfast, UK). The spectral pixel dispersion of the imaging spectrometer is 0.7 nm. A C-mount lens is placed in front of the camera. Three different lenses were used for the experiments: two standard lenses (Cinegon 8 mm f/1.4 and Xenoplan 28 mm f/1.4, Jos. Schneider Optische Werke, Germany) and a telecentric lens with fixed working distance of 86 mm (S5LPJ2426, Sill Optics, Wendelstein, Germany) for small samples. To eliminate the excitation wavelength of the laser, a 532 nm long-pass filter is mounted before the lens (LP03-532RE, Semrock, USA).

The imaging spectrometer is built in the Offner design and optimized for a wavelength range from 400 to 1000 nm. The spectrometer can be equipped with optical slits of different widths ranging from 25 μm to 60 μm. The Si-EMCCD-detector has 1004×1002 pixels with a pixel size of 8×8 μm and 14-bit depth. The rows and columns of the sensor can be binned together up to $8\times$ to increase the signal-to-noise ratio of the measurement, sacrificing resolution. The sensor is cooled to -20 °C and can be connected and controlled using USB. To reduce the read-out noise of the camera and to increase the signal-to-noise ratio (SNR), an electron multiplication gain between $2\times$ and $200\times$ can be activated.

For sample movement, a linear motion unit (VT 80, PI Micos, Eschbach, Germany) is positioned underneath the hyperspectral camera. The linear stage has a travel length of 200 mm and a step size of 0.5 μm. The whole system is housed with blackboards to minimize ambient light.

The camera, the linear motion unit, and the laser can be controlled using the imanto®pro software package (version 3.7, Fraunhofer IWS, Dresden, Germany). This program allows the configuration of the camera, the movement of the linear motion unit, and the image acquisition, as well as the display of the live image from the camera and the current measurement. In addition, the generated spectral images can be viewed, pre-processed and saved as .envi or .hsi.jpg formats. The latter is a new data format for hyperspectral data, which includes metadata and a preview picture from the measurement in the file. Furthermore, it is possible to integrate machine learning models in the software. In this work, most data evaluation was done using Matlab® version R2017a (Mathworks, Natick, MA, USA).

The Laser-HSI system operates as a pushbroom imager. This means that the hypercube is generated line by line and, therefore, the sample is moved underneath the camera system during the measurement using the linear motion unit. To achieve a sufficient Raman and fluorescence intensity, the imaging plane of the camera has to be aligned to the excitation laser line. Therefore, the dichroitic mirror has to be adjusted prior to the measurement.

2.2. Spectral Calibration and Spectral Resolution

The aim of the spectral calibration of the Laser-HSI is to assign correct absolute wavelength or relative Raman shifts to the rows of the detector array. For the absolute wavelength calibration, a neon lamp calibration source with characteristic emission lines between 550 nm and 750 nm (Renishaw, Wotton-under-Edge, UK) is used. The calibration source is placed underneath the camera and the spectrum is averaged 200 times using slit widths of 25, 40 and 60 µm, the Cinegon lens and no binning. For each measurement, the positions of the measured lines are used to fit a linear function between the known wavelengths of the calibration source and the row indices of the detector array. The calibration can be done automatically using an own-written Matlab® script.

To obtain the spectral resolution of the imaging system, two sharp peaks of the neon lamp (583.4 nm and 701.6 nm) were used. In theory, these peaks should cover only one pixel of the detector. Due to the point spread function of the imaging system, the measured peaks are broadened. The resolution of the imaging system can therefore be calculated using the full width at half maximum (FWHM) of the measured peaks.

2.3. Spatial Resolution

The spatial resolution perpendicular to the measurement line depends on the speed of the moving stage and the frame-rate of the camera, and can, therefore, be easily controlled and adjusted to achieve squared pixels. The spatial resolution parallel to the measurement line depends on several factors, namely the working distance, the focal length of the used lens, and the number of detector pixels in the spatial dimension of the camera. The fixed structure of the spectrograph and camera allows the spatial resolution required for the application to be adjusted with a lens with an appropriate focal length, which simultaneously results in a certain field of view (FOV). To measure the spatial resolution, a calibration target (R2L2S1P1, Thorlabs, Newton, MA, USA) with black and white lines is placed underneath the camera and illuminated with halogen light. One line of the target is measured 200 times and averaged. The measured black and white transitions of the sample at a wavelength of 600 nm are differentiated and the FWHM of the obtained peaks is taken as the spatial resolution. The spatial resolution for the three different lenses was determined with working distances of 330 mm and 82.7 mm and FOVs of 386 mm, 104 mm and 4 mm. A slit width of 40 µm was used for the measurements.

2.4. Example Measurements

To evaluate the developed Laser-HSI, four samples from quite different fields of application were measured: (a) a mixture of sulphur and naphthalene; (b) a carrot slice; (c) black polymer particles; and, (d) a lead zirconate titanate (PZT) piezoelectric actuator with contaminations. The specific experimental conditions are shown in Table 1.

Table 1. Experimental conditions.

Experiment	Lens	Laser Power (mW)	Integration Time (ms)	Speed (mm·s^{-1})	Frame rate (Hz)	Binning (Lateral × Spectral)	Total Time (s)	Total Area (cm^2)
(a) sulphur and naphthalene	Xenoplan	150	25	2.4	24	2 × 1	20	50
(b) carrot slice	Xenoplan	300	100	0.8	8	2 × 1	50	42
(c) black polymers	Xenoplan	150	25	2.4	24	2 × 1	42	105
(d) PZT actuator	Sill	300	100	0.013	8	4 × 1	170	2.3

For referencing the experiments (a) and (d), a Raman microscope (inVia, Renishaw GmbH, Wotton-under-Edge, UK) with a 514.5 nm argon ion laser source was used. In case of (a), a 100× magnification lens was used and the spectra were accumulated for 10 s and measured between 150 cm^{-1} and 3500 cm^{-1}.

In case of (d), areas with suspected contaminations were mapped using a 50× magnification lens and a spatial resolution of approximately 5 μm. The integration time was 0.1 s and the spectra were measured between 150 cm^{-1} and 2000 cm^{-1}.

For referencing the experiment (b), the same Raman microscope with a 785 nm diode laser (to reduce fluorescence) and 50× magnification lens was used. The spectra were measured between 800 cm^{-1} and 1800 cm^{-1} and accumulated ten times for 1 s.

2.5. Classification Experiments

To show that the system can be used for process monitoring and sorting tasks, classification experiments were performed for the experiments (a), (c), and (d). Therefore, 500 spectra per class were selected at random and a random forest [23] model was trained using the fitcensemble function of Matlab® using default hyperparameters. Prior to the training, a principal component analysis was conducted and only the scores of the first five principal components were used. To estimate the accuracy of the trained model, 10-fold cross validation was used and the cross-validated accuracy was used as a measure of the quality of the model. Finally, the trained model was applied to all spectra in the hyperspectral measurement to generate a classification image of the sample.

3. Results

3.1. System Design

The individual components of the Laser-HSI were selected and optimized with regard to the specific application requirements. With a wavelength of 532 nm, the frequency-doubled Nd:YAG laser used offers a good compromise between the expected intensity of the Raman scattering and the fluorescence intensity. Furthermore, the wavelength of the light emitted by the samples is in the range of the optimum quantum efficiency of the used silicon detector. If Raman measurements are made, this wavelength is particularly suitable for inorganic or weakly fluorescent samples. A near-infrared laser (e.g., 785 nm) would be more suitable for organic or highly fluorescent samples, but this would also further decrease the Raman intensity. Compared to argon ion lasers, which emit a similar wavelength, the Nd:YAG laser offers low power consumption, easy handling and a compact design, which facilitates industrial use (e.g., in recycling plants).

To generate the laser line, a scanning mirror was used. Alternatively, one could use a cylindrical or Powell lens [24] to generate the excitation line. Both approaches have advantages and disadvantages. If a cylindrical lens is used, no moving parts are required and a compact design can be achieved. On the other hand, a cylinder lens produces a non-uniform intensity distribution along the excitation line. This can be reduced by using a Powell lens, but these lenses must be optimized for a certain beam diameter and fan angle, which reduces the flexibility of the system. A disadvantage of the scanning mirror is an increased intensity at the reversal points of the mirror (at the edges of the excitation line) and the limited scanning speed, which can lead to problems using short integration times for the detector. Because the setup presented here should be as flexible as possible, a scanning mirror was used. For later industrial applications, the use of cylindrical or Powell lenses can be considered.

The used imaging spectrometer is optimized for spectroscopy in the visible and near-infrared wavelength range (400–1000 nm). Therefore, the spectral resolution is quite low for Raman spectroscopy (see Section 3.2). This has a negative effect on the usability of the device for tasks where a high spectral resolution is required. On the other hand, it is possible to measure a much larger wavelength range than with a Raman spectrometer. This could be useful for some questions; for example, when fluorescence signals are to be measured, which often extend over a wider wavelength range (see experiment (c)). In addition, the used imaging spectrometer is cheaper and more readily available then a Raman imaging spectrometer.

The EMCCD detector used enables a reduction of the readout noise due to the built-in amplification and thus enables the measurement of weak Raman and fluorescence signals.

The disadvantage is the relatively low frame rate of ~12 Hz using the full detector, which can, however, be increased using binning.

The C-mount connection allows flexibility regarding the FOV and the spatial resolution of the measurement. In summary, due to the modular design of the Laser-HSI system an individual adaptation to the respectively specific measuring application is possible.

3.2. Spectral Calibration and Spectral and Spatial Resolution

The results of the spectral calibration for different slit widths are shown in Table 2. For all three slit widths, a similar covered wavelength range of approximately 330 nm to just over 1100 nm is obtained. A wavelength interval of about 0.8 nm per pixel is determined for all three slits. The quality of the spectral calibration can be determined from the mean absolute error of the line positions to the reference line positions. The mean absolute error is relatively small for all three measurements, but increases with wider slits. The reason for this is the decreasing spectral resolution, which leads to broader lines and makes the automated peak selection more difficult, resulting in larger calibration errors. The spectral resolution of the system was measured for the peaks at 583.4 nm and 701.6 nm, and ranges from about 2 nm for 25 μm slit width, to over 4 nm for 60 μm slit width. It can also be seen that the spectral resolution decreases slightly for larger slit widths.

Table 2. Results of the spectral calibration of the Laser-HSI.

Slit Width (μm)	Wavelength Interval (nm)	Mean Absolute Error (nm)	FWHM (583 nm) (nm)	FWHM (702 nm) (nm)	FWHM (583 nm) (cm^{-1})	FWHM (702 nm) (cm^{-1})
25	0.79	0.25	1.8	2.0	52.9	40.6
40	0.8	0.58	2.8	3.1	82.3	63.0
60	0.79	1.28	4.1	4.5	120.5	91.4

Because the used spectrometer is optimized for visible and near-infrared hyperspectral imaging, the spectral resolution (in nanometers) decreases inherently due to the equidistant wavelength-resolution of the detector. In addition, it is possible to cover a much wider spectral range, from 535 nm up to 1000 nm, compared to a classic Raman spectrometer. Because of the low spectral resolution of the system, the typical calibration for Raman spectrometer, using reference substances like sulphur or naphthalene, could not be applied. Sometimes it was not possible to clearly assign the measured signals to the bands of the reference spectra. Therefore, the typical calibration could not be performed and the Raman shifts in the measurements were only calculated using the wavelengths obtained by the absolute calibration with the neon lamp calibration source.

The results for the determination of the spatial resolution for different lenses and FOVs are shown in Table 3. The measured resolutions are 1.31 mm for the Cinegon lens, 0.41 mm for the Xenoplan and 17 μm for the Sill telecentric lens with FOVs of 386 mm, 104 mm, and 4 mm, respectively. This shows the versatility of the developed laser-hyperspectral imager in terms of resolution and measurement area. It can be used to measure small samples with high resolution or to measure larger samples; e.g., in a conveyer belt application for sorting and recycling tasks. The FOV and the achievable resolution mostly depend on the used optics. Nevertheless, it has to be taken into account that with increasing FOV and working distance, more laser power is needed to maintain the signal intensity.

Table 3. Results of the spatial calibration of the Laser-HSI.

Lens	Working Distance (mm)	FOV (mm)	FWHM (pixel)	Spatial Resolution (mm)
Cinegon, f/1.4, 8 mm	330	386	3.49	1.34
Xenoplan, f/1.4, 28 mm	330	104	3.97	0.41
Sill, S5LPJ2426	85.7	4	4.18	0.02

3.3. Example Measurements

3.3.1. Spatial Distribution of Naphthalene Granules in a Sulphur Matrix (a)

A mixture of sulphur and naphthalene was measured using the Laser-HSI and the results are summarized in Figure 2. A picture of the sample (Figure 2a,c) shows a false color image of the measured intensity at ~180 cm^{-1}. At this wavenumber there is a slight overlap between the peaks of the two materials, but the intensity of sulphur is much higher than the intensity of naphthalene, so the position of the naphthalene can be clearly identified in the false color image. Despite the low resolution of the spectrograph, the characteristic peaks of both sulphur and naphthalene are clearly visible and the spectra show good agreement with reference spectra (Figure 2b). Fine spectral features like the sulphur peaks at 154 cm^{-1} and 219 cm^{-1} are not resolved. In addition, a precise determination of the peak position is not possible due to the low spectral resolution. Nonetheless, it is possible to determine the spatial distribution of the two materials in the sample. The measurement of the sample area of ~16 cm^2 only took around 20 s with a spatial resolution of approximately 0.4 mm. The acquired spectra were used to train a classification model to discriminate between sulphur and naphthalene. The resulting model has a high cross-validation accuracy of 99.9%. The classification image is shown in Figure 2d.

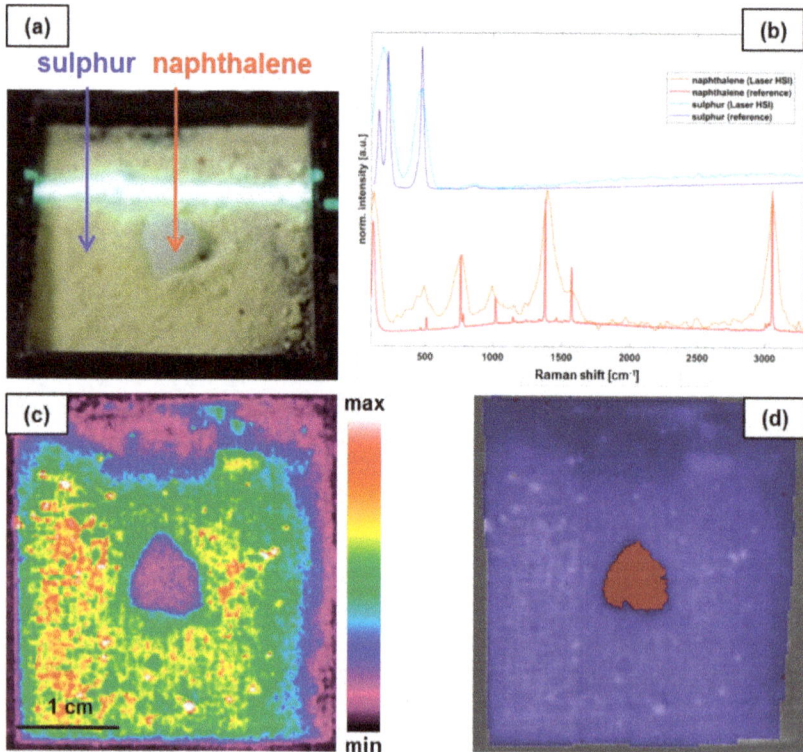

Figure 2. Laser-HSI of mixture of sulphur and naphthalene. (**a**) Picture of the sample with laser excitation line. (**b**) Normalized example spectra of sulphur and naphthalene with reference spectra acquired with the Raman microscope. (**c**) False color image of the measured intensity at ~180 cm^{-1} (sulphur). (**d**) Results of the Laser-HSI classification (blue = sulphur; red = naphthalene).

3.3.2. Carotenoid Distribution in a Carrot Slice (b)

A carrot slice was measured with the Laser-HSI and the results are shown in Figure 3. A picture of the sample (Figure 3a,c) shows a false color image of the measured intensity at ~1000 cm^{-1} before any pre-processing. This image, therefore, shows the fluorescence of the carrot slice. Figure 3b shows an example spectrum of the carrot slice scan after fluorescence background removal with two characteristic carotenoid peaks at 1156 cm^{-1} and 1521 cm^{-1}, which originate from C=C stretching vibrations of the carotenoids [25]. The background was removed fitting a polynomial of fourth degree to the spectra using a non-quadratic cost function [26]. The peaks are also in good agreement with the peaks from the reference spectra acquired with the Raman microscope. The intensity of the peaks is proportional to the carotenoid concentration. The band at 1521 cm^{-1} was used to calculate the distribution of the relative carotenoid concentration (Figure 3d). The image shows an uneven distribution of the carotenoid in the carrot slice. The concentration in the xylem is relatively low and increases towards the cambium. Towards the outer areas of the slice (phloem), the concentration decreases again. This is in good agreement with previous findings and shows the potential of the Laser-HSI in the area of food surveillance, quality control, and research [25]. When comparing images Figure 3c and d, it can be seen that there is a correlation between the distribution of the fluorescence and the carotenoid distribution in the inner part of the slice. However, at the edge of the slice, the carotenoid concentration is very low, but a strong fluorescence can be observed. This shows that it can be useful to measure Raman and fluorescence signals at the same time to get the information with only one measurement, which could be especially useful for time-critical tasks.

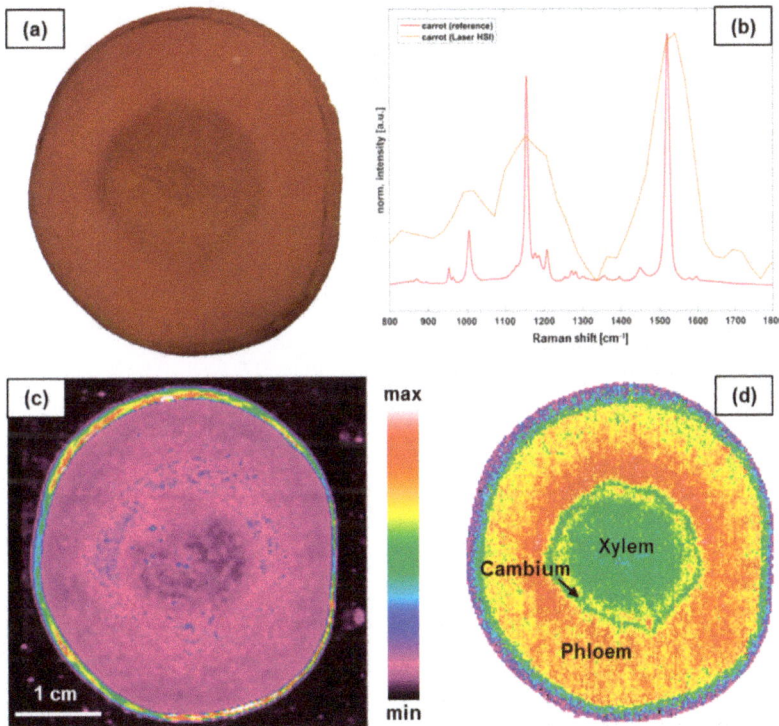

Figure 3. Laser-HSI of the carrot slice. (**a**) Picture of the carrot slice. (**b**) Normalized example spectra of the carrot slice from the phloem and reference spectra acquired with the Raman microscope. (**c**) False color image of the measured intensity at ~1000 cm^{-1} before any pre-processing of the spectra. (**d**) Color-coded relative intensity of the carotenoid peak at 1521 cm^{-1}.

3.3.3. Black Polymer Sorting (c)

The appearance of black polymers is achieved by adding filling materials like carbon black. The recycling of black polymers to high purity recycled materials is difficult because sorting techniques fail due to the low recognition rates of common sensor techniques like near-infrared spectroscopy. Laser-HSI is a promising technology to solve this problem because many black polymers show characteristic fluorescence signals. Therefore, three different kind of black polymer particles, namely polyamide 6 (PA6), polyamide 66 (PA66), and thermoplastic polyurethane (TPU) were imaged using the Laser-HSI. A picture of the three black polymers (Figure 4a,b) shows a false color image of the measured intensity at 580 nm. Figure 4d shows some Laser-HSI example spectra of the polymers. It is evident that there are clear differences in the spectral signatures which may be used to classify and, in consequence, sort the polymer particles.

The obtained classification model for the polymers shows a cross validation accuracy of 99.6%, which is very promising for the use of the Laser-HSI for sorting black plastics for recycling. The trained model was applied to all spectra in the hyperspectral measurement to generate a classification image of the sample (Figure 4c).

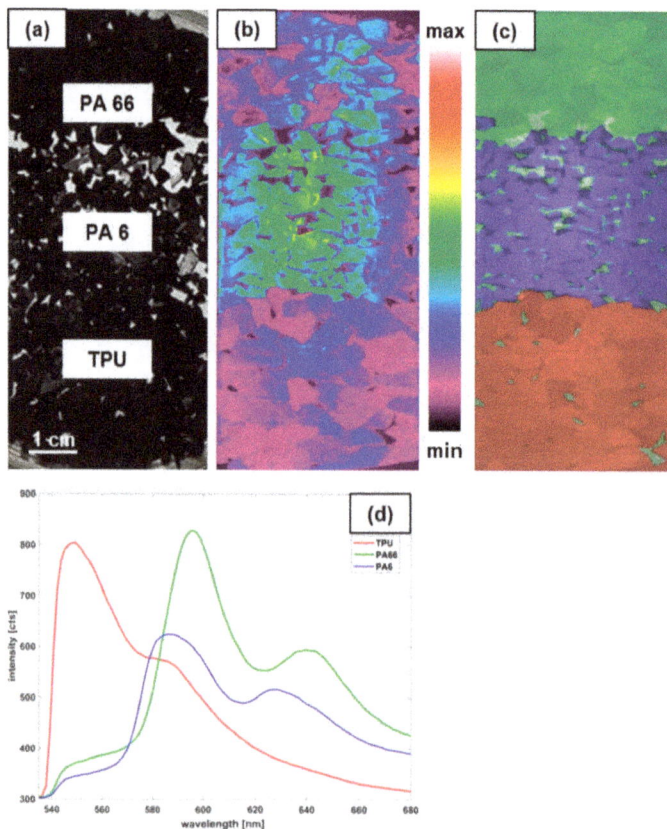

Figure 4. Laser-HSI measurement of the black polymer samples. (**a**) Picture of the sample showing the different kind of black polymer particles. (**b**) False color image of the measured intensity at 580 nm. (**c**) Classification result of the Laser-HSI measurement (green = PA66 (polyamide 66); blue = PA6 (polyamide 6); red = TPU (thermoplastic polyurethane). (**d**) Example spectra of the three black polymers.

3.3.4. Contaminations on PZT Piezoelectric Actuators (d)

For the last experiment, a PZT piezoelectric actuator was investigated with a high-resolution Laser-HSI measurement. The actuator is contaminated with organic substances of unknown origin and the aim of the experiment is to locate these impurities, using the Laser-HSI. Figure 5a shows a picture of the PZT piezoelectric actuator and Figure 5b shows a section of a measurement of a contaminated sample acquired with the Laser-HSI. Figure 5c shows a section of the same area, which was measured with the Raman microscope. Both figures show a false color representation of the Raman intensity at 230 cm^{-1}. Figure 5d shows spectra of the PZT ceramic and the contamination, measured with the Laser-HSI and the Raman microscope. The PZT spectra shows the typical bands of the PZT ceramic [27]. In the areas of contamination, strong fluorescence caused by the organic impurities can be seen. The shape and dimensions of the contamination show a good agreement for both measurements. The trained classification model shows a cross validation accuracy of 100% and the classification image is shown in Figure 5e. If needed, the peaks of the PZT ceramic bands could be used to find, for example, defects in the layers of the piezoelectric actuator.

This experiment shows that the developed Laser-HSI provides equivalent results to a Raman microscope for this task. While the Laser-HSI measurement takes about three min per sample, a complete mapping with the Raman microscope would take several hours. Therefore, the Laser-HSI could be used as a fast tool for the high-resolution quality control of small components.

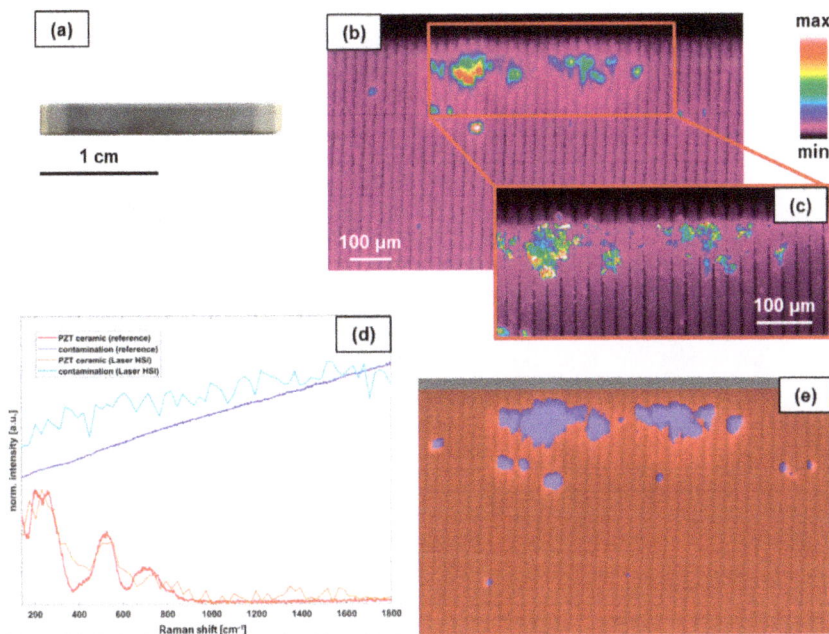

Figure 5. Results of the fourth experiment. (**a**) Picture of the measured PZT piezoelectric actuator. (**b**) False color image of the measured intensity at ~230 cm^{-1} measured with the Laser-HSI. (**c**) False color image of the measured intensity at ~230 cm^{-1} measured with the Raman microscope. (**d**) Normalized example spectra of the PZT ceramic and the contamination and reference spectra acquired with the Raman microscope. (**e**) Classification result of the Laser-HSI measurement (blue = contamination; red = PZT ceramic).

4. Conclusions

In this paper a fast, line-scanning hyperspectral imaging system with laser excitation using fluorescence or/and Raman scattered light for imaging analysis was introduced.

Four different application examples demonstrate the wide range of possible applications. The experiments carried out show that the described Laser-HSI system makes it possible to measure the Raman or fluorescence spectra of relatively large sample areas, with a high spatial resolution and in a relatively short amount of time. This opens up interesting application possibilities in many areas, from quality control in the food industry to surface inspection and recycling. The modular design of the system makes it possible to adapt the measuring range and spatial resolution to the problem at hand. In the experiments, measuring widths between 38 cm and 4 mm, with resolutions between 1.3 mm and 20 mm, were achieved. Furthermore, the system was spectrally calibrated and it could be shown that the spectra obtained for a number of test substances are in good agreement with literature or reference measurements. The minimal spectral resolution of approximately 1.8 nm at 583 nm is below the resolution of a conventional Raman spectrometer. On the other hand, the spectral range covered is much larger. This will be used in future attempts to obtain further information about the samples; for example, by measuring reflective properties in the spectral ranges in which no Raman or fluorescence signals are expected. Future developments will focus, in particular, on the investigation of the application for the recycling of black polymers. In addition, it will be investigated how measurements can be further accelerated. Other hyperspectral cameras and stronger lasers with different wavelengths could be used for this purpose.

5. Patents

In reference to the work presented here, two patents with the patent numbers DE102018210019.5 and DE102018210015.2 have been filed.

Author Contributions: F.G. conducted the research presented in this study and wrote the paper. P.W., W.G., and S.K. contributed to the development of the overall research design, provided guidance along the way, and aided in writing of the paper.

Funding: This research was funded by the scholarship program of the Deutsche Bundesstiftung Umwelt grant number 20016/421.

Acknowledgments: Special thanks go to Beate Leupoldt and Oliver Throl.

Conflicts of Interest: The authors declare no conflict of interest.

References

1. Borengasser, M.; Hungate, W.S. *Hyperspectral Remote Sensing. Principles and Applications*; CRC Press: Boca Raton, FL, USA, 2007.
2. Dale, L.M.; Thewis, A.; Boudry, C.; Rotar, I.; Dardenne, P.; Baeten, V.; Pierna, J.A.F. Hyperspectral imaging applications in agriculture and agro-food product quality and safety control: A review. *Appl. Spectrosc. Rev.* **2013**, *48*, 142–159. [CrossRef]
3. Elmasry, G.; Kamruzzaman, M.; Sun, D.W.; Allen, P. Principles and applications of hyperspectral imaging in quality evaluation of agro-food products: A review. *Crit. Rev. Food Sci. Nutr.* **2012**, *52*, 999–1023. [CrossRef] [PubMed]
4. Gowen, A.A.; O'Donnell, C.P.; Cullen, P.J.; Bell, S.E.J. Recent applications of chemical imaging to pharmaceutical process monitoring and quality control. *Eur. J. Pharm. Biopharma.* **2008**, *69*, 10–22. [CrossRef] [PubMed]
5. Gendrin, C.; Roggo, Y.; Collet, C. Pharmaceutical applications of vibrational chemical imaging and chemometrics: A review. *J. Pharm. Biomed. Anal.* **2008**, *48*, 533–553. [CrossRef] [PubMed]
6. Calin, M.A.; Parasca, S.V.; Savastru, D.; Manea, D. Hyperspectral imaging in the medical field: present and future. *Appl. Spectrosc. Rev.* **2013**, *49*, 435–447. [CrossRef]
7. Lu, G.; Fei, B. Medical hyperspectral imaging: A review. *J. Biomed. Opt.* **2014**, *19*, 10901. [CrossRef] [PubMed]

8.	Tilling, A.K.; O'Leary, G.J.; Ferwerda, J.G.; Jones, S.D.; Fitzgerald, G.J.; Rodriguez, D.; Belford, R. Remote sensing of nitrogen and water stress in wheat. *Field Crop. Res.* **2007**, *104*, 77–85. [CrossRef]

9.	Lewis, E.N.; Kidder, L.H.; Lee, E. NIR chemical imaging—Near infrared spectroscopy on steroids. *NIR News* **2005**, *16*, 2–4. [CrossRef]

10.	Martin, M.E.; Wabuyele, M.B.; Chen, K.; Kasili, P.; Panjehpour, M.; Phan, M.; Overholt, B.; Cunningham, G.; Wilson, D.; DeNovo, R.C.; et al. Development of an Advanced Hyperspectral Imaging (HSI) System with Applications for Cancer Detection. *Ann. Biomed. Eng.* **2006**, *34*, 1061–1068. [CrossRef] [PubMed]

11.	Gowen, A.; Odonell, C.; Cullen, P.; Downey, G.; Frias, J. Hyperspectral imaging—An emerging process analytical tool for food quality and safety control. *Trends Food Sci. Tech.* **2007**, *18*, 590–598. [CrossRef]

12.	Boldrini, B.; Kessler, W.; Rebner, K.; Kessler, R.W. Hyperspectral imaging: A review of best practice, performance and pitfalls for in-line and on-line applications. *J. Near Infrared Spec.* **2012**, *20*, 483–508. [CrossRef]

13.	Stewart, S.; Priore, R.J.; Nelson, M.P.; Treado, P.J. Raman imaging. *Ann. Rev. Anal. Chem.* **2012**, *5*, 337–360. [CrossRef] [PubMed]

14.	Févotte, G. In situ Raman spectroscopy for in-line control of pharmaceutical crystallization and solids elaboration processes: A review. *Chem. Eng. Res. Des.* **2007**, *85*, 906–920. [CrossRef]

15.	Adar, F.; Geiger, R.; Noonan, J. Raman spectroscopy for process/quality control. *Appl. Spectrosc. Rev.* **1997**, *32*, 45–101. [CrossRef]

16.	Lakowicz, J.R. *Principles of Fluorescence Spectroscopy*; Springer: New York, NY, USA, 2006.

17.	Zimmermann, T.; Rietdorf, J.; Pepperkok, R. Spectral imaging and its applications in live cell microscopy. *FEBS Lett.* **2003**, *546*, 87–92. [CrossRef]

18.	El-Mashtoly, S.F.; Petersen, D.; Yosef, H.K.; Mosig, A.; Reinacher-Schick, A.; Kötting, C.; Gerwert, K. Label-free imaging of drug distribution and metabolism in colon cancer cells by Raman microscopy. *Analyst* **2014**, *139*, 1155–1161. [CrossRef] [PubMed]

19.	Hartschuh, A.; Sánchez, E.J.; Xie, X.S.; Novotny, L. High-resolution near-field Raman microscopy of single-walled carbon nanotubes. *Phys. Rev. Lett.* **2003**, *90*, 95503. [CrossRef] [PubMed]

20.	McCreery, R.L. *Raman Spectroscopy for Chemical Analysis*; Wiley: New York, NY, USA, 2000.

21.	Qin, J.; Chao, K.; Kim, M.S. High-throughput Raman chemical imaging for evaluating food safety and quality. In Proceedings of the Sensing for Agriculture and Food Quality and Safety VI, San Francisco, CA, USA, 1–6 February 2014; p. 91080F.

22.	Qin, J.; Kim, M.S.; Chao, K.; Schmidt, W.F.; Cho, B.K.; Delwiche, S.R. Line-scan Raman imaging and spectroscopy platform for surface and subsurface evaluation of food safety and quality. *J. Food Eng.* **2017**, *198*, 17–27. [CrossRef]

23.	Breiman, L. Random forests. *Mach. Learn.* **2001**, *45*, 5–32. [CrossRef]

24.	Palonpon, A.F.; Ando, J.; Yamakoshi, H.; Dodo, K.; Sodeoka, M.; Kawata, S.; Fujita, K. Raman and SERS microscopy for molecular imaging of live cells. *Nat. Protoc.* **2013**, *8*, 677–692. [CrossRef] [PubMed]

25.	Baranska, M.; Baranski, R.; Schulz, H.; Nothnagel, T. Tissue-specific accumulation of carotenoids in carrot roots. *Planta* **2006**, *224*, 1028–1037. [CrossRef] [PubMed]

26.	Mazet, V.; Carteret, C.; Brie, D.; Idier, J.; Humbert, B. Background removal from spectra by designing and minimising a non-quadratic cost function. *Chemometr. Intell. Lab.* **2005**, *76*, 121–133. [CrossRef]

27.	Camargo, E.R.; Frantti, J.; Kakihana, M. Low-temperature chemical synthesis of lead zirconate titanate (PZT) powders free from halides and organics. *J. Mater. Chem.* **2001**, *11*, 1875–1879. [CrossRef]

Journal of
Imaging

MDPI

Article

Spatial Referencing of Hyperspectral Images for Tracing of Plant Disease Symptoms

Jan Behmann [1,*], David Bohnenkamp [1], Stefan Paulus [2] and Anne-Katrin Mahlein [2]

[1]   INRES Plant Diseases and Plant Protection, University of Bonn, 53115 Bonn, Germany; davidb@uni-bonn.de
[2]   Institute for Sugar Beet Research (IFZ), 37079 Göttingen, Germany; paulus@ifz-goettingen.de (S.P.);
       mahlein@ifz-goettingen.de (A.-K.M.)
*   Correspondence: jbehmann@uni-bonn.de; Tel.: +49-228-73-4998

Received: 5 November 2018; Accepted: 2 December 2018; Published: 4 December 2018

Abstract: The characterization of plant disease symptoms by hyperspectral imaging is often limited by the missing ability to investigate early, still invisible states. Automatically tracing the symptom position on the leaf back in time could be a promising approach to overcome this limitation. Therefore we present a method to spatially reference time series of close range hyperspectral images. Based on reference points, a robust method is presented to derive a suitable transformation model for each observation within a time series experiment. A non-linear 2D polynomial transformation model has been selected to cope with the specific structure and growth processes of wheat leaves. The potential of the method is outlined by an improved labeling procedure for very early symptoms and by extracting spectral characteristics of single symptoms represented by Vegetation Indices over time. The characteristics are extracted for brown rust and septoria tritici blotch on wheat, based on time series observations using a VISNIR (400–1000 nm) hyperspectral camera.

Keywords: hyperspectral imaging; plant phenotyping; disease detection; spectral tracking; time series

---

1. Introduction

Hyperspectral images of plants are suitable to assess the health and vitality state of plants [1,2]. Leaf diseases show characteristic symptoms, allowing a hyperspectral characterization of symptom development [1,3]. The spectral dynamic of symptom development during pathogenesis has been described for numerous plant-pathogen systems [4]. Therefore, hyperspectral imaging has been applied on multiple scales from the leaf level via full plants up to the field and landscape scale [5–8]. Platforms, microscope stands, laboratory systems, high-throughput facilities, as well as Unmanned Aerial Vehicles (UAVs), planes, and satellites are used [5–7,9,10].

This publication focuses on tracking leaf diseases of wheat at the leaf scale. Wheat (*Triticum aestivum*) is the second most cultivated crop worldwide which is threatened by various pathogens infecting root, stem, leaves, and ears [11]. At the leaf scale, a limiting factor is the natural variability in the spatial and temporal development of disease symptoms [12]. The exact position of symptom appearance and dynamics of development are bound to multiple parameters and, to a certain extent, unpredictable [13,14]. Therefore, symptoms of different development steps are present at a certain point in time, hampering the clear extraction of the typical symptom development.

At best, each symptom has to be traced during the different observation days on its own to have a clearer look on the different steps of pathogenesis. Performing this task is extremely labor intensive and in many situations not feasible, e.g., for the early parts of the pathogenesis before expression of a visible symptom. Few studies have focused this task [5,6], but are restricted to the characterization of single symptoms instead of extracting a representative description of the pathogenesis.

A further advantage of spatially referenced hyperspectral time series is the generation of large amounts of training data with high quality annotation for the training of machine learning models [15].

Even very early effects of the disease could be included as positions on the leaves showing symptoms at a later point in time are known. The underlying assumption is that if disease symptoms are visible, the first changes could most probably be recorded by the hyperspectral camera a few days earlier.

Prerequisite for these applications is a common measurement coordinate system for every image, but its generation on leaf scale is challenging due to leaf movements and growing. The spatial assignment of two images is a common task in computer vision and addressed by the terms image matching or image referencing [16–18].

In remote sensing a joint spatial reference system for multiple observations is often provided by the data distributor. Images are georeferenced by an automatic process to localize reflectance characteristics and perform multi-temporal analyses based on repetitive observations of a location. Space and airborne images are georeferenced either by ground control points with known coordinates or by additional sensors determining the location and orientation of the sensor platform. Usually Global Navigation Satellite System (GNSS) receiver are combined with Inertial Measurement Systems (IMS) to reach sufficient global accuracy as well as local continuity [19,20]. Ortho-rectified images are generated by projecting the reflectance information obtained from the 3D earth surface on a 2D reference plane using the determined camera models. By this approach, spatial distortion within the images can be removed [20].

However, in close range scenarios with plants, the image referencing relies typically on the image content instead of external sensors as the measured object cannot be assumed to be solid. Therefore the joint coordinate system is a new concept on the leaf scale. Most of the approaches of extracting geometrical features aim at the classification of plant species based on the shape of its leaf or organs [21–23] but the generated features can be transferred to the image matching problem. Gupta et al. [24] investigated the growing characteristics of multiple species using a dense grid of ink markers on the leaves. Based on reference points on a separate reference object, camera models have been further determined for the combination of reflectance and 3D surface data of plants [25,26].

Generally, three method groups to establish correspondence between RGB images can be identified: Point approaches [27], area/contour approaches [28,29], and global approaches in pixel [30] or frequency space [31]. The point approaches detect relevant suitable points within the images and describe their local neighborhood by robust descriptors (e.g., Scale-invariant feature transform, SIFT [32]) allowing an assignment of points from different images. Such methods are rarely used for plant leaves (e.g., [27]) but are the standard procedure for RGB image matching for 3D reconstruction. Area and contour approaches at first extract the leaf within the image and perform an assignment based on the shape or texture characteristics of the leaf surface [29]. Yin et al. [33] used chamfer matching to perform the assignment of Arabidopsis leaves based on their shape. Bar-Sinai et al. [28] used a matching of the local graph of leaf veins to investigate the response of the leaf to mechanical stress. An approach for the multi-modal registration of RGB images and thermal images of plants based on extracted silhouettes in both image types has been developed [34]. There, a non-rigid spline model was used to match the silhouettes and generate the data for the detection of diseased plant tissue [35].

These methods for spatial image matching rely on characteristic shapes or textures. In the present case of parts of wheat leaves none of that is informative. In monocot plants the leaf veins are arranged in parallel and provide no intersections to detect. Moreover, the low spatial resolution of hyperspectral images compared to RGB images prevents the extraction of detailed surface texture information. Disease symptoms may be suitable patterns at later time points but are not present at the earlier days. The silhouette of the fixated leaf parts matches approximately a rectangle, especially for larger leaves where the leaf tip is not captured.

To handle this challenge, artificial reference points of white color have been applied on the leaf surface to allow the image referencing within this study. Leaves are fixated to reduce the complexity to a 2D case approximating the leaf surface by a horizontal 2D plane. In such a case, transformations between different image coordinate systems can be performed by 2D transformation like Affine,

Similarity, or Polynomial transformation models [36]. The optimal choice depends on the required model flexibility.

We introduce an approach to spatially reference multiple hyperspectral image cubes of time series experiments. These spatially referenced images form a new 4 dimensional data type with two spatial axis (x and y), one spectral axis (λ), and a fourth temporal axis (t). Within this new data type disease symptoms can easily be traced back in time, even to the point when no symptom is visible for the human eye.

The referencing is performed using an algorithm robust against missing or non-stable reference points by including the RanSaC algorithms [37] and multiple 2D geometric transformations [36] in combination with a well-defined set of control points. As data, time series measurements of wheat leaves fixated in a grid frame assessed by a VISNIR hyperspectral pushbrom camera sensible in the visible (400–680 nm) and near-infrared part of the electro-magnetic spectrum (680–1000 nm) were used. Two relevant diseases of wheat with different symptom expressions were covered: Septoria tritici blotch and brown rust (Figure 1).

Figure 1. RGB images of the symptoms of (**A**) brown rust caused by *Puccinia triticina* and (**B**) septoria tritici blotch caused by *Zymoseptoria tritici*. Brown rust symptoms are dominated by reddish spore stocks whereas necrotic lesions are characteristic for septoria tritici blotch.

2. Materials and Methods

2.1. Plant Material & Fungal Pathogens

2.1.1. Plant Material

Wheat plants (triticum aestivum), cv. Taifun and cv. Chamsin (KWS SAAT SE, Einbeck, Germany) were grown in plastic pots in substrate ED73 (Balster Erdenwerk GmbH, Sinntal-Altengronau, Germany). The plants were grown under greenhouse conditions with 22/20 °C day/night temperature, 60 ± 10% RH and a photoperiod of 16 h per day. Plants were inoculated after reaching BBCH growth stage 30 (sprouting) [38].

2.1.2. Fungal Pathogens

Inoculations of wheat plants were done with isolates of *Zymoseptoria tritici* and *Puccinia triticina*. The isolate of *Z. tritici* was cultured on artificial ISP2 media. As obligate biotrophic pathogen, *P. triticina* was maintained on living plants.

2.2. Hyperspectral Measurements

A hyperspectral line scanning spectrograph (ImSpector V10E, Spectral Imaging Ltd., Oulu, Finland), covering the VISNIR spectral range from 400–1000 nm has been used for image assessment [6]. Images had a spectral resolution of up to 2.8 nm and a spatial resolution of 0.14 mm per pixel. This results in a hyperspectral datacube with 211 bands and 1600 px per image line. A homogeneous illumination was ensured by using six ASD-Pro-Lamps (Analytical Spectral Devices Inc., Boulder, CO, USA). Camera and illumination were installed on a motorized line scanner (Spectral Imaging Ltd.).

Camera settings and motor control were adapted using the SpectralCube software (Version 3.62, 2000, Spectral Imaging Ltd.).

Leaves have been horizontally fixed in a tray using strings allowing imaging of around 20 cm of each leaf. Multiple leaves have been placed side by side within a single image. White color spots are applied as reference points on the leaves to allow image referencing. As shown in Figure 2, six spots in two rows are applied to each leaf section of approximately 4 cm. Observing five leaf sections limited by the fixating strings results in 30 white color spots on each leaf.

As background material, blue cardboard has been selected as it supports background segmentation. For brown rust, time series measurements from 2 to 12 days after inoculation (dai) have been captured. Contrarily, for septoria tritici blotch 15 to 27 dai were covered due to the deviating process of pathogenesis. The data has been normalized, i.e., reflectance was calculated relative to a barium sulphate white reference (Spectral Imaging Ltd.) and a dark current measurement. Normalization has been performed following [39] using ENVI 4.6 + IDL 7.0 (EXELIS Visual Information Solutions, Boulder, CO, USA).

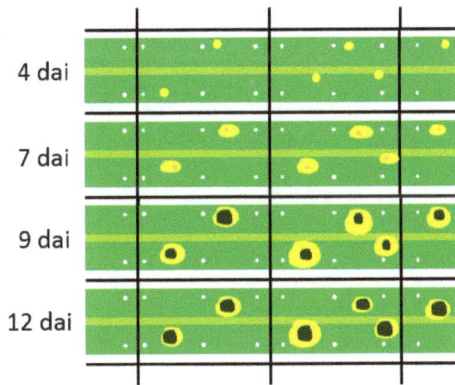

Figure 2. Scheme of the referenced hyperspectral time series of a single leaf observed at four exemplary days. Reference points (white spots), fixating strings (in black), and the developing disease symptoms are included.

2.3. Algorithm for Hyperspectral Image Referencing

The proposed algorithm is divided into four steps: 1. background segmentation, 2. reference point detection, 3. matching of reference points, and 4. the spatial transformation (Figure 3).

2.3.1. Background Segmentation

The background segmentation relies on the classification method Random Forest algorithm [40] to separate leaf regions and background. The model was trained by manual annotation of a single hyperspectral image. As training data 1000 pixels of the blue background, 1000 pixels of healthy leaf tissue, 1000 pixels of chlorotic leaf tissue, and 500 pixels of spore stocks were randomly sampled from the annotation in which the human expert has tried to represent the class variability. Remaining artifacts of misclassified pixels causing very small regions and holes within large regions were corrected using connected components approach. Leaf regions were extracted and identified by corresponding leaf number based on the position within the image.

2.3.2. Reference Point Extraction

For the selection of the reference points, the Random Forest algorithm was applied as well. By manual annotation 100 pixels of the white reference points and 1000 pixels of plant material (balanced mixture of healthy, chlorotic, and spore stock tissue) were selected and used to train the model.

Classified regions within a size range of 3 to 40 pixels were regarded and the center of gravity was extracted as the pixel position. To exclude mixed pixels, the reference point region has been extended by 3 pixel and removed from the leaf regions.

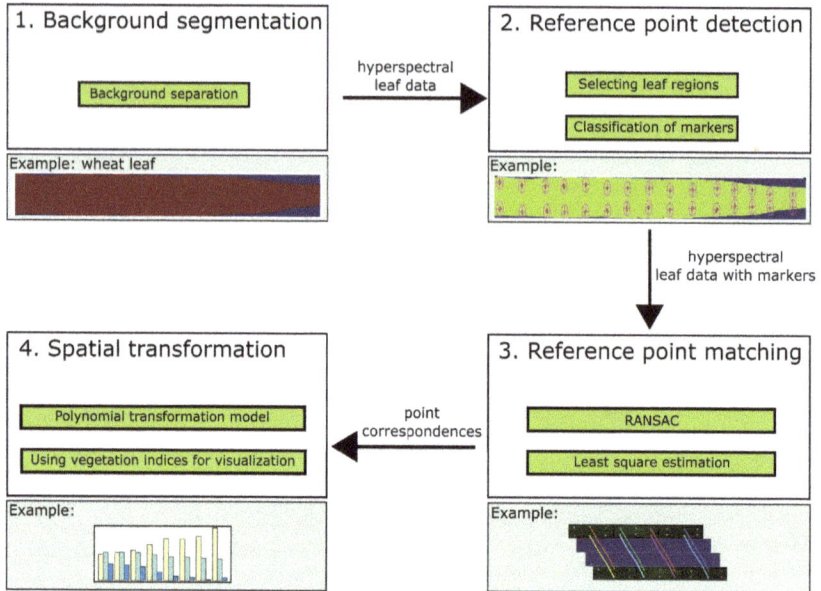

Figure 3. Dataflow of the proposed algorithm for geometric referencing of hyperspectral images.

2.3.3. Assignment of Reference Points

Point correspondence was used to derive the geometric transformation model between two images of a leaf recorded at different observation days. Following, the assignment of single reference points between the different observation dates is a prerequisite for image transformation. In our approach, the Random Sample Consensus (RanSaC) algorithm [37] determines a preliminary nonreflective similarity transformation by assigning two random reference points in the base image to two random points in the image to reference. The models were evaluated by projecting each reference point of the origin image to the target image and assess the distance to the closest reference points of the target image. Reference points within a distance of 20 px are assumed to be correct and, therefore, support the model. By repeating this process and selecting the transformation model with the maximum number of supporting reference points, a preliminary referencing was performed. Using nearest neighbor assignment with a distance threshold of 20 px, reference points were assigned and reference points without counterpart were excluded from the further process.

2.3.4. Spatial Image Transformation Models

A transformation model was derived using the corresponding reference points. It is used to reference all images of a time series to the first observation day. The type and flexibility need to be adapted to the specific task. We compared different transformation types [36]: Nonreflective Similarity, Affine, Projective, Polynomial, and Local Weighted Mean (LWM).

Nonreflective Similarity transformation is defined by rotation, translation, and scaling. Adding a shearing parameter and another scale factor results in an Affine transformation. The Projective transformation represents a central-projective transformation between the two image coordinate systems and is defined by eight coefficients [36]. The Polynomial model relates the coordinates within the two images by a mathematical description based on two 2D polynomials. We selected polynomials of order 3 as they are able to represent the assumed leaf movements. The LWM model differs from the mentioned transformation due to its local character. The image is divided into regions in which a local polynomial transformation model is applied [41].

All steps of the algorithm have been performed using Matlab 2013a (The Mathworks, Natick, MA, USA) and the corresponding Image Processing Toolbox.

2.3.5. Evaluation of Transformation Accuracy

Evaluation of the transformation accuracy was performed by the Root Mean Square Error (RMSE) on Euclidean deviations of n reference points

$$RMSE = \sqrt{\frac{\sum_i \sqrt{(x_i - x_{ref})^2 + (y_i - y_{ref})^2}^2}{n}}. \tag{1}$$

To evaluate the transformation quality, assessing the mean accuracy of projected reference points is not sufficient. Large distortions or missing image information can occur in parts of the leaf without affecting the RMSE parameter. Therefore, two further quality parameters were introduced: Stability and Extrapolation. The first is defined as the RMSE of an inner point, if it is not included into the transformation model. This approximates the transformation quality of arbitrary points of the leaf. The extrapolation parameter approximates the transformation quality at the leaf border by the transformation accuracy of the four outer points point in the first and last point columns at the leaf base and the leaf tip (cv. Figure 2). The evaluation has been performed on the full time series (11 images) of a representative leaf showing brown rust symptoms. For the accuracy measurement $11 \times 30 = 330$, for the extrapolation $11 \times 4 = 44$, and for the stability $11 \times 5 = 55$ reference points have been used. The hold out reference points for stability calculation have been evenly distributed within the inner points.

2.4. Vegetation Indices

For characterization of the spectral development of symptomatic areas, Vegetation Indices based on selected spectral bands (λ_i) have been used. The Normalized Difference Vegetation Index (NDVI) uses a combination of a red band (670 nm) and a NIR band (800 nm) according to formula 2 to extract information about plant vitality and Chlorophyll content [10]. In addition, the Photochemical Reflectance Index (PRI) was calculated, using the difference between two bands in the green color range (531 nm and 570 nm) [42] as well as the Anthocyanin Reflectance Index (ARI) that uses a green (550 nm) and a NIR (700 nm) band which is sensitive to changes in carotenoid pigments [43].

$$NDVI = \frac{\lambda_{670} - \lambda_{800}}{\lambda_{670} + \lambda_{800}} \tag{2}$$

$$PRI = \frac{\lambda_{531} - \lambda_{570}}{\lambda_{531} + \lambda_{570}} \tag{3}$$

$$ARI = \frac{1}{\lambda_{550}} - \frac{1}{\lambda_{700}} \tag{4}$$

The selected Vegetation Indices are correlated to different plant-physiological parameters (chlorophyll content, photochemical activity, anthocyanin content) which are significantly influenced during disease development.

2.5. Presymptomatic Labeling

To demonstrate the advantages of a fully referenced data set, we used spatial referencing to move the border of symptoms that can be annotated retrospectively regarding the infection time. At first a supervised classification model (Random Forest algorithm [40]) has been derived on the full spectral information based on manual annotations of vital plant material, chlorotic regions, and spore stocks of brown rust (training data composition given in Section 2.3.1). Such models reproduce the visual annotation with good accuracy but are not able to detect invisible effect.

Here, spectra of pixels that were two days later detected as diseased were include in the training data and the Random Forest model is retrained. Furthermore, only pixels were included that are observed at every observation day to guarantee continuous time series observation for every point on the leaf surface within the data set. The extended annotations are used to retrain the classification model.

3. Results

This section presents the obtained results of the spatial referencing algorithm for hyperspectral images and the proposed applications tracing of symptoms and advanced labeling. The referencing approach has been applied to time series observations of two different diseases: Septoria tritici blotch and brown rust, each represented by twelve leaves. Figure 4 shows the effect of referencing by the RGB visualization of hyperspectral images showing a progressing brown rust infection 2–12 dai.

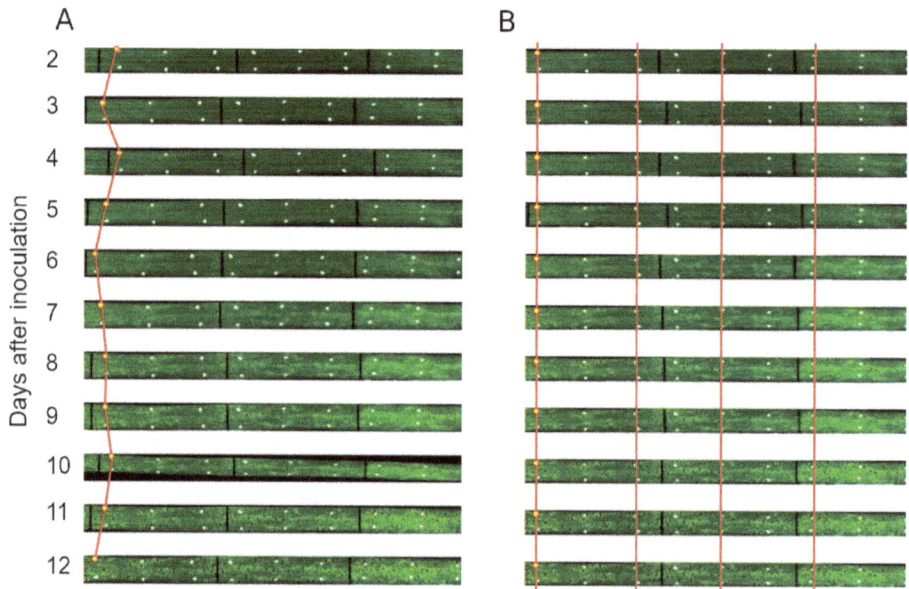

Figure 4. Effect of the spatial referencing. (**A**) shows the time series of unaligned hyperspectral images and (**B**) the same time series after the application of the proposed algorithm. To illustrate the result, identical points within the images were connected by red lines.

Figure 5 shows the results of tracing mature symptoms back in time. For brown rust and septoria tritici blotch, a continuous transition starting from healthy tissue has been extracted. The final state represents the deviating symptom appearance of the diseases (cv. Figure 1) .

3.1. Background Segmentation and Reference Point Detection

Parts of the algorithm are common and well understood steps in many image analysis pipelines. 1. background segmentation and 2. reference point detection did not limit the accuracy or performance of the algorithm. The used Random Forest classifier was trained on manually annotated but representative training data and reached on this data an accuracy of more than 99%. The derived class images "background vs. leaf" and "leaf vs. reference point" showed a high level of concordance with the visual impression. In transfer regions, e.g., the unsharp transition of leaf tissue to the white color of the reference points, a true region boundary is not defined. Consequences are neglectable as the center of gravity of the reference point region showed a high level of reproducibility within the different images.

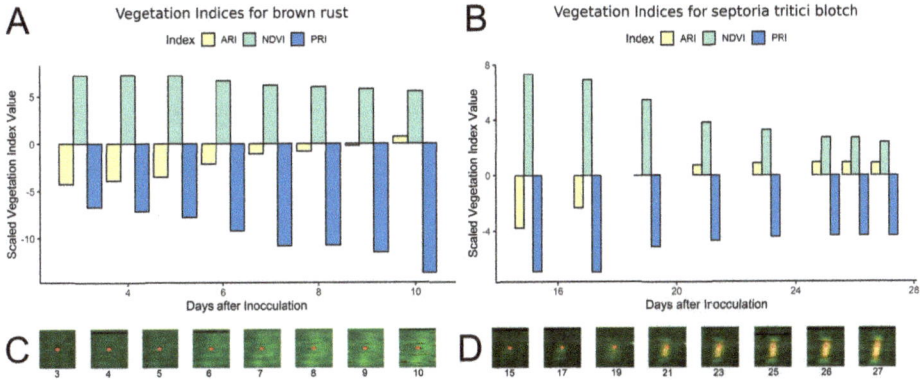

Figure 5. Visualization of the tracing results by using RGB visualizations of a symptom and the corresponding Vegetation indices anthocyanin reflectance index (ARI), normalized difference vegetation index (NDVI), and photochemical reflectance index (PRI) for brown rust (**A,C**) and septoria tritici blotch (**B,D**). For visualization purposes the NDVI was multiplied by 10 and the PRI by 40. Timeseries of spectral characteristics are derived to uncover the deviating spectral dynamics of the diseases.

In contrast, the matching of the reference points to the base image of day 1 is challenging if larger movement had occur or missing points complicate the assignment process. The RanSaC algorithm determines a preliminary non-reflective similarity transformation and allows a nearest neighbor assignment. More flexible transformations tend to degenerated cases assigning multiple reference point to a single base point. Using the RanSaC with 10,000 iteration, an optimum has been found in each case, whereas using only 1000 iteration led to a suboptimal result in around 10% of the runs.

3.2. Transformation Model

The flexibility and robustness of a transformation model type determines the suitability of the model for a specific task. Average quality parameters of the different models for the full time series of a representative leaf of the brown rust data set are given in Table 1.

Table 1. Performance parameters (accuracy, stability, and extrapolation) of the five transformation model types assessing the suitability for the spatial referencing of wheat leaves. The accuracy is measured by the reprojection error of used reference points. The stability is measured by the reproduction error of unused control points within the leaf and the ability to extrapolate is measured by the root mean square error (RMSE) of unused control points at the base and the tip of the leaf. Displayed are the mean results during the whole time series of eleven days (standard deviation in brackets).

	Similarity	Affine	Projective	Polynomial	LWM
Accuracy (px)	1.24 (0.75)	1.23 (0.76)	1.09 (0.75)	0.26 (0.092)	0.19 (0.07)
Stability (px)	1.06 (0.62)	1.08 (0.61)	0.97 (0.60)	0.47 (0.17)	0.36 (0.15)
Extrapolation (px)	2.61 (1.57)	2.61 (1.59)	2.37 (1.59)	0.83 (0.31)	0.76 (0.30)

The extracted quality parameters indicate advantages of using transformation functions with higher flexibility and more model parameters. In each of the three performance parameters accuracy, stability, and extrapolation, the LWM transformations reached the lowest error rates whereas Similarity and Affine transformation obtained the highest error rates.

3.3. Presymptomatic Labeling

The effect of the extended labeling is shown in Figure 6. The manual annotation allows to detect the spore stocks with good accordance with the visual impression. The highly chlorotic tissue at the later observation data was also selected whereas the transition areas to the vital plant tissue is mostly assigned to the healthy class. The extended labelling allows to move this border between vital and chlorotic tissue. The transition areas are now separated from the vital area and regions showing no visual symptoms are detected.

Non-continuously observed points of the leaf surface were excluded causing that the covered area by the fixation strings are widened as it is summed up for each day.

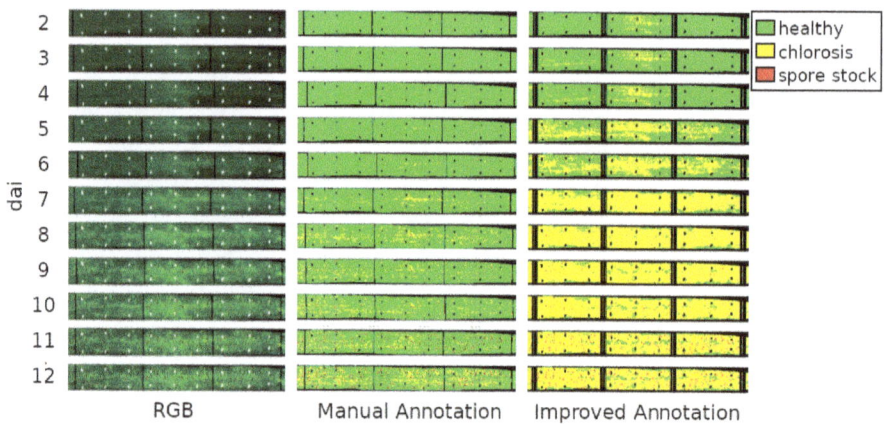

Figure 6. RGB visualization and corresponding classification results of a brown rust time series. Compared are the classification results based on a manual annotation and an improved annotation including the data two days before detection.

4. Discussion

The results show that automated referencing of hyperspectral images is possible. The shown approach enables the tracking of spatial symptoms regarding size and reflection in particular for the spectral area between 400–1000 nm. Furthermore, tracking of the spectral characteristics of diseases over time gives new insights for a biological interpretation and an improved detection. This has been shown for brown rust for periods between 2–10 days after inoculation and for 15–30 days for septoria tritici blotch (Figure 5). The time series uncover similarities as well as differences within the type of effect and its dynamics. For both diseases the NDVI is reduced and the ARI increased indicating the degradation of chlorphylls and the production of anthocyanins, but the change by septoria tritici blotch is much sharper. The PRI shows contrary effects if the two disease: Brown rust induces a decrease whereas septoria tritici blotch induces an increase. PRI is related to the photochemical activity meaning the productivity of photosynthesis [42]. This is surprising as the brown rust permits vital leaf tissue whereas septoria tritici blotch causes necrotic tissue. Such response may be explained by the pigments of the produced brown rust spores interfering with the used bands of PRI [44].

The extended labeling showed high potential to train machine learning models with higher sensitivity even in very early symptom stages (Figure 6). To the best of our knowledge, this is a new approach in hyperspectral close range imaging. In non-imaging setups the early symptoms has been classified [45]. In such scenario a spatial referencing is not required, however, this neglects the high sensitivity for small scale symptoms [6].

Conditions for tracing are proper measurement setup and suitable background material allowing a clear background separation. The background material has to differ significantly from the plant material, wilted leaves, reference points, and disease symptoms. The selected blue paper material fulfills these demands. Minor errors occur at reflecting metal edges of the tray but these can be filtered out by a minimal region size. Same applies for the reference points. They need to be durable enough over the experimental period and, furthermore, need to differ significantly from plant material, wilted leaves, and disease symptoms.

Critical parts include the selection of the transformation model as it is always a compromise between robustness and accuracy. The LWM model provides the highest accuracy (Table 1). Nevertheless, the Polynomial transformation model was selected as it reached very similar results compared to the LWM model which tends to produce distorted image areas due to its local character. As local transformation models depend always on a small part of the information, they are less stable but have the ability to represent also local changes due to wilting. In the present data set, the resulting distortions and twisting effects could not be represented by any of the models. Advantages of the local model were therefore very rare whereas negative effects occur much more frequent especially at the leaf borders.

One disadvantage of the showed method is the use of markers on the leaf tissue. Effects of the marker material on the underlying leaf tissue cannot be completely excluded as well as a possible change of the disease development. However, differences were not observed between marked leaves and unmarked control plants.

Limitations for the method are given by the size of the reference points. Using the shortwave-infrared camera (1000–2500 nm; Specim Spectral Imaging Ltd., Oulu, Finland) of the measurement platform, it was not possible to detect them with the required accuracy. The spatial resolution of the shortwave-infrared camera is by a factor of 10 lower than the spatial resolution of the VISNIR camera [25]. Upcoming hyperspectral cameras may be able to allow the spectral referencing also within different spectral regions.

Further research will focus on the use of interest operators on image data as they are used for motion tracking. In particular the use of the SIFT [32] operator for tracking within RGB Image sequences could be an applicable alternative. This could be adapted to the needs of hyperspectral image sequences using different bands and their spectral relation.

Spatial tracking of markers is a key capability for transferring the findings of the shown experiments to the high throughput greenhouse scale (e.g., [46]). Tracking is needed due to leaf movement (external) and plant growth (internal) which leads to a complex transformation of the complete plant. At the moment, the method is limited to 2D leaves due to the integrated geometrical transformations. Even since the data fusion of hyperspectral images and 3D models is possible [25,26], the referencing has many more degrees of freedom. Changes in the shape of the leaf over time has to be represented within the transformation model having the potential to result in a runaway model complexity. Furthermore, reference points within the 3D model has to be selected, described (e.g., by RIFT descriptor [47]), and assigned to image locations. However, extensive studies are required to represent leaf growing and leaf movements by a compact and applicable model. Modelling the complete development of not only geometry, but also nutrient supply and changes in reflectance is possible when using L-Systems [48] or FSP models (FSPM–functional-structural-plant-models) [49] which use substitution in a grammar structure to model the plant development. By this, the continuous geometric referencing of hyperspectral images provides valuable input data for modelling plant growing and development.

5. Conclusions

This publication introduces a method for referencing of hyperspectral images. Field of application is the improved tracking of hyperspectral information of disease symptoms and their development over time. Results have been shown for symptoms and their development of septoria tritici blotch and brown rust, using a VISNIR camera measuring between 400 and 1000 nm. The potential of the method

J. Imaging **2018**, *4*, 143

has been demonstrated by extracting the dynamic of spectral indices of a single symptom over time. Furthermore, the possibility to annotate invisible symptoms by tracing visible symptoms back in time to the invisible phase of pathogenesis. The concept of spectral tracking over time can contribute to a more dynamic research of disease development instead of focusing to mature symptoms and their appearing in the visible bands.

Author Contributions: J.B., A.K.M. and D.B. conceived and designed the experiments; J.B. and D.B. performed the experiments and analyzed the data; J.B., D.B. and S.P. contributed Figures; J.B., S.P., D.B. and A.K.M. wrote the paper.

Funding: This work was financially supported by BASF Digital Farming.

Acknowledgments: We acknowledge the detailed and constructive remarks of the reviewers.

Conflicts of Interest: The authors declare no conflict of interest.

Abbreviations

ARI	Anthocyanin Reflectance Index
LWM	Local Weighted Mean
NDVI	Normalized Difference Vegetation Index
PRI	Photochemical Reflectance Index
RMSE	Root Mean Square Error
UAV	Unmanned Aerial Vehicle
VISNIR	Visual-nearinfrared

References

1. Mahlein, A.K.; Kuska, M.; Behmann, J.; Polder, G.; Walter, A. Hyperspectral sensors and imaging technologies in phytopathology: State of the art. *Annu. Rev. Phytopathol.* **2018**, *56*, 535–558. [CrossRef] [PubMed]
2. Blackburn, G.A. Hyperspectral remote sensing of plant pigments. *J. Exp. Bot.* **2006**, *58*, 855–867. [CrossRef] [PubMed]
3. Bock, C.; Poole, G.; Parker, P.; Gottwald, T. Plant disease severity estimated visually, by digital photography and image analysis, and by hyperspectral imaging. *Crit. Rev. Plant Sci.* **2010**, *29*, 59–107. [CrossRef]
4. Mahlein, A.K. Plant disease detection by imaging sensors–parallels and specific demands for precision agriculture and plant phenotyping. *Plant Dis.* **2016**, *100*, 241–251. [CrossRef]
5. Kuska, M.; Wahabzada, M.; Leucker, M.; Dehne, H.W.; Kersting, K.; Oerke, E.C.; Steiner, U.; Mahlein, A.K. Hyperspectral phenotyping on the microscopic scale: Towards automated characterization of plant-pathogen interactions. *Plant Methods* **2015**, *11*, 28. [CrossRef] [PubMed]
6. Mahlein, A.K.; Rumpf, T.; Welke, P.; Dehne, H.W.; Plümer, L.; Steiner, U.; Oerke, E.C. Development of spectral indices for detecting and identifying plant diseases. *Remote Sens. Environ.* **2013**, *128*, 21–30, doi:10.1016/j.rse.2012.09.019. [CrossRef]
7. Aasen, H.; Burkart, A.; Bolten, A.; Bareth, G. Generating 3D hyperspectral information with lightweight UAV snapshot cameras for vegetation monitoring: From camera calibration to quality assurance. *ISPRS J. Photogramm. Remote Sens.* **2015**, *108*, 245–259. [CrossRef]
8. Honkavaara, E.; Rosnell, T.; Oliveira, R.; Tommaselli, A. Band registration of tuneable frame format hyperspectral UAV imagers in complex scenes. *ISPRS J. Photogramm. Remote Sens.* **2017**, *134*, 96–109. [CrossRef]
9. Burkart, A.; Aasen, H.; Alonso, L.; Menz, G.; Bareth, G.; Rascher, U. Angular dependency of hyperspectral measurements over wheat characterized by a novel UAV based goniometer. *Remote Sens.* **2015**, *7*, 725–746. [CrossRef]
10. Rouse, J., Jr.; Haas, R.; Schell, J.; Deering, D. Monitoring vegetation systems in the Great Plains with ERTS. In Proceedings of the Third Earth Resources Technology Satellite-1 Symposium, Washington, DC, USA, 10–14 December 1973.
11. Bockus, W.W.; Bowden, R.; Hunger, R.; Murray, T.; Smiley, R. *Compendium of Wheat Diseases and Pests*, 3rd ed.; American Phytopathological Society (APS Press): Sao Paulo, MN, USA, 2010.

12. Camargo, A.; Smith, J. An image-processing based algorithm to automatically identify plant disease visual symptoms. *Biosyst. Eng.* **2009**, *102*, 9–21. [CrossRef]
13. West, J.S.; Bravo, C.; Oberti, R.; Lemaire, D.; Moshou, D.; McCartney, H.A. The potential of optical canopy measurement for targeted control of field crop diseases. *Annu. Rev. Phytopathol.* **2003**, *41*, 593–614. [CrossRef] [PubMed]
14. Bravo, C.; Moshou, D.; West, J.; McCartney, A.; Ramon, H. Early Disease Detection in Wheat Fields using Spectral Reflectance. *Biosyst. Eng.* **2003**, *84*, 137–145, doi:10.1016/s1537-5110(02)00269-6. [CrossRef]
15. Behmann, J.; Mahlein, A.K.; Rumpf, T.; Römer, C.; Plümer, L. A review of advanced machine learning methods for the detection of biotic stress in precision crop protection. *Precis. Agric.* **2015**, *16*, 239–260. [CrossRef]
16. Hartley, R.; Zisserman, A. *Multiple View Geometry in Computer Vision*; Cambridge University Press: Cambridge, UK, 2003.
17. Zitova, B.; Flusser, J. Image registration methods: A survey. *Image Vis. Ccomput.* **2003**, *21*, 977–1000. [CrossRef]
18. Salvi, J.; Matabosch, C.; Fofi, D.; Forest, J. A review of recent range image registration methods with accuracy evaluation. *Image Vis. Comput.* **2007**, *25*, 578–596. [CrossRef]
19. Eling, C.; Klingbeil, L.; Kuhlmann, H. Real-time single-frequency GPS/MEMS-IMU attitude determination of lightweight UAVs. *Sensors* **2015**, *15*, 26212–26235. [CrossRef] [PubMed]
20. Toutin, T. Geometric processing of remote sensing images: Models, algorithms and methods. *Int. J. Remote Sens.* **2004**, *25*, 1893–1924. [CrossRef]
21. Gwo, C.Y.; Wei, C.H. Plant identification through images: Using feature extraction of key points on leaf contours1. *Appl. Plant Sci.* **2013**, *1*, 1200005. [CrossRef]
22. Mouine, S.; Yahiaoui, I.; Verroust-Blondet, A. Combining leaf salient points and leaf contour descriptions for plant species recognition. In Proceedings of the International Conference Image Analysis and Recognition, Povoa do Varzim, Portugal, 26–28 June 2013; Springer: Berlin, Germany, 2013; pp. 205–214.
23. Kolivand, H.; Fern, B.M.; Rahim, M.S.M.; Sulong, G.; Baker, T.; Tully, D. An expert botanical feature extraction technique based on phenetic features for identifying plant species. *PLoS ONE* **2018**, *13*, e0191447. [CrossRef]
24. Gupta, M.D.; Nath, U. Divergence in patterns of leaf growth polarity is associated with the expression divergence of miR396. *Plant Cell* **2015**. [CrossRef]
25. Behmann, J.; Mahlein, A.K.; Paulus, S.; Kuhlmann, H.; Oerke, E.C.; Plümer, L. Calibration of hyperspectral close-range pushbroom cameras for plant phenotyping. *ISPRS J. Photogramm. Remote Sens.* **2015**, *106*, 172–182. [CrossRef]
26. Behmann, J.; Mahlein, A.K.; Paulus, S.; Dupuis, J.; Kuhlmann, H.; Oerke, E.C.; Plümer, L. Generation and application of hyperspectral 3D plant models: Methods and challenges. *Mach. Vis. Appl.* **2016**, *27*, 611–624. [CrossRef]
27. De Vylder, J.; Douterloigne, K.; Vandenbussche, F.; Van Der Straeten, D.; Philips, W. A non-rigid registration method for multispectral imaging of plants. In Proceedings of the 2012 SPIE Defense, Security, and Sensing, Baltimore, MD, USA, 23–27 April 2012; Volume 8369, p. 836907.
28. Bar-Sinai, Y.; Julien, J.D.; Sharon, E.; Armon, S.; Nakayama, N.; Adda-Bedia, M.; Boudaoud, A. Mechanical stress induces remodeling of vascular networks in growing leaves. *PLoS Comput. Boil.* **2016**, *12*, e1004819. [CrossRef] [PubMed]
29. Balduzzi, M.; Binder, B.M.; Bucksch, A.; Chang, C.; Hong, L.; Iyer-Pascuzzi, A.S.; Pradal, C.; Sparks, E.E. Reshaping plant biology: Qualitative and quantitative descriptors for plant morphology. *Front. Plant Sci.* **2017**, *8*, 117. [CrossRef] [PubMed]
30. Wang, X.; Yang, W.; Wheaton, A.; Cooley, N.; Moran, B. Efficient registration of optical and IR images for automatic plant water stress assessment. *Comput. Electron. Agric.* **2010**, *74*, 230–237. [CrossRef]
31. Henke, M.; Junker, A.; Neumann, K.; Altmann, T.; Gladilin, E. Automated alignment of multi-modal plant images using integrative phase correlation approach. *Front. Plant Sci.* **2018**, *9*, 1519. [CrossRef] [PubMed]
32. Lowe, D.G. Distinctive image features from scale-invariant keypoints. *Int. J. Comput. Vis.* **2004**, *60*, 91–110, doi:10.1023/b:visi.0000029664.99615.94. [CrossRef]

33. Yin, X.; Liu, X.; Chen, J.; Kramer, D.M.; others. Multi-leaf alignment from fluorescence plant images. In Proceedings of the IEEE 2014 IEEE Winter Conference on Applications of Computer Vision (WACV), Steamboat Springs, CO, USA, 24–26 March 2014; pp. 437–444.

34. Raza, S.E.A.; Sanchez, V.; Prince, G.; Clarkson, J.P.; Rajpoot, N.M. Registration of thermal and visible light images of diseased plants using silhouette extraction in the wavelet domain. *Pattern Recognit.* **2015**, *48*, 2119–2128. [CrossRef]

35. Raza, S.E.A.; Prince, G.; Clarkson, J.P.; Rajpoot, N.M. Automatic detection of diseased tomato plants using thermal and stereo visible light images. *PLoS ONE* **2015**, *10*, e0123262. [CrossRef]

36. Luhmann, T.; Robson, S.; Kyle, S.; Harley, I. *Close Range Photogrammetry: Principles, Techniques and Applications*; Whittles: Dunbeath, UK, 2006.

37. Fischler, M.A.; Bolles, R.C. Random sample consensus: A paradigm for model fitting with applications to image analysis and automated cartography. *Commun. ACM* **1981**, *24*, 381–395. [CrossRef]

38. Meier, U. *Growth Stages of Mono-and Dicotyledonous Plants*; Blackwell, Wissenschafts-Verlag: Berlin, Germany, 1997.

39. Grahn, H.; Geladi, P. *Techniques and Applications of Hyperspectral Image Analysis*; John Wiley Sons: Hoboken, NJ, USA, 2007.

40. Breiman, L. Random forests. *Mach. Learn.* **2001**, *45*, 5–32. [CrossRef]

41. Goshtasby, A. Image registration by local approximation methods. *Image Vis. Comput.* **1988**, *6*, 255–261. [CrossRef]

42. Gamon, J.; Penuelas, J.; Field, C. A narrow-waveband spectral index that tracks diurnal changes in photosynthetic efficiency. *Remote Sens. Environ.* **1992**, *41*, 35–44. [CrossRef]

43. Gitelson, A.A.; Merzlyak, M.N.; Chivkunova, O.B. Optical properties and nondestructive estimation of anthocyanin content in plant leaves. *Photochem. Photobiol.* **2001**, *74*, 38, doi:10.1562/0031-8655(2001)074<0038:opaneo>2.0.co;2. [CrossRef]

44. Wang, E.; Dong, C.; Park, R.F.; Roberts, T.H. Carotenoid pigments in rust fungi: Extraction, separation, quantification and characterisation. *Fungal Boil. Rev.* **2018**, *32*, 166–180. [CrossRef]

45. Rumpf, T.; Mahlein, A.K.; Steiner, U.; Oerke, E.C.; Dehne, H.W.; Plümer, L. Early detection and classification of plant diseases with Support Vector Machines based on hyperspectral reflectance. *Comput. Electron. Agric.* **2010**, *74*, 91–99. [CrossRef]

46. Kuska, M.T.; Behmann, J.; Grosskinsky, D.K.; Roitsch, T.; Mahlein, A.K. Screening of barley resistance against powdery mildew by simultaneous high-throughput enzyme activity signature profiling and multispectral imaging. *Front. Plant Sci.* **2018**, *9*, 1074. [CrossRef]

47. Lazebnik, S.; Schmid, C.; Ponce, J. A sparse texture representation using local affine regions. *IEEE Trans. Pattern Anal. Mach. Intell.* **2005**, *27*, 1265–1278. [CrossRef] [PubMed]

48. Prusinkiewicz, P.; Lindenmayer, A. *The Algorithmic Beauty of Plants*; Springer: Berlin/Heidelberg, Germany, 1996.

49. Vos, J.; Evers, J.B.; Buck-Sorlin, G.H.; Andrieu, B.; Chelle, M.; de Visser, P.H.B. Functional-structural plant modelling: A new versatile tool in crop science. *J. Exp. Bot.* **2009**, *61*, 2101–2115, doi:10.1093/jxb/erp345. [CrossRef]

Journal of
Imaging

MDPI

Article

A Low-Rate Video Approach to Hyperspectral Imaging of Dynamic Scenes

Charles M. Bachmann [1,*], Rehman S. Eon [1], Christopher S. Lapszynski [1], Gregory P. Badura [1], Anthony Vodacek [1], Matthew J. Hoffman [2], Donald McKeown [1], Robert L. Kremens [1], Michael Richardson [1,†], Timothy Bauch [1] and Mark Foote [1]

1   Chester F. Carlson Center for Imaging Science, Rochester Institute of Technology, Rochester, NY 14623-5603, USA; rse4949@rit.edu (R.S.E.); csl3172@rit.edu (C.S.L.); gpb6751@rit.edu (G.P.B.); vodacek@cis.rit.edu (A.V.); mckeown@cis.rit.edu (D.M.); kremens@cis.rit.edu (R.L.K.); richardson@cis.rit.edu (M.R.); tdbpci@cis.rit.edu (T.B.); mlf9871@rit.edu (M.F.)
2   School of Mathematical Sciences, Rochester Institute of Technology, Rochester, NY 14623-5603, USA; mjhsma@rit.edu
*   Correspondence: bachmann@cis.rit.edu; Tel.: +1-585-475-7238
†   Current address: Harris Corporation, Rochester, NY 14623, USA.

Received: 10 November 2018; Accepted: 26 December 2018; Published: 31 December 2018

Abstract: The increased sensitivity of modern hyperspectral line-scanning systems has led to the development of imaging systems that can acquire each line of hyperspectral pixels at very high data rates (in the 200–400 Hz range). These data acquisition rates present an opportunity to acquire full hyperspectral scenes at rapid rates, enabling the use of traditional push-broom imaging systems as low-rate video hyperspectral imaging systems. This paper provides an overview of the design of an integrated system that produces low-rate video hyperspectral image sequences by merging a hyperspectral line scanner, operating in the visible and near infra-red, with a high-speed pan-tilt system and an integrated IMU-GPS that provides system pointing. The integrated unit is operated from atop a telescopic mast, which also allows imaging of the same surface area or objects from multiple view zenith directions, useful for bi-directional reflectance data acquisition and analysis. The telescopic mast platform also enables stereo hyperspectral image acquisition, and therefore, the ability to construct a digital elevation model of the surface. Imaging near the shoreline in a coastal setting, we provide an example of hyperspectral imagery time series acquired during a field experiment in July 2017 with our integrated system, which produced hyperspectral image sequences with 371 spectral bands, spatial dimensions of 1600 × 212, and 16 bits per pixel, every 0.67 s. A second example times series acquired during a rooftop experiment conducted on the Rochester Institute of Technology campus in August 2017 illustrates a second application, moving vehicle imaging, with 371 spectral bands, 16 bit dynamic range, and 1600 × 300 spatial dimensions every second.

Keywords: hyperspectral; video; imaging; coastal dynamics; moving vehicle imaging; bi-directional reflectance distribution function (BRDF); hemispherical conical reflectance factor (HCRF); stereo imaging; digital elevation model; Virginia Coast Reserve Long Term Ecological Research (VCR LTER)

---

1. Introduction

Hyperspectral imaging has been a powerful tool for identifying the composition of materials in scene pixels. Over the years, a large number of applications have been considered, ranging from environmental remote sensing to identification of man-made objects [1–10]. Some applications involve dynamic scenes which naturally would be well addressed by hyperspectral imaging systems operated at very high data rates. For example, coastal regions with rapidly changing conditions due to the persistent action of tides provide an example of a dynamic landscape where both the water and the land near shore change from moment to moment. Similarly, imaging of moving vehicles provides a challenging but different set of demands which would benefit from a system which can image rapidly.

The coastal zone, in particular, offers a range of important applications where hyperspectral imaging at video rates can have an impact. A wide variety of imaging systems have been used to study near-shore dynamics [11–13]. Considerable effort also has been made to develop hydrodynamic models that attempt to capture the dynamics of flowing sediment in the littoral zone [14], and models of flowing sediment are critical to understanding erosion and accretion processes. At the shoreline, modeling sediment transport and in particular accurately characterizing frictional effects is challenging due to the complicated dynamics as waves break on shore and then retreat [15].

Imaging of the coastal zone has taken many forms. For example, multi-spectral and hyperspectral imaging systems have been used to characterize in water constituents, bottom type, and bathymetry using radiative transfer models [7,16]. In addition, video imaging has been used to estimate flow of the water column and its constituents using video imaging (monochromatic and 3-band multi-spectral) through particle imaging velocimetry (PIV) [17–19]. Limitations of these past approaches are that traditional airborne and satellite remote sensing, while providing important details about sediment concentrations near shore, have produced essentially an instantaneous look at what is in fact a dynamical system. At the same time, while video systems have been used to image the water column and model its flow, the limited number of bands has meant that little information is available from these systems regarding in water constituents or bottom properties.

Relatively recently, the commercial marketplace has begun to deliver sensors which are advancing toward the long-term goal of high-frame rate hyperspectral imagery. Several different approaches have been taken, which include the use of so-called "snapshot" imaging systems [20,21]. Some hyperspectral imaging systems that fall into this category use a Fabry–Perot design [22]. However, the signal-to-noise ratio (SNR) that has been achieved by existing systems or those under development [23] is typically lower than that obtained with conventional hyperspectral imaging systems. In some cases, other trade-offs must be made to obtain comparable performance such as using fewer spectral bands or spatial pixels. Recently reported results using a snapshot imaging system based on a Fabry–Perot design indicate other issues such as mis-registration of band images during airborne data acquisition since whole band images are acquired sequentially in time [24]; this same system delivers image cubes that have 1025 × 648 spatial pixels with only 23 spectral bands and 12 bits per pixel acquired within 0.76 s, meaning that the data volumes recorded are only 9% of the data rates achieved below by our approach in comparable time (0.67 s). For our applications and the scientific goals described later in this Section, the 12-bit dynamic range found in the system described in [24] and in other snapshot hyperspectral imaging systems [25] is too limited, and this is one of the motivating factors for our having designed an overall system uses a hyperspectral line scanner with 16-bit dynamic range. On the other hand, progress has been made in co-registration of the mis-aligned band images captured via snapshot hyperspectral imaging, with one recent work demonstrating mis-registration errors of ≤0.5 pixels [26]. Some designs have included much smaller conventional hyperspectral imaging arrays that have been resampled then to a panchromatic image acquired simultaneously [27]. Of course, the potential limitation for these systems is that they may not produce hyperspectral image sequences that are truly representative of what would be recorded in an actual full-resolution imaging spectrometer.

At the same time, among conventional imaging spectrometer designs, the commercial marketplace, driven by consumer demand for portable imaging technologies as well as unmanned aerial systems (UAS) in a number of important commercial application areas, has led to cost-effective improvements to spectrographs with progressively greater sensitivity, and this in combination with improvements to data capture capabilities have together led to hyperspectral imaging systems that can frame at high rates while maintaining high data quality (low aberration) as well as excellent spectral and spatial resolution [28,29]. The current generation of conventional imaging systems requires shorter integration times, and therefore can record a line of hyperspectral pixels at much higher rates, in the 200–400 Hz range. Most of these spectrometers are incorporated in systems that operate as line scanners. The hyperspectral line-scanner incorporated in the system described in this paper is an example of a system operating with data rates in this range [30]. Line scanners such as this have been the norm in so called "push-broom" imaging system design, used in hyperspectral imaging from space- and air-borne systems [31–39], where the motion of the platform produces one spatial dimension, the along-track spatial dimension, of the image data cube.

In developing a system such as the the one described here in this paper, we had several specific objectives. For coastal applications, our objectives included: (1) to be able to acquire short-time-interval hyperspectral imagery time series to support long-term goals of modeling both dynamics of the near-shore water column including the mapping of in water constituents (suspended sediments, color-dissolved organic matter (CDOM), chlorophyll, etc.) and their transport, (2) to capture near-shore land characteristics (sediments and vegetation) and in particular change in sediments on short time scales due to the influence of waves and tides, (3) more broadly to be able to image from a variety of geometries from the same location in order to obtain multi-view imagery from samples of the bi-directional reflectance distribution function for use in retrieval of geophysical parameters of the surface through inversion of radiative transfer models [40] and for construction of digital surface models (DSM) to enhance derived products and contribute to validation, (4) to image at very fine-scale spatial resolutions (mm to cm in the near range) in order to derive water-column and land surface products as just described on scales where variation might occur and to consider how products derived at these resolutions then scale up to more traditional scales so often used in remote sensing from airborne and satellite platforms where resolutions have often been measured in meters, to tens of meters, or greater; and (5) to acquire imagery for all of these purposes with a hyperspectral imager with sufficient dynamic range that retrieval in both the water column and on land would be possible.

Our objectives for moving vehicle applications overlap a number of those just described for the water column, especially goals (1), (3) and (5) listed above. Our objectives here were: (a) to obtain short-time interval hyperspectral imagery, which is critical to identification and tracking of moving vehicles while minimizing distortion due to vehicle movement; (b) to be able to characterize BRDF effects for moving vehicles, and (c) to ensure that shadows due to occlusions and nearby structures in the vicinity of moving vehicles could be better characterized.

Our approach in this paper uses the very high data rates found in modern hyperspectral line scanners to achieve a low-rate hyperspectral video acquisition system. Our overall system design incorporates a modern hyperspectral imaging spectrometer integrated into a high-speed pan tilt system with onboard Inertial Measurement Unit Global Positioning System (IMU-GPS) for pointing and data time synchronization. In field settings, the system is deployed from a telescopic mast, meeting our objectives (3) and (4). By combining these components into one integrated system, we describe how a time sequence of hyperspectral images can be acquired at ~1.5 Hz, thus operating as a low-rate hyperspectral video of dynamic scenes and satisfying objectives (1), (2), and (a). In order to meet objectives (5) and (c) above, the imaging system that we selected had 16-bit dynamic range. Traditional video systems have been used to examine coastal regions in the past, however, these have been primarily monochromatic [15] or multi-spectral imaging systems [17] with a very limited number of spectral bands (typically 3 bands); these systems provide a more limited understanding of the dynamics of the littoral zone and have been primarily used to estimate current flow vectors. Similarly, previous studies have recognized the potential of spectral information to improve persistent vehicle tracking, but most tracking studies have used panchromatic or RGB imaging due to the cost and availability of spectral imaging equipment [41–47].

2. Approach

2.1. Low-Rate Hyperspectral Video System

At the heart of our approach is a state-of-the-art Headwall micro High Efficiency (HE) Hyperspec [30]. This system is advertised to achieve "frame rates" of up to 250 Hz. Here the term "frame rate" refers to the rate at which a line of hyperspectral pixels can be acquired and stored in a data capture unit. Our Headwall micro HE Hyperspec E-Series is a hyperspectral line scanner with 1600 across-track spatial pixels and 371 spectral pixels, with 16-bit dynamic range. Headwall currently manufactures both visible and near infrared (VNIR) as well as short-wave infrared (SWIR) versions of the Hyperspec. This paper describes an overall system design in which a VNIR Hyperspec is the imaging unit of the system.

Our design integrates (Figure 1) a Headwall Hyperspec into a high-speed maritime-rated General Dynamics Vector 20 pan-tilt unit [48]. Along-track motion of the Headwall Hyperspec line-scanner is accomplished by nodding of the pan-tilt unit. A Vectornav VN-300 IMU-GPS [49] is also integrated to provide pointing information for the system as well as GPS time-stamps for acquired hyperspectral data. In field settings, we mount the integrated Headwall Hyperspec and General Dynamics pan-tilt and the Headwall compact data unit atop a BlueSky AL-3 telescopic mast [50] which can raise the system from 1.5–15 m above the ground. Integration of the Hyperspec, General Dyanmics pan-tilt, and Vectornav VN-300 GPS-IMU was accomplished by Headwall under contract to RIT, and under the same contract, Headwall modified their Hyperspec data acquisition software to meet our RIT data acquisition specifications. The key data acquisition features allow direct user control of camera parameters such as integration time as well as rates of azimuthal slewing and nodding in the zenith direction of the pan-tilt system. Additional engineering, including development of custom mounting plate for the General Dynamics pan-tilt containing the Headwall Hyperspec and the Vectornav GPS-IMU components, as well as development of a field portable power supply to provide power to all components, was undertaken at RIT to further integrate the Headwall/General-Dynamics/Vectornav configuration onto the BlueSky AL-3 telescopic mast to make the final configuration field-ready.

The control software allows a variety of scan sequences to be implemented. This includes nodding at the same azimuthal orientation, typical for hyperspectral video modes, where bi-directional scanning is used to maximize hyperspectral data acquisition rates, as well as scanning sequences that step in azimuth between image frames, in combination with the normal zenith nodding mode used to produce the along track motion for each full hyperspectral image frame.

Our current instrument configuration incorporates a 12 mm lens on the Headwall Hyperspec imaging system. When operated from our telescopic mast with this lens, the Headwall system provides very fine scale hyperspectral imagery with a GSD in the millimeter to centimeter range. A table of GSD values obtainable with our system at various mast heights and distances from the mast appears in Table 1.

We note that in its current configuration, when the General Dynamics pan-tilt housing is leveled, it has a maximum deflection angle above or below the horizontal of 34°. This, however, is not a permanent limitation as the addition of a rotational stage in future planned upgrades will allow the system to reach and measure hyperspectral data from a much broader range of zenith angles.

2.2. System Calibration

To characterize the system, we used our calibration facility, which includes a LabSphere Helios (Labsphere, North Sutton, NH, USA) 0.5 m diameter integrating sphere [51] paired with a calibrated spectrometer to collect radiance data in order to derive system calibration curves, signal-to-noise ratio (SNR), and noise-equivalent spectral radiance (NESR). In the examples provided here, we show results for an integration time of 2.5 ms, which is the integration time used in the surf zone hyperspectral imagery time series example provided below.

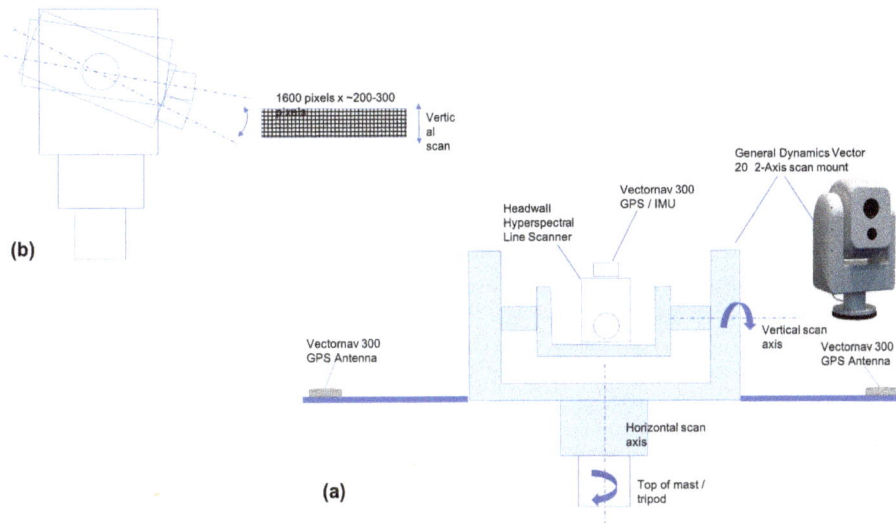

Figure 1. Hyperspectral video imaging concept. (a) Headwall Hyperspec HE E-Series hyperspectral line scanner and Vectornav 300 GPS/IMU integrated into the General Dynamics Vector 20 high-speed pan-tilt unit. (b) Nodding motion of the pan-tilt provides the along-track motion normally produced by movement of an aircraft when these types of imaging systems are used in an airborne platform.

Table 1. GSD (m).

Height of Camera (m)	1	2	3	4	5	6	7	8	9	10	15
Distance from Mast (m)											
1	0.0008	0.0012	0.0017	0.0022	0.0028	0.0033	0.0038	0.0044	0.0049	0.0054	0.0081
2	0.0012	0.0015	0.0020	0.0024	0.0029	0.0034	0.0039	0.0045	0.0050	0.0055	0.0082
5	0.0028	0.0029	0.0032	0.0035	0.0038	0.0042	0.0047	0.0051	0.0056	0.0061	0.0086
10	0.0054	0.0055	0.0057	0.0058	0.0061	0.0063	0.0066	0.0069	0.0073	0.0077	0.0098
15	0.0081	0.0082	0.0083	0.0084	0.0086	0.0088	0.0090	0.0092	0.0095	0.0098	0.0115
20	0.0109	0.0109	0.0110	0.0111	0.0112	0.0113	0.0115	0.0117	0.0119	0.0121	0.0136
25	0.0136	0.0136	0.0136	0.0137	0.0138	0.0139	0.0141	0.0142	0.0144	0.0146	0.0158
30	0.0163	0.0163	0.0163	0.0164	0.0165	0.0166	0.0167	0.0168	0.0170	0.0171	0.0182
35	0.0190	0.0190	0.0190	0.0191	0.0192	0.0192	0.0193	0.0195	0.0196	0.0197	0.0206
40	0.0217	0.0217	0.0217	0.0218	0.0218	0.0219	0.0220	0.0221	0.0222	0.0223	0.0232
45	0.0244	0.0244	0.0244	0.0245	0.0245	0.0246	0.0247	0.0248	0.0249	0.0250	0.0257
50	0.0271	0.0271	0.0271	0.0272	0.0272	0.0273	0.0274	0.0274	0.0275	0.0276	0.0283
55	0.0298	0.0298	0.0299	0.0299	0.0299	0.0300	0.0301	0.0301	0.0302	0.0303	0.0309
60	0.0325	0.0325	0.0326	0.0326	0.0326	0.0327	0.0327	0.0328	0.0329	0.0330	0.0335

Through various system ports, our integrating sphere is configured with three different light sources (a quartz tungsten halogen (QTH) bulb and two highly stable xenon plasma arc lamps), a VNIR spectrometer (Ocean Optics, Largo, FL, USA), and two single point broad band detectors, one silicon detector (Hamamatsu Photonics, Hamamatsu City, Japan) measuring in the visible portion of the spectrum, and an Indium Gallium Arsenide (InGaAs) detector (Teledyne Judson, Montgomeryville, PA, USA) measuring the total energy in the shortwave infrared. The external illumination sources allow the instrument to be utilized as a source capable of outputting a constant illumination across the entire 0.2 m exit port, and the radiometrically calibrated detectors are capable of measuring internal illumination conditions. For calibration purposes, the sphere operates as a source, utilizing the two high intensity plasma lamps (Labsphere, North Sutton, NH, USA), which are capable of producing almost full daylight illumination conditions through the exit port.

In our calibration, we use typically in the range of 10–30 different illumination levels, and we average 255 scans at the desired integration time. The Xenon plasma lamps provide a highly stable illumination source for the measurements, however, in order to minimize any residual instrument drift, we use two sets of dark current measurements, one before and one after imaging system measurements for the various illumination levels provided by the integrating sphere. According to manufacturer specifications each lamp maintains an approximate correlated color temperature of 5100 K $\pm$ 200 K with rated lifetime of 30,000 h. Because the Plasma External Lamps (PEL) are microwave induced sources the emitter requires feedback to maintain desired light levels. This fluctuation results in a 0.1 Hz sawtooth shaped waveform. In rest mode short term stability is $\pm$3% from peak to peak (P-P) resulting in 6% change in magnitude of desired output. To further reduce error, Labsphere has implemented a Test Mode, during which the short term drift is $\pm$0.5% P-P, (0.6% magnitude). Test Mode can only be maintained for a 30 min period, after which the system requires a minimum of 5 min before the next activation cycle. The long term stability reported by Labsphere for every 100 h is less than 1%. Correlated Color Temperature (CCT) change for the same time period was reported to be <100 K. Lastly observed spectral stability had fluctuations <0.5 nm for every 10 h. The quoted stability values provided here are manufacturer specifications, indicating expected performance. To obtain the results provided below, we used the Test Mode during data collection.

To develop calibrations for each wavelength, we perform a linear regression between the NIST-traceable light levels (radiance) as recorded by the onboard spectrometer attached to our integrating sphere and the recorded radiance at each wavelength in our Headwall Hyperspec imaging system. We measure system dark current by blocking the entrance aperture with the lens cap in the dark room of the calibration facility. Figure 2 shows the noise equivalent spectral radiance (NESR) and signal-to-noise ratio (SNR) for the 2.5 ms integration time used by our Headwall system during the acquisition of the hyperspectral imagery time series of the surf zone described later in this paper. The curves correspond to different illumination levels, varying from the base noise of the system to just below the saturation limit of the detector at 30,000 electrons. The SNR curves in Figure 2 show that at near full daylight levels, the peak SNR in the visible part of the spectrum is around 150, while at 0.9 μm in the near infra-red, the SNR drops to around 40. Note that the spatial resolution of our system is usually quite high (mm to cm range, as shown in Table 1, depending on the height of the mast and proximity of the ground element to the sensor). Thus, if higher SNR is desired, spatial binning by even a modest amount can provide significant enhancements; for example a 3 × 3 spatial window would provide a peak SNR of 450 at the peak in the visible and 120 at 0.9 μm.

2.3. Imaging the Dynamics of the Surf Zone

Imaging of the coastal zone has taken many forms. For example, multi-spectral and hyperspectral imaging systems have been used to quantify concentrations of water constituents and characterize bottom type and bathymetry using radiative transfer models [7,52]. In addition, video imaging has been used to estimate flow of the water column and its constituents using video imaging (monochromatic and 3-band multi-spectral) through particle imaging velocimetry (PIV) [18,19]. Limitations of these past approaches are that traditional airborne and satellite remote sensing, while providing important details about sediment concentrations near shore, have produced essentially an instantaneous look at what is in fact a dynamical system. At the same time, while video systems have been used to image the water column and model its flow, the limited number of bands has meant that little information is available from these systems regarding in water constituents or bottom properties.

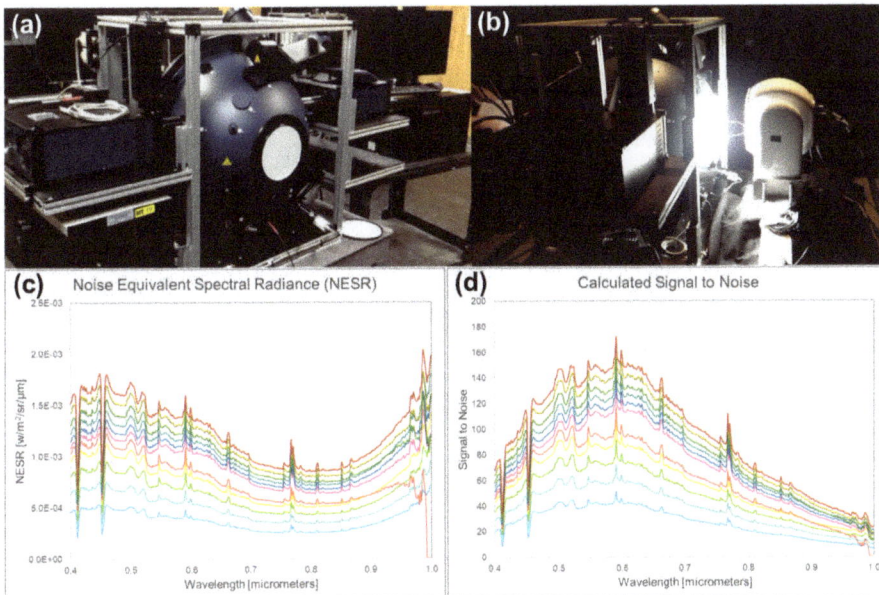

Figure 2. (**a**) Labsphere Helios 0.5 m diameter integrating sphere in our Rochester Institute of Technology (RIT) calibration laboratory. Plasma lamps, attached to the sphere, are visible on the top shelves to the left and right. (**b**) Our Headwall imaging system in the pan-tilt unit in front of the sphere during calibration. (**c**) typical NESR curves for 10 light levels up to the maximum output of the two plasma lamps, near daylight levels. (**d**) Typical SNR obtained over the same 30 light levels for a 2.5×10^{-3} s integration time. Hyperspectral video sequences shown in this paper used either a 2.5×10^{-3} s or 3.0×10^{-3} s integration time.

The dynamics of sediment flow in coastal settings plays a significant role in the evolution of shorelines, determining processes such as erosion and accretion. As sea levels continue to rise, improved modeling of the evolution of coastal regions is a priority for environmental stewards, natural resource managers, urban planners, and decision makers. Understanding the details of this evolution is critical and improved knowledge of the dynamics of flowing sediment near shore can contribute significantly to hydrodynamic models that ultimately predict the future of coastal regions. Imaging systems have been used to acquire snapshots of the coastal zone from airborne and satellite platforms. Multi-spectral and especially hyperspectral imaging systems can provide an instantaneous look at the distribution of in-water constituents, bottom-type, and depth; however, these have not produced a continuous time series that looks at the short time scale dynamics of the flowing sediment. Video systems have also been used to examine coastal regions, however, these have been primarily monochromatic [15] or multi-spectral imaging systems [17] with a very limited number of spectral bands (typically 3 bands); these systems provide a more limited understanding of the dynamics of the littoral zone and have been primarily used to estimate current flow vectors without the ability to determine local particle densities. In the Results section below, we demonstrate a low-rate hyperspectral video time series. This imaging demonstration offers the advantage of bringing together the power of spectral imaging to estimate in water constituents and bottom type along with low-rate video that offers the potential to track the movement of these in-water constituents on very short time scales.

2.4. Real-Time Vehicle Tracking Using Hyperspectral Imagery

In recent years, vehicle detection and tracking has become important in a number of applications, including analyzing traffic flow, monitoring accidents, navigation for autonomous vehicles, and surveillance [41,47,53–55]. Most traffic monitoring and vehicle movement applications use relatively high-resolution video and have a high number of pixels on each vehicle, which allows tracking algorithms to rely on appearance features in the spatial domain for detection and identification. Tracking from airborne imaging platforms, on the other hand, poses several unique challenges. Airborne imaging systems typically have fewer pixels representing each vehicle within the scene due to the longer viewing distance, as well as being prone to blur or smear due to the relative motion of the sensor and the object and parallax error [41,56]. Beyond imaging system limitations, vehicle tracking/detection algorithms also must be able to handle complex, cluttered scenes that include traffic congestion and occlusions from the environment [41,47]. Occlusions are more common in airborne images and are particularly challenging for persistent vehicle tracking. When a tracked vehicle is obscured by a tree or a building, it is common for it to be assigned a new label once it reemerges. This can be avoided if the vehicle can be uniquely identified, however while traditional tracking methods can rely on high-resolution spatial features, the low resolution of airborne imagery, where the object is only represented by 100–200 or fewer pixels, makes reidentification by spatial features difficult and can lead to the tracker following a different vehicle or dropping the track entirely [41,56].

Compared to panchromatic or RGB systems, hyperspectral sensors can more effectively identify different materials based on their spectral signature and can thereby provide additional spectral features that can reidentify vehicles. Vodacek et al. [47] and Uzkent et al. [56] suggested using a multi-modal sensor design consisting of a wide field of view (FOV) panchromatic system alongside a narrow FOV hyperspectral sensor for real-time vehicle tracking and developed a tracking method leveraging the spectral information. Due to the lack of hyperspectral data, the method—along with subsequent additions to the tracking system—has only been tested on synthetic hyperspectral images generated by the Digital Imaging and Remote Sensing Image Generation model [41,56]. Results using the synthetic data have demonstrated that the spectral signatures can provide the necessary information to isolate targets of interest (TOI) in occluded backgrounds. This can be especially important when tracking vehicles in highly congested traffic or in the presence of dense buildings or trees within the scene. Experiments in cluttered synthetic scenes have shown that utilizing the spectral data outperforms other algorithms for persistent airborne tracking [56]. In addition, there has been increased recent interest in using advanced computer vision and machine learning algorithms to efficiently exploit the large amount of information contained in hyperspectral video, but no data sets currently exist with which to train—much less validate—a neural network model.

3. Results

3.1. Hyperspectral Data Collection Experiment

The first demonstration of the hyperspectral low-rate video imaging concept that we have described took place during an RIT experiment on Hog Island, VA, a barrier island which is part of the Virginia Coast Reserve (VCR) [57], a National Science Foundation Long-Term Ecological Research (LTER) site [58]. Over an 11-day period, the imaging system was used repeatedly from atop the BlueSky telescopic mast system (Figure 3) to acquire a wide variety of hyperspectral imagery of the island. By integrating the system onto the telescopic mast, the system is also able to acquire imagery from the same region on the ground, or of the water, from multiple viewing geometries, allowing the bi-directional reflectance distribution function (BRDF) of the surface to be sampled in collected imagery (Figure 4). For field data collections such as these, the term hemispherical conical reflectance factor (HCRF) is also sometimes used as a descriptor since: (a) the sediment radiance is compared with the radiance of a Lambertian standard reference (Spectralon panel), (b) the sensor has a finite aperture, and (c) the primary illumination source is not a single point source but contains both direct (solar) and indirect sources of illumination (skylight and adjacency effects) [59–61]. In our experiment, described in greater detail below, we deployed our hyperspectral field-portable goniometer system, the Goniometer of the Rochester Institute of Technology-Two (GRIT-T) [62] for direct comparison with hyperspectral imagery acquired from our Headwall integrated hyperspectral imaging system at varying heights on the telescopic mast. HCRF of the surface provides information on the geophysical state of the surface, such as the fill factor, which can be inferred by inverting radiative transfer models [40], which has been done previously using GRIT-T hyperspectral multi-angular data [63].

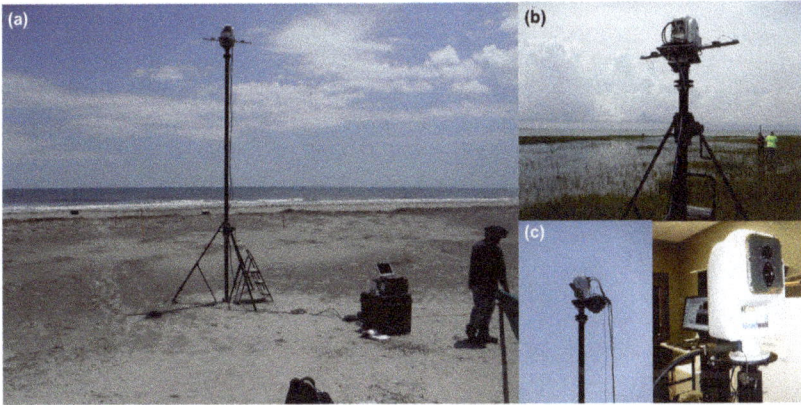

Figure 3. Data collection with the integrated imaging system: Headwall Hyperspec imaging system, General Dynamics maritime pan-tilt, and Vectornav-300 GPS IMU atop a BlueSky AL-3 telescopic mast. (**a**) on the western shore of Hog Island, VA while imaging littoral zone dynamics; (**b**) on the eastern shore imaging coastal wetlands; (**c**) close-ups of the imaging system while in operation at Hog Island and in the lab during testing. Closer to the shoreline in (**a**), white Spectralon calibration panels are deployed; also visible are various fiducials (orange stakes) used for image registration and geo-referencing. Fiducials were surveyed with real-time kinematic GPS.

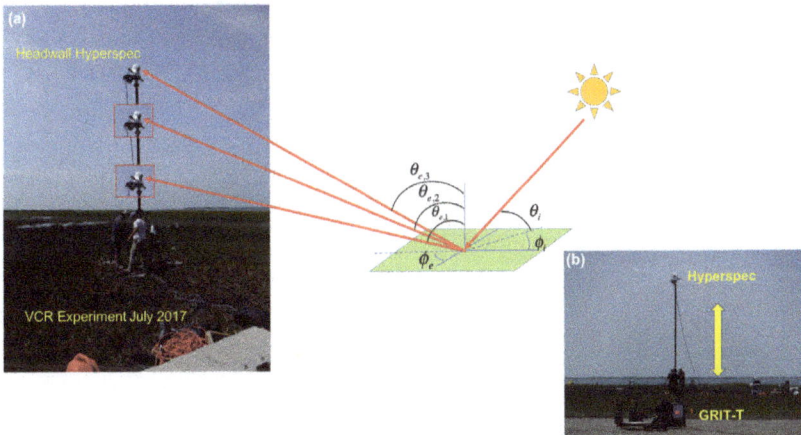

Figure 4. (**a**) Hyperspectral HCRF imagery sequences from our integrated hyperspectral Hyperspec imaging system atop a telescopic mast. Mast height determines view zenith angle. (**b**) The Hyperspec imaging a salt panne region during the July 2017 experiment on Hog Island while the Goniometer of the Rochester Institute of Technology-Two (GRIT) [62] records HCRF from the surface.

3.2. Digital Elevation Model and HCRF from Hyperspectral Stereo Imagery

The hyperspectral imagery acquired from differing viewing geometries of the same surface also allows us to develop digital elevation models (DEMs) of the surface which can be merged with the hyperspectral imagery and used in modeling and retrieval of surface properties. An example DEM-derived from the multi-view hyperspectral imagery that we acquired at our study site on Hog Island in July 2017 appears in Figure 5, which shows the resulting DEM from a set of fourteen hyperspectral scenes acquired from our mast-mounted system. DEM construction used the structure from motion (SFM) algorithm PhotoScan developed by AgiSoft LLC [64]. Similar results have been obtained from stereo views of a surface from unmanned aerial system (UAS) platforms [38] as well as from a ground-based hyperspectral imaging system [65], although the latter result was obtained from significantly longer distances, requiring the use of atmospheric correction algorithms. The example provided in Figure 5 was for stand-off distances significantly less than those for which atmospheric correction would be necessary.

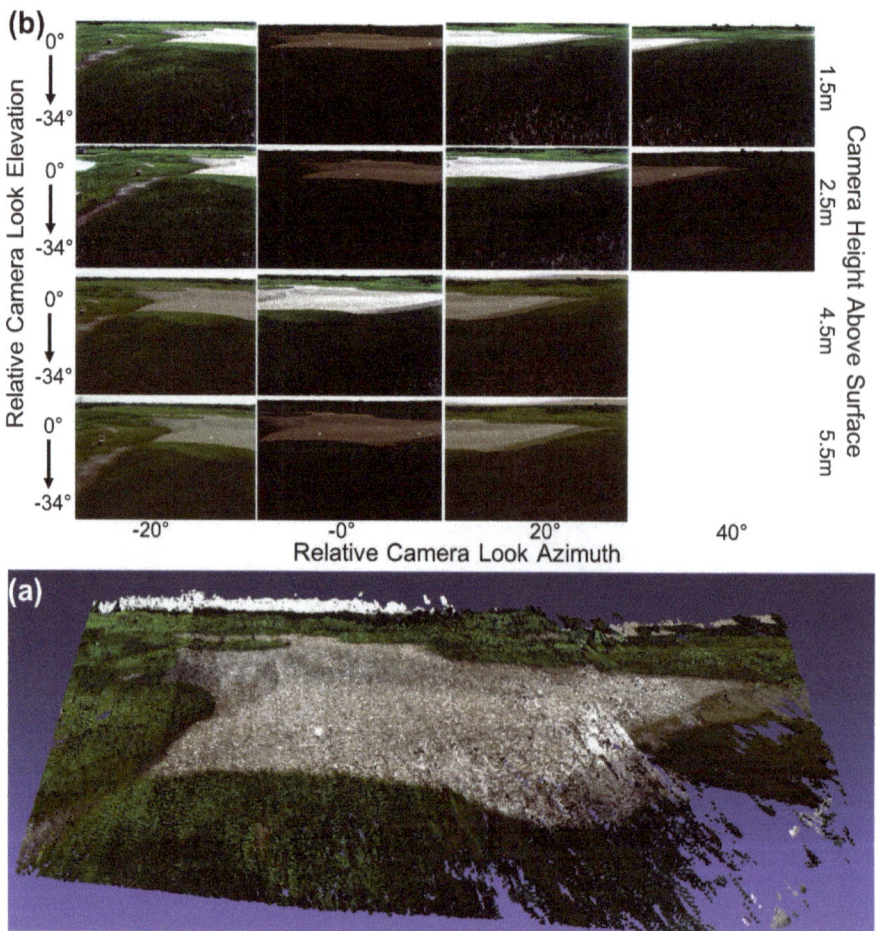

Figure 5. (a) DEM-derived from multi-view imagery from our mast-mounted hyperspectral system. (b) the fourteen hyperspectral scenes used as input to a Structure-from-Motion (SFM) algorithm. These hyperspectral scenes had spatial dimensions 1600 × 971 with 371 spectral bands.

Each image of the set of 14 used in creating the DEM is a hyperspectral scene acquired with the full 371 spectral bands from 0.4–1.0 μm, 1600 across track spatial pixels, and 971 spatial pixels in the second spatial dimension produced by the nodding of the pan-tilt. Each row was produced from a series of scans that overlapped in azimuth and were acquired at different mast heights. In these examples, the height of the hyperspectral imager above the surface during image acquisition was respectively 1.5 m, 2.5 m, 4.5m, and 5.5 m. Each scene shows a salt panne surrounded by coastal salt marsh vegetation, predominantly *Spartina alterniflora*.

Having the ability to produce a DEM as part of the data collection workflow has potential advantages. Lorenz et al. [65] used this information to correct for variations in illumination over rocky outcrops by determining the true angle of the sun to the surface normal derived from the DEM. For our own workflow, which is focused on problems such as inversion of radiative transfer models to retrieve geophysical properties of the surface [40,63], we require both the true viewing zenith and azimuth angles of our imaging system in the reference frame of the tilted surface normal as well as the incident zenith and azimuth angles of solar illumination within this tilted coordinate system. The onboard GPS-IMU of our mast-mounted imaging system, provides pointing (view orientation) and timestamps which together with the DEM allow the calculation of these angles. Fiducials placed in the scene enhance the overall accuracy of these angle calculations.

The fourteen scenes portrayed in Figure 5 also represent another important aspect of our overall approach described earlier in Section 3.1: the acquisition of multi-view imagery that sub-sample the HCRF distributions that form the core of inversion of radiative transfer models to retrieve geophysical properties of the surface [40,63] and satisfy goal (3) stated in the Introduction. Figure 6 shows examples of the spectral reflectance derived from 4 of the 14 scenes acquired from the salt panne at different mast heights, which as Figure 4 illustrates, provide us with a sub-sample of the HCRF. We have previously demonstrated an approach to retrieving sediment fill factor from laboratory bi-conical reflectance factor (BCRF) measurements [63] and then extended this to retrieval from multi-view hyperspectral time-series imagery acquired by NASA G-LiHT and multi-spectral time series imagery from GOES-R [40]; in each case, these retrievals represented a more restricted sub-sample of points from the HCRF distribution. Imagery from our mast-mounted system can allow us to more completely validate the inversion of this modified radiative transfer model to retrieve and map sediment fill factor from imagery and in particular help in assessing how many and which views of the surface are most critical for successful inversion.

3.3. Low-Rate Hyperspectral Video Image Sequence of the Surf Zone

On 14 July 2017, our integrated system was used for the first time in the low-rate video mode to acquire imagery of the surf zone on the eastern shore of Hog Island, VA. One such image sequence is shown in Figure 7, which shows a subset of a longer sequence of images acquired every 0.67 s. Each image in the scene is 1600 across-track pixels (horizontal dimension) with 371 spectral bands each by 212 along-track pixels (vertical dimension produced by the nodding motion of the General Dynamics pan-tilt unit). The integration time for each line of 1600 across-track spatial pixels with 371 spectral pixels each was approximately 2.5 ms. Once other latencies in data acquisition are accounted for and the necessary time is allowed for the pan-tilt to reverse direction, the acquisition rate of 0.67 s for the full hyperspectral scene is achieved. We emphasize that there is no specific limitation of the system that prevents longer integration times and/or slower slewing rates from being used, and other data collected during the experiment did use slower scan rates to obtain larger scenes (see for example the 14 hyperspectral scenes in Figure 5 and the long scan in the lower right portion of Figure 7 which shows a scene with 1600 × 2111 spatial pixels and 371 spectral bands acquired over a 12-s interval during a very slow scan with a longer integration time). The latter hyperspectral scene, in particular, represents our goal (4) stated in the Introduction of being able to produce mm- to cm-scale imagery in the near range to better characterize the land surface at scales typical of the variation found near the waterline. However, in the hyperspectral low-rate video mode, image frames of the size shown in Figure 7 are typical. The quality of the spectra obtained in the imagery is indicated by the spectral time series derived from a small 5 × 5 window near the shoreline over time. A well-known local minimum in the liquid water absorption spectrum [66] normally appears in very shallow waters as a peak in the reflectance spectrum around 810 nm. This peak is well correlated with shallow water bathymetry typically in depths that are ≤1 m , and this feature was previously used in a shallow water bathymetry retrieval algorithm and demonstration which compared favorably with bathymetry directly measured in situ [67]. Obtaining spectral data of sufficient quality is important to the success of retrievals based on spectral features, such as the 810 nm feature just described, band combinations and regressions based on band combinations [68,69], "semi-analytical" models [70,71], or inversion of forward-modeled look-up tables generated from radiative transfer models such as Hydrolight [7], which rely on the spectral and radiometric accuracy of the hyperspectral data. These short-time-scale hyperspectral imagery sequences satisfy our stated goals (1) and (2).

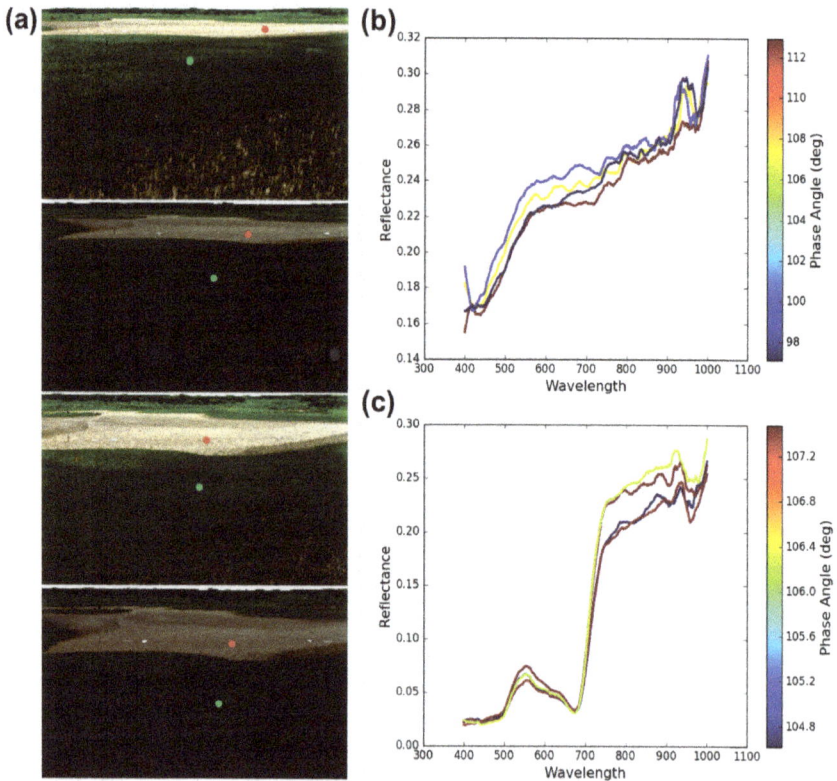

Figure 6. (a) Enlargement of four hyperspectral scenes acquired from four mast heights in the set of fourteen shown in Figure 5. (b,c) Set of spectra from the same location in each of the four scenes for (b) a position (red dot) in the salt panne , and (c) a position (green dot) in the salt marsh vegetation.

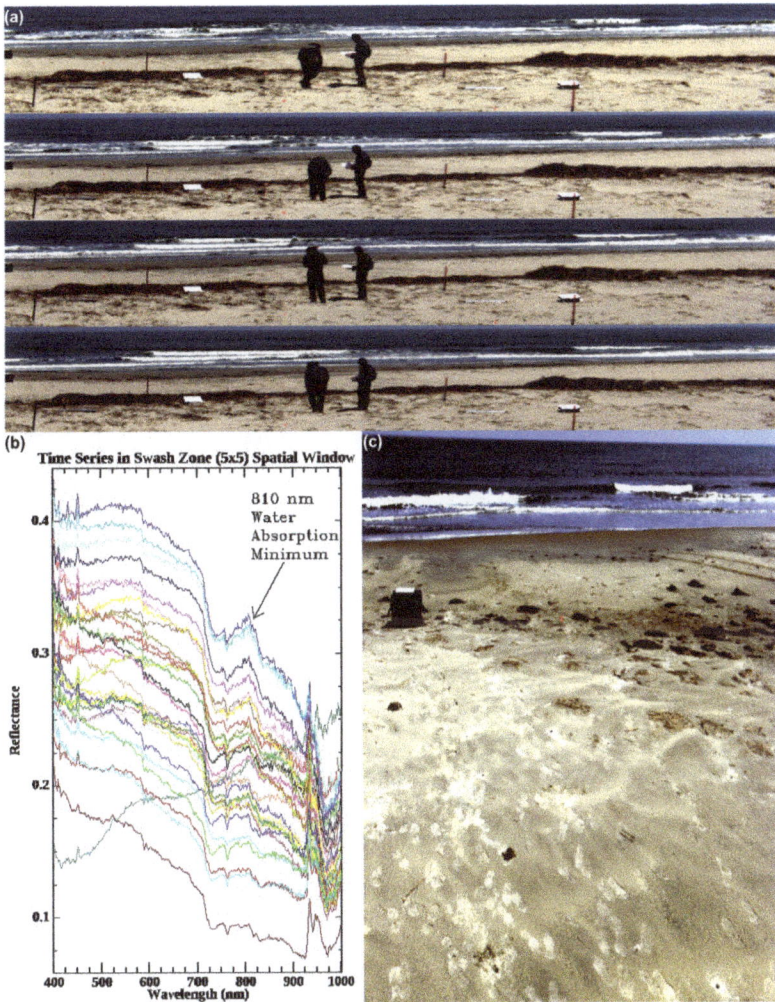

Figure 7. (**a**) Hyperspectral video image sequence using our integrated Headwall micro-HE VNIR hyperspectral imaging system on Hog Island, VA on 14 July 2017. The representative sequence subset (from a time series of 30 images) shown here contains hyperspectral image frames with spatial dimensions 1600 × 212 each with 371 spectral bands. Each hyperspectral scene was acquired approximately once every 0.67 s. Two Spectralon reference panels used in reflectance calculations and several orange fiducial stakes used in geo-referencing are also visible. (**b**) Spectral reflectance captured by the integrated system for a co-registered pixel in the swash zone of the hyperspectral video image sequence. The spectral reflectance, for a 5 × 5 spatial window, is shown over all 30 hyperspectral images, which were acquired once every 0.67 s. The 810 nm peak indicated corresponds to a well-known minimum in the water absorption spectrum [66], well-correlated with shallow water bathymetry [67]. (**c**) Slow scan/longer integration time hyperspectral scene with 1600 × 2111 spatial pixels and 371 spectral channels acquired closer to the waterline with the system deployed at 1.5 m height, showing the mm- to cm-scale resolution possible: the details of footprints can be clearly seen.

3.4. Time Series Hyperspectral Imagery of Moving Vehicles

A second test of the video capabilities of our hyperspectral imaging system was performed on 9 August 2017 at RIT. For this experiment, imaging of moving vehicles was the primary focus. The same instrument configuration was used from atop the Chester F. Carlson Center for Imaging Science at RIT (Figure 8). Slewing rates of the pan-tilt as well as integration time of the Headwall micro HE were adjusted to achieve a larger image in the along-track dimension (zenith or nodding dimension). The integration time for this data collection was 3.0 ms, and the images in the sequence were acquired once every second. These images have 1600 across-track pixels and 299 along-track pixels with 371 spectral bands.

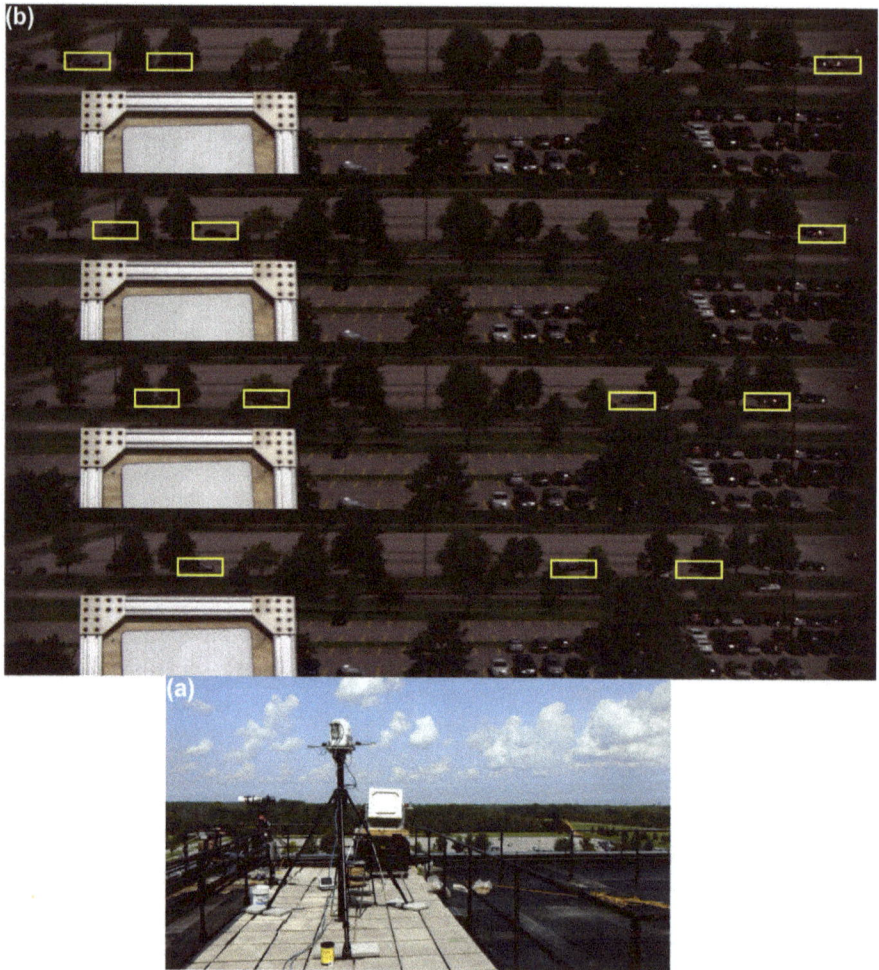

Figure 8. (a) Our integrated system deployed on the roof of the Chester F. Carlson Center for Imaging Science on the RIT campus on 9 August 2017 during a second test focused on imaging of moving vehicles. Shown also is a Spectralon panel deployed on the roof and elevated to be within the field of view of the imaging system. (b) Hyperspectral image time series (**top to bottom**) of the RIT parking lot showing five moving vehicles in a cluttered environment. The yellow boxes outline the positions of the test vehicles over time, but a box is drawn only when a test vehicle is clearly visible.

The objective of the data collection was to obtain hyperspectral image sequences that would be useful for studies of the detection and tracking of vehicles driving through the parking lot at ∼2.75 mps and passing behind various occlusions within the scene, such as the trees in the background and parked cars. An example sequence of hyperspectral image frames from the experiment appear in Figure 8. Each image contains the calibration panel. Note the partly cloudy conditions leading to potential rapid changes to the illumination state. Yellow boxes are drawn around the test vehicles controlled for the experiment. Note that vehicles are occluded at times so the number of boxes drawn can change from image to image. The figure illustrates the complexity of vehicle tracking when a large number of occlusions are present. Video sequences such as this will be produced in future experiments but with a longer duration and with coincident intensive reference data collection to serve as community resources. Such data sequences of short-time interval data of moving vehicles are especially useful for modeling purposes to address the challenge of occlusions (our objective (c) stated in the Introduction) and mixtures that appear in the spectral imagery. Similarly, as vehicles move through the scene, the imaging geometry changes significantly leading to BRDF effects that must be properly modeled for successful extraction and tracking of moving vehicles. The imagery shown, therefore, is important to be able to meet objectives (a) and (b) described in the Introduction.

4. Conclusions

We have described an approach to acquiring full hyperspectral data cubes at low video rates. Our approach integrated a state-of-the-art hyperspectral line scanner capable of high data acquisition rates into a high speed maritime pan-tilt unit. The system also included an integrated GPS/IMU to provide position and pointing information. The entire system is integrated onto a telescopic mast system that allows us to acquire hyperspectral time series imagery from multiple vantage points. This feature also makes possible the creation of a DEM from the resulting stereo hyperspectral views, an approach which was illustrated in this study. Similarly, the multi-view capability also allows the system to sample the bi-directional reflectance distribution function. We provided two examples of the low-rate hyperspectral video approach, showing hyperspectral imagery time series acquired in two different settings for very different applications: imaging of the dynamics of the surf zone in a coastal setting and moving vehicle imaging in the presence of many occlusions. We evaluated SNR and NESR and found values within acceptable limits for the data rates and integration times used in the examples. We noted that SNR could be further improved by spatial binning, an acceptable trade-off in some applications given the very high spatial resolution that we obtain with this system. Within the present system architecture, we noted that further improvements in hyperspectral image acquisition rates could be achieved by reducing the size of the across-track spatial dimension. Our particular system does not allow spectral binning on-chip, however, such capabilities do exist in commercially available systems and could be used to further accelerate image acquisition rates.

Author Contributions: C.M.B. developed the hyperspectral video imaging concept and approach, and C.M.B., D.M., M.R. and T.B. designed the overall hardware implementation, which Headwall Photonics integrated under contract to RIT (Rochester Institute of Technology). R.L.K. designed the system power supply and cable system. T.B. and M.F. performed additional engineering for platform integration. C.M.B., R.S.E., C.S.L., and G.P.B. conceived the VCR LTER experiment, and C.M.B., R.S.E., C.S.L., G.P.B, and M.F. carried it out and processed the data. C.M.B., R.S.E., C.S.L., and G.P.B. also conducted the measurements for the RIT moving vehicle experiment, and processed the data. A.V., M.J.H., M.R., and D.M. conceived the RIT moving vehicle experiment and participated in the execution of the experiment. C.M.B., R.S.E., C.S.L., A.V., and M.J.H. wrote the article.

Funding: This research is funded by an academic grant from the National Geospatial-Intelligence Agency (Award No. #HM0476-17-1-2001, Project Title: Hyperspectral Video Imaging and Mapping of Littoral Conditions), Approved for public release, 19-083. The development of this instrument was made possible by a grant from the Air Force Office of Scientific Research (AFOSR) Defense University Research Instrumentation Program (DURIP). System integration was performed by Headwall Photonics under contract to RIT, as well as by co-authors identified under Author Contributions.

Conflicts of Interest: The authors declare no conflict of interest.

References

1. Dutta, D.; Goodwell, A.E.; Kumar, P.; Garvey, J.E.; Darmody, R.G.; Berretta, D.P.; Greenberg, J.A. On the feasibility of characterizing soil properties from aviris data. *IEEE Trans. Geosci. Remote Sens.* **2015**, *53*, 5133–5147. [CrossRef]

2. Huesca, M.; García, M.; Roth, K.L.; Casas, A.; Ustin, S.L. Canopy structural attributes derived from AVIRIS imaging spectroscopy data in a mixed broadleaf/conifer forest. *Remote Sens. Environ.* **2016**, *182*, 208–226. [CrossRef]

3. Schlerf, M.; Atzberger, C.; Hill, J. Remote sensing of forest biophysical variables using HyMap imaging spectrometer data. *Remote Sens. Environ.* **2005**, *95*, 177–194. [CrossRef]

4. Giardino, C.; Brando, V.E.; Dekker, A.G.; Strömbeck, N.; Candiani, G. Assessment of water quality in Lake Garda (Italy) using Hyperion. *Remote Sens. Environ.* **2007**, *109*, 183–195. [CrossRef]

5. Haboudane, D.; Miller, J.R.; Pattey, E.; Zarco-Tejada, P.J.; Strachan, I.B. Hyperspectral vegetation indices and novel algorithms for predicting green LAI of crop canopies: Modeling and validation in the context of precision agriculture. *Remote Sens. Environ.* **2004**, *90*, 337–352. [CrossRef]

6. Garcia, R.A.; Lee, Z.; Hochberg, E.J. Hyperspectral Shallow-Water Remote Sensing with an Enhanced Benthic Classifier. *Remote Sens.* **2018**, *10*, 147. [CrossRef]

7. Mobley, C.D.; Sundman, L.K.; Davis, C.O.; Bowles, J.H.; Downes, T.V.; Leathers, R.A.; Montes, M.J.; Bissett, W.P.; Kohler, D.D.; Reid, R.P.; et al. Interpretation of hyperspectral remote-sensing imagery by spectrum matching and look-up tables. *Appl. Opt.* **2005**, *44*, 3576–3592. [CrossRef]

8. Liu, Y.; Gao, G.; Gu, Y. Tensor matched subspace detector for hyperspectral target detection. *IEEE Trans. Geosci. Remote Sens.* **2017**, *55*, 1967–1974. [CrossRef]

9. Prasad, S.; Bruce, L.M. Decision fusion with confidence-based weight assignment for hyperspectral target recognition. *IEEE Trans. Geosci. Remote Sens.* **2008**, *46*, 1448–1456. [CrossRef]

10. Chang, C.I.; Jiao, X.; Wu, C.C.; Du, Y.; Chang, M.L. A review of unsupervised spectral target analysis for hyperspectral imagery. *EURASIP J. Adv. Signal Process.* **2010**, *2010*, 503752. [CrossRef]

11. Power, H.; Holman, R.; Baldock, T. Swash zone boundary conditions derived from optical remote sensing of swash zone flow patterns. *J. Geophys. Res. Oceans* **2011**. [CrossRef]

12. Lu, J.; Chen, X.; Zhang, P.; Huang, J. Evaluation of spatiotemporal differences in suspended sediment concentration derived from remote sensing and numerical simulation for coastal waters. *J. Coast. Conserv.* **2017**, *21*, 197–207. [CrossRef]

13. Dorji, P.; Fearns, P. Impact of the spatial resolution of satellite remote sensing sensors in the quantification of total suspended sediment concentration: A case study in turbid waters of Northern Western Australia. *PLoS ONE* **2017**, *12*, e0175042. [CrossRef] [PubMed]

14. Kamel, A.; El Serafy, G.; Bhattacharya, B.; Van Kessel, T.; Solomatine, D. Using remote sensing to enhance modelling of fine sediment dynamics in the Dutch coastal zone. *J. Hydroinform.* **2014**, *16*, 458–476. [CrossRef]

15. Puleo, J.A.; Holland, K.T. Estimating swash zone friction coefficients on a sandy beach. *Coast. Eng.* **2001**, *43*, 25–40. [CrossRef]

16. Concha, J.A.; Schott, J.R. Retrieval of color producing agents in Case 2 waters using Landsat 8. *Remote Sens. Environ.* **2016**, *185*, 95–107. [CrossRef]

17. Puleo, J.A.; Farquharson, G.; Frasier, S.J.; Holland, K.T. Comparison of optical and radar measurements of surf and swash zone velocity fields. *J. Geophys. Res. Oceans* **2003**. [CrossRef]

18. Muste, M.; Fujita, I.; Hauet, A. Large-scale particle image velocimetry for measurements in riverine environments. *Water Resour. Res.* **2008**. [CrossRef]

19. Puleo, J.A.; McKenna, T.E.; Holland, K.T.; Calantoni, J. Quantifying riverine surface currents from time sequences of thermal infrared imagery. *Water Resour. Res.* **2012**. [CrossRef]

20. Gao, L.; Wang, L.V. A review of snapshot multidimensional optical imaging: Measuring photon tags in parallel. *Phys. Rep.* **2016**, *616*, 1–37. [CrossRef]

21. Hagen, N.A.; Gao, L.S.; Tkaczyk, T.S.; Kester, R.T. Snapshot advantage: A review of the light collection improvement for parallel high-dimensional measurement systems. *Opt. Eng.* **2012**, *51*, 111702. [CrossRef] [PubMed]

22. Saari, H.; Aallos, V.V.; Akujärvi, A.; Antila, T.; Holmlund, C.; Kantojärvi, U.; Mäkynen, J.; Ollila, J. Novel miniaturized hyperspectral sensor for UAV and space applications. In Proceedings of the International

Society for Optics and Photonics, Sensors, Systems, and Next-Generation Satellites XIII, Berlin, Germany, 31 August–3 September 2009; Volume 7474, p. 74741M.

23. Pichette, J.; Charle, W.; Lambrechts, A. Fast and compact internal scanning CMOS-based hyperspectral camera: The Snapscan. In Proceedings of the International Society for Optics and Photonics, Instrumentation Engineering IV, San Francisco, CA, USA, 28 January–2 February 2017; Volume 10110, p. 1011014.

24. de Oliveira, R.A.; Tommaselli, A.M.; Honkavaara, E. Geometric calibration of a hyperspectral frame camera. *Photogramm. Rec.* **2016**, *31*, 325–347. [CrossRef]

25. Aasen, H.; Burkart, A.; Bolten, A.; Bareth, G. Generating 3D hyperspectral information with lightweight UAV snapshot cameras for vegetation monitoring: From camera calibration to quality assurance. *ISPRS J. Photogramm. Remote Sens.* **2015**, *108*, 245–259. [CrossRef]

26. Honkavaara, E.; Rosnell, T.; Oliveira, R.; Tommaselli, A. Band registration of tuneable frame format hyperspectral UAV imagers in complex scenes. *ISPRS J. Photogramm. Remote Sens.* **2017**, *134*, 96–109. [CrossRef]

27. Aasen, H.; Bendig, J.; Bolten, A.; Bennertz, S.; Willkomm, M.; Bareth, G. *Introduction and Preliminary Results of a Calibration for Full-Frame Hyperspectral Cameras to Monitor Agricultural Crops with UAVs*; ISPRS Technical Commission VII Symposium; Copernicus GmbH: Istanbul, Turkey, 29 September–2 October 2014; Volume 40, pp. 1–8.

28. Hill, S.L.; Clemens, P. Miniaturization of high spectral spatial resolution hyperspectral imagers on unmanned aerial systems. In Proceedings of the International Society for Optics and Photonics, Next-Generation Spectroscopic Technologies VIII, Baltimore, MD, USA, 3 June 2015; Volume 9482, p. 94821E.

29. Warren, C.P.; Even, D.M.; Pfister, W.R.; Nakanishi, K.; Velasco, A.; Breitwieser, D.S.; Yee, S.M.; Naungayan, J. Miniaturized visible near-infrared hyperspectral imager for remote-sensing applications. *Opt. Eng.* **2012**, *51*, 111720. [CrossRef]

30. Headwall E-Series Specifications. Available online: https://cdn2.hubspot.net/hubfs/145999/June%202018%20Collateral/MicroHyperspec0418.pdf (accessed on 28 December 2018).

31. Greeg, R.O.; Eastwood, M.L.; Sarture, C.M.; Chrien, T.G.; Aronsson, M.; Chippendale, B.J.; Faust, J.A.; Pavri, B.E.; Chovit, C.J.; Solis, M.; et al. Imaging Spectroscopy and the Airborne Visible/Infrared Imaging Spectrometer (AVIRIS). *Remote Sens. Environ.* **2013**, *65*, 227–248.

32. Pearlman, J.S.; Barry, P.S.; Segal, C.C.; Shepanski, J.; Beiso, D.; Carman, S.L. Hyperion, a space-based imaging spectrometer. *IEEE Trans. Geosci. Remote Sens.* **2003**, *41*, 1160–1173. [CrossRef]

33. Diner, D.J.; Beckert, J.C.; Reilly, T.H.; Bruegge, C.J.; Conel, J.E.; Kahn, R.A.; Martonchik, J.V.; Ackerman, T.P.; Davies, R.; Gerstl, S.A.; et al. Multi-angle Imaging SpectroRadiometer (MISR) instrument description and experiment overview. *IEEE Trans. Geosci. Remote Sens.* **1998**, *36*, 1072–1087. [CrossRef]

34. Guanter, L.; Kaufmann, H.; Segl, K.; Foerster, S.; Rogass, C.; Chabrillat, S.; Kuester, T.; Hollstein, A.; Rossner, G.; Chlebek, C.; et al. The EnMAP spaceborne imaging spectroscopy mission for earth observation. *Remote Sens.* **2015**, *7*, 8830–8857. [CrossRef]

35. Babey, S.; Anger, C. A compact airborne spectrographic imager (CASI). In Proceedings of IGARSS '89 and Canadian Symposium on Remote Sensing: Quantitative Remote Sensing: An Economic Tool for the Nineties, Vancouver, BC, Canada, 10–14 July 1989; Volume 1, pp. 1028–1031.

36. Johnson, W.R.; Hook, S.J.; Mouroulis, P.; Wilson, D.W.; Gunapala, S.D.; Realmuto, V.; Lamborn, A.; Paine, C.; Mumolo, J.M.; Eng, B.T. HyTES: Thermal imaging spectrometer development. In Proceedings of the 2011 IEEE Aerospace Conference, Big Sky, MT, USA, 5–12 March 2011; pp. 1–8.

37. Rickard, L.J.; Basedow, R.W.; Zalewski, E.F.; Silverglate, P.R.; Landers, M. HYDICE: An airborne system for hyperspectral imaging. In Proceedings of the International Society for Optics and Photonics, Imaging Spectrometry of the Terrestrial Environment, Orlando, FL, USA, 11–16 April 1993; Volume 1937; pp. 173–180.

38. Lucieer, A.; Malenovský, Z.; Veness, T.; Wallace, L. HyperUAS—Imaging spectroscopy from a multirotor unmanned aircraft system. *J. Field Robot.* **2014**, *31*, 571–590. [CrossRef]

39. Cocks, T.; Jenssen, R.; Stewart, A.; Wilson, I.; Shields, T. The HyMap^{TM} airborne hyperspectral sensor: The system, calibration and performance. In Proceedings of the 1st EARSeL Workshop on Imaging Spectroscopy, EARSeL, Zurich, Switzerland, 6–8 October 1998; pp. 37–42.

40. Eon, R.; Bachmann, C.; Gerace, A. Retrieval of Sediment Fill Factor by Inversion of a Modified Hapke Model Applied to Sampled HCRF from Airborne and Satellite Imagery. *Remote Sens.* **2018**, *10*, 1758. [CrossRef]

41. Uzkent, B.; Rangnekar, A.; Hoffman, M.J. Aerial vehicle tracking by adaptive fusion of hyperspectral likelihood maps. In Proceedings of the 2017 IEEE Conference on Computer Vision and Pattern Recognition Workshops (CVPRW), Honolulu, HI, USA, 21–26 July 2017; pp. 233–242.

42. Bhattacharya, S.; Idrees, H.; Saleemi, I.; Ali, S.; Shah, M. Moving object detection and tracking in forward looking infra-red aerial imagery. In *Machine Vision beyond Visible Spectrum*; Springer: New York, NY, USA, 2011; pp. 221–252.

43. Cao, Y.; Wang, G.; Yan, D.; Zhao, Z. Two algorithms for the detection and tracking of moving vehicle targets in aerial infrared image sequences. *Remote Sens.* **2015**, *8*, 28. [CrossRef]

44. Teutsch, M.; Grinberg, M. Robust detection of moving vehicles in wide area motion imagery. In Proceedings of the IEEE Conference on Computer Vision and Pattern Recognition Workshops, Las Vegas, NV, USA, 26 June–1 July 2016; pp. 27–35.

45. Cormier, M.; Sommer, L.W.; Teutsch, M. Low resolution vehicle re-identification based on appearance features for wide area motion imagery. In Proceedings of the 2016 IEEE Applications of Computer Vision Workshops (WACVW), Lake Placid, NY, USA, 10 March 2016; pp. 1–7.

46. Tuermer, S.; Kurz, F.; Reinartz, P.; Stilla, U. Airborne vehicle detection in dense urban areas using HoG features and disparity maps. *IEEE J. Sel. To. Appl. Earth Obs. Remote Sens.* **2013**, *6*, 2327–2337. [CrossRef]

47. Vodacek, A.; Kerekes, J.P.; Hoffman, M.J. Adaptive Optical Sensing in an Object Tracking DDDAS. *Procedia Comput. Sci.* **2012**, *9*, 1159–1166. [CrossRef]

48. General Dynamics Vector 20 Maritime Pan Tilt. Available online: https://www.gd-ots.com/wp-content/uploads/2017/11/Vector-20-Stabilized-Maritime-Pan-Tilt-System-1.pdf (accessed on 8 October 2018).

49. Vectornav VN-300 GPS IMU. Available online: https://www.vectornav.com/products/vn-300 (accessed on 8 October 2018).

50. BlueSky AL-3 (15m) Telescopic Mast. Available online: http://blueskymast.com/product/bsm3-w-l315-al3-000/ (accessed on 9 October 2018).

51. Labsphere Helios 0.5 m Diameter Integrating Sphere. Available online: https://www.labsphere.com/labsphere-products-solutions/remote-sensing/helios/ (accessed on 9 October 2018).

52. Mobley, C.D. *Light and Water: Radiative Transfer in Natural Waters*; Academic Press: Cambridge, MA, USA, 1994.

53. Coifman, B.; Beymer, D.; McLauchlan, P.; Malik, J. A real-time computer vision system for vehicle tracking and traffic surveillance. *Transp. Res. Part C Emerg. Technol.* **1998**, *6*, 271–288. [CrossRef]

54. Hsieh, J.W.; Yu, S.H.; Chen, Y.S.; Hu, W.F. Automatic traffic surveillance system for vehicle tracking and classification. *IEEE Trans. Intell. Transp. Syst.* **2006**, *7*, 175–187. [CrossRef]

55. Kim, Z.; Malik, J. Fast vehicle detection with probabilistic feature grouping and its application to vehicle tracking. In Proceedings of the Ninth IEEE International Conference on Computer Vision, Nice, France, 13–16 October 2003; p. 524.

56. Uzkent, B.; Hoffman, M.J.; Vodacek, A. Real-time vehicle tracking in aerial video using hyperspectral features. In Proceedings of the IEEE Conference on Computer Vision and Pattern Recognition Workshops, Las Vegas, NV, USA, 26 June–1 July 2016; pp. 36–44.

57. Virginia Coast Reserve Long Term Ecological Research. Available online: https://www.vcrlter.virginia.edu/home2/ (accessed on 10 November 2018).

58. National Science Foundation Long Term Ecological Research Network. Available online: https://lternet.edu/ (accessed on 10 November 2018).

59. Nicodemus, F.E.; Richmond, J.; Hsia, J.J. *Geometrical Considerations and Nomenclature for Reflectance*; US Department of Commerce, National Bureau of Standards: Washington, DC, USA, 1977; Volume 160.

60. Schaepman-Strub, G.; Schaepman, M.; Painter, T.H.; Dangel, S.; Martonchik, J.V. Reflectance quantities in optical remote sensing—Definitions and case studies. *Remote Sens. Environ.* **2006**, *103*, 27–42. [CrossRef]

61. Bachmann, C.M.; Abelev, A.; Montes, M.J.; Philpot, W.; Gray, D.; Doctor, K.Z.; Fusina, R.A.; Mattis, G.; Chen, W.; Noble, S.D.; et al. Flexible field goniometer system: The goniometer for outdoor portable hyperspectral earth reflectance. *J. Appl. Remote Sens.* **2016**, *10*, 036012. [CrossRef]

62. Harms, J.D.; Bachmann, C.M.; Ambeau, B.L.; Faulring, J.W.; Torres, A.J.R.; Badura, G.; Myers, E. Fully automated laboratory and field-portable goniometer used for performing accurate and precise multiangular reflectance measurements. *J. Appl. Remote Sens.* **2017**, *11*, 046014. [CrossRef]

63. Bachmann, C.M.; Eon, R.S.; Ambeau, B.; Harms, J.; Badura, G.; Griffo, C. Modeling and intercomparison of field and laboratory hyperspectral goniometer measurements with G-LiHT imagery of the Algodones Dunes. *J. Appl. Remote Sens.* **2017**, *12*, 012005. [CrossRef]

64. AgiSoft PhotoScan Professional (Version 1.2.6) (Software). 2016. Available online: http://www.agisoft.com/downloads/installer/ (accessed on 10 November 2018).

65. Lorenz, S.; Salehi, S.; Kirsch, M.; Zimmermann, R.; Unger, G.; Vest Sørensen, E.; Gloaguen, R. Radiometric correction and 3D integration of long-range ground-based hyperspectral imagery for mineral exploration of vertical outcrops. *Remote Sens.* **2018**, *10*, 176. [CrossRef]

66. Curcio, J.A.; Petty, C.C. The near infrared absorption spectrum of liquid water. *JOSA* **1951**, *41*, 302–304. [CrossRef]

67. Bachmann, C.M.; Montes, M.J.; Fusina, R.A.; Parrish, C.; Sellars, J.; Weidemann, A.; Goode, W.; Nichols, C.R.; Woodward, P.; McIlhany, K.; Hill, V.; Zimmerman, R.; Korwan, D.; Truitt, B.; Schwarzschild, A. Bathymetry retrieval from hyperspectral imagery in the very shallow water limit: A case study from the 2007 virginia coast reserve (VCR'07) multi-sensor campaign. *Marine Geodesy* **2010**, *33*, 53–75. [CrossRef]

68. Pacheco, A.; Horta, J.; Loureiro, C.; Ferreira, Ó. Retrieval of nearshore bathymetry from Landsat 8 images: A tool for coastal monitoring in shallow waters. *Remote Sens. Environ.* **2015**, *159*, 102–116. [CrossRef]

69. Lyzenga, D.R.; Malinas, N.P.; Tanis, F.J. Multispectral bathymetry using a simple physically based algorithm. *IEEE Trans. Geosci. Remote Sens.* **2006**, *44*, 2251–2259. [CrossRef]

70. Brando, V.E.; Anstee, J.M.; Wettle, M.; Dekker, A.G.; Phinn, S.R.; Roelfsema, C. A physics based retrieval and quality assessment of bathymetry from suboptimal hyperspectral data. *Remote Sens. Environ.* **2009**, *113*, 755–770. [CrossRef]

71. Lee, Z.; Weidemann, A.; Arnone, R. Combined Effect of reduced band number and increased bandwidth on shallow water remote sensing: The case of worldview 2. *IEEE Trans. Geosci. Remote Sens.* **2013**, *51*, 2577–2586. [CrossRef]

Journal of
Imaging

|MDPI|

Article

Fusing Multiple Multiband Images

Reza Arablouei

Commonwealth Scientific and Industrial Research Organisation (CSIRO), Pullenvale QLD 4069, Australia;
reza.arablouei@csiro.au

Received: 21 August 2018; Accepted: 8 October 2018; Published: 12 October 2018

Abstract: High-resolution hyperspectral images are in great demand but hard to acquire due to several existing fundamental and technical limitations. A practical way around this is to fuse multiple multiband images of the same scene with complementary spatial and spectral resolutions. We propose an algorithm for fusing an arbitrary number of coregistered multiband, i.e., panchromatic, multispectral, or hyperspectral, images through estimating the endmember and their abundances in the fused image. To this end, we use the forward observation and linear mixture models and formulate an appropriate maximum-likelihood estimation problem. Then, we regularize the problem via a vector total-variation penalty and the non-negativity/sum-to-one constraints on the endmember abundances and solve it using the alternating direction method of multipliers. The regularization facilitates exploiting the prior knowledge that natural images are mostly composed of piecewise smooth regions with limited abrupt changes, i.e., edges, as well as coping with potential ill-posedness of the fusion problem. Experiments with multiband images constructed from real-world hyperspectral images reveal the superior performance of the proposed algorithm in comparison with the state-of-the-art algorithms, which need to be used in tandem to fuse more than two multiband images.

Keywords: alternating direction method of multipliers; Cramer–Rao lower bound; forward observation model; linear mixture model; maximum likelihood; multiband image fusion; total variation

1. Introduction

The wealth of spectroscopic information provided by hyperspectral images containing hundreds or even thousands of contiguous bands can immensely benefit many remote sensing and computer vision applications, such as source/target detection [1–3], object recognition [4], change/anomaly detection [5,6], material classification [7], and spectral unmixing [8,9], commonly encountered in environmental monitoring, resource location, weather or natural disaster forecasting, etc. Therefore, finely-resolved hyperspectral images are in great demand [10–14]. However, limitations in light intensity as well as efficiency of the current sensors impose an inexorable trade-off between the spatial resolution, spectral sensitivity, and the signal-to-noise ratio (SNR) of existing spectral imagers [15]. As a results, typical spectral imaging systems can capture multiband images of high spatial resolution at a small number of spectral bands or multiband images of high spectral resolution with a reduced spatial resolution. For example, imaging devices onboard Pleiades or IKONOS satellites [16] provide single-band panchromatic images with spatial resolutions of less than a meter and multispectral images with a few bands and spatial resolutions of a few meters while NASA's airborne visible/infrared imaging spectrometer (AVIRIS) [17] provides hyperspectral images with more than 200 bands but with a spatial resolution of several ten meters.

One way to surmount the abovementioned technological limitation of acquiring high-resolution hyperspectral images is to capture multiple multiband images of the same scene with practical spatial and spectral resolutions, then fuse them together in a synergistic manner. Fusing multiband images combines their complementary information obtained through multiple sensors that may have different spatial and spectral resolutions and cover different spectral ranges.

Initial multiband image fusion algorithms were developed to fuse a panchromatic image with a multispectral image and the associated inverse problem was dubbed pansharpening [18–22]. Many pansharpening algorithms are based on either of the two popular pansharpening strategies: component substitution (CS) and multiresolution analysis (MRA). The CS-based algorithms substitute a component of the multispectral image obtained through a suitable transformation by the panchromatic image. The MRA-based algorithms inject the spatial detail of the panchromatic image obtained by a multiscale decomposition, e.g., using wavelets [23], into the multispectral image. There also exist hybrid methods that use both CS and MRA. Some of the algorithms originally proposed for pansharpening have been successfully extended to be used for fusing a panchromatic image with a hyperspectral image, a problem that is called hyperspectral pansharpening [21].

Recently, significant research effort has been expended to solve the problem of fusing a multispectral image with a hyperspectral one. This inverse problem is essentially different from the pansharpening and hyperspectral pansharpening problems since a multispectral image has multiple bands that are intricately related to the bands of its corresponding hyperspectral image. Unlike a panchromatic image that contains only one band of reflectance data usually covering parts of the visible and near-infrared spectral ranges, a multispectral image contains multiple bands each covering a smaller spectral range, some being in the shortwave-infrared (SWIR) region. Therefore, extending the pansharpening techniques so that they can be used to inject the spatial details of a multispectral image into a hyperspectral image is not straightforward. Nonetheless, an effort towards this end has led to the development of a framework called hypersharpening, which is based on adapting the MRA-based pansharpening methods to multispectral–hyperspectral image fusion. The main idea is to synthesize a high-spatial-resolution image for each band of the hyperspectral image by linearly combining the bands of the multispectral image using linear regression [24].

In some works on multispectral–hyperspectral image fusion, it is assumed that each pixel on the hyperspectral image, which has a lower spatial resolution than the target image, is the average of the pixels of the same area on the target image [25–29]. Clearly, the size of this area depends on the downsampling ratio. Based on this pixel-aggregation assumption, one can divide the problem of fusing two multiband images into subproblems dealing with smaller blocks and hence significantly decrease the complexity of the overall process. However, it is more realistic to allow the area on the target image corresponding to a pixel of the hyperspectral image to span as many pixels as determined by the point-spread function of the sensor, which induces spatial blurring. The downsampling ratio generally depends on the physical and optical characteristics of a sensor and is usually fixed. Therefore, spatial blurring and downsampling can be expressed as two separate linear operations. The spectral degradation of a panchromatic or multispectral image with respect to the target image can also be modeled as a linear transformation. Articulating the spatial and spectral degradations in terms of linear operations forms a realistic and convenient forward observation model to relate the observed multiband images to the target image.

Hyperspectral image data is generally known to have a low-rank structure and reside in a subspace that usually has a dimension much smaller than the number of the spectral bands [8,30–33]. This is mainly due to correlations among the spectral bands and the fact that the spectrum of each pixel can often be represented as a linear combination of a relatively few spectral signatures. These signatures, called endmembers, may be the spectra of the material present at the scene. Consequently, a hyperspectral image can be linearly decomposed into its constituent endmembers and the fractional abundances of the endmembers for each pixel. This linear decomposition is called spectral unmixing and the corresponding data model is called the linear mixture model. Other linear decompositions that can be used to reduce the dimensionality of a hyperspectral image in the spectral domain are dictionary-learning-based sparse representation and principle-component analysis.

Many recent works on multiband image fusion, which mostly deal with fusing a multispectral image with a hyperspectral image of the same scene, employ the abovementioned forward observation model and a form of linear spectral decomposition. They mostly extract the endmembers or the spectral dictionary from the hyperspectral image. Some of the works use the extracted endmember or dictionary matrix to reconstruct the multispectral image via sparse regression and calculate the endmember abundances or the representation coefficients [34]. Others cast the multiband image fusion problem as reconstructing a high-spatial-resolution hyperspectral datacube from two datacubes degraded according to the mentioned forward observation model. When the number of spectral bands in the multispectral image is smaller than the number of endmembers or dictionary atoms, the linear inverse problem associated with the multispectral–hyperspectral fusion problem is ill-posed and needs be regularized to have a meaningful solution. Any prior knowledge about the target image can be used for regularization. Natural images are known to mostly consist of smooth segments with few abrupt changes corresponding to the edges and object boundaries [35–37]. Therefore, penalizing the total-variation [38–40] and sparse (low-rank) representation in the spatial domain [41–44] are two popular approaches to regularizing the multiband image fusion problems. Some algorithms, developed within the framework of the Bayesian estimation, incorporate the prior knowledge or conjecture about the probability distribution of the target image into the fusion problem [45–47]. The work of [48] obviates the need for regularization by dividing the observed multiband images into small spatial patches for spectral unmixing and fusion under the assumption that the target image is locally low-rank.

When the endmembers or dictionary atoms are induced from an observed hyperspectral image, the problem of fusing the hyperspectral image with a multispectral image boils down to estimating the endmember abundances or representation coefficients of the target image, a problem that is often tractable (due to being a convex optimization problem) and has a manageable size and complexity. The estimate of the target image is then obtained by mixing the induced endmembers/dictionary and the estimated abundances/coefficients. It is also possible to jointly estimate the endmembers/dictionary and the abundances/coefficients from the available multiband data. This joint estimation problem is usually formulated as a non-convex optimization problem of non-negative matrix factorization, which can be solved approximately using block coordinate-descent iterations [49–52].

To the best of our knowledge, all existing multiband image fusion algorithms are designed to fuse a pair of multiband images with complementary spatial and spectral resolutions. Therefore, fusing more than two multiband images using the existing algorithms can only be realized by performing a hierarchical procedure that combines multiple fusion processes possibly implemented via different algorithms as, for example, in [53,54]. In addition, there are potentially various ways to arrange the pairings and often it is not possible to know beforehand which way will provide the best overall fusion result. For instance, in order to fuse a panchromatic, a multispectral, and a hyperspectral image of a scene, one can first fuse the panchromatic and multispectral images, then fuse the resultant pansharpened multispectral image with the hyperspectral image. Another way would be to first fuse the multispectral and hyperspectral images, then pansharpen the resultant hyperspectral image with the panchromatic image. Apart from the said ambiguity of choice, such combined pair-wise fusions can be slow and inaccurate since they may require several runs of different algorithms and may suffer from propagation and accumulation of errors. Therefore, the increasing availability of multiband images with complementary characteristics captured by modern spectral imaging devices has brought about the demand for efficient and accurate fusion techniques that can handle multiple multiband images simultaneously.

In this paper, we propose an algorithm that can simultaneously fuse an arbitrary number of multiband images. We utilize the forward observation and linear mixture models to effectively model the data and reduce the dimensionality of the problem. Assuming matrix normal distribution for the observation noise, we derive the likelihood function as well as the Fisher information matrix

(FIM) associated with the problem of recovering the endmember abundance matrix of the target image from the observations. We study the properties of the FIM and the conditions for existence of a unique maximum-likelihood estimate and the associated Cramer–Rao lower bound. We regularize the problem of maximum-likelihood estimation of the endmember abundances by adding a vector total-variation penalty term to the cost function and constraining the abundances to be non-negative and add up to one for each pixel. The total-variation penalty serves two major purposes. First, it helps us cope with the likely ill-posedness of the maximum-likelihood estimation problem. Second, it allows us to take into account the spatial characteristics of natural images that is they mostly consist of piecewise plane regions with few sharp variations. Regularization with a vector total-variation penalty can effectively advocate this desired feature by promoting sparsity in the image gradient, i.e., local differences between adjacent pixels, while encourages the local differences to be spatially aligned across different bands [37]. The non-negativity and sum-to-one constraints on the endmember abundances ensure that the abundances have practical values. They also implicitly promote sparsity in the estimated endmember abundances.

We solve the resultant constrained optimization problem using the alternating direction method of multipliers (ADMM) [55–60]. Simulation results indicate that the proposed algorithm outperforms several combinations of the state-of-the-art algorithms, which need be cascaded to carry out fusion of multiple (more than two) multiband images.

2. Data Model

2.1. Forward Observation Model

Let us denote the target multiband image by $\mathbf{X} \in \mathbb{R}^{L \times N}$ where L is the number of spectral bands and N is the number of pixels in the image. We wish to recover $\mathbf{X}$ from K observed multiband images $\mathbf{Y}_k \in \mathbb{R}^{L_k \times N_k}$, $k = 1, \ldots, K$, that are spatially or spectrally downgraded and degraded versions of $\mathbf{X}$. We assume that these multiband images are geometrically coregistered and are related to $\mathbf{X}$ via the following forward observation model

$$\mathbf{Y}_k = \mathbf{R}_k \mathbf{X} \mathbf{B}_k \mathbf{S}_k + \mathbf{P}_k \tag{1}$$

where

$L_k \leq L$ and $N_k = N/D_k^2$ with D_k being the spatial downsampling ratio of the kth image;

$\mathbf{R}_k \in \mathbb{R}^{L_k \times N}$ is the spectral response of the sensor producing $\mathbf{Y}_k$;

$\mathbf{B}_k \in \mathbb{R}^{N \times N}$ is a band-independent spatial blurring matrix that represents a two-dimensional convolution with a blur kernel corresponding to the point-spread function of the sensor producing $\mathbf{Y}_k$;

$\mathbf{S}_k \in \mathbb{R}^{N \times N_k}$ is a sparse matrix with N_k ones and zeros elsewhere that implements a two-dimensional uniform downsampling of ratio D_k on both spatial dimensions and satisfies $\mathbf{S}_k^\top \mathbf{S}_k = \mathbf{I}_N$;

$\mathbf{P}_k \in \mathbb{R}^{L_k \times N_k}$ is an additive perturbation representing the noise or error associated with the observation of $\mathbf{Y}_k$.

We assume that the perturbations $\mathbf{P}_k$, $k = 1, \ldots, K$, are independent of each other and have matrix normal distributions expressed by

$$\mathbf{P}_k \sim \mathcal{MN}_{L_k \times N_k}\left(\mathbf{0}_{L_k \times N_k}, \mathbf{\Sigma}_k, \mathbf{I}_{N_k}\right) \tag{2}$$

where $\mathbf{0}_{L_k \times N_k}$ is the $L_k \times N_k$ zero matrix, $\mathbf{I}_{N_k}$ is the $N_k \times N_k$ identity matrix, and $\mathbf{\Sigma}_k \in \mathbb{R}^{L_k \times L_k}$ is a diagonal matrix that represents the correlation among rows of $\mathbf{P}_k$, which correspond to different spectral bands. Note that we consider the column-covariance matrices to be identity assuming that the perturbations are independent and identically-distributed in the spatial domain. However, by considering diagonal row-covariance matrices, we assume that the perturbations are independent in the spectral domain but may have nonidentical variances at different bands. Moreover, the instrument noise of an optoelectronic device can also have a multiplicative nature. A prominent example is the shot noise that is generally modeled using the Poisson distribution. By virtue of the central limit theorem and since a Poisson distribution with a reasonably large mean can be well approximated by a Gaussian distribution, our assumption of additive Gaussian perturbation for the acquisition noise/error is a sensible working hypothesis given that the SNRs are adequately high.

Note that $\mathbf{Y}_k$, $k = 1, \ldots, K$, in (1) contain the corrected (preprocessed) spectral values, not the raw measurements produced by the spectral imagers. The preprocessing usually involves several steps including radiometric calibration, geometric correction, and atmospheric compensation [61]. The radiometric calibration is generally performed to obtain radiance values at the sensor. It converts the sensor measurements in digital numbers into physical units of radiance. The reflected sunlight passing through the atmosphere is partially absorbed and scattered through a complex interaction between the light and various parts of the atmosphere. The atmospheric compensation counters these effects and converts the radiance values into ground-leaving radiance or surface reflectance values. To obtain accurate reflectance values, one additionally has to account for the effects of the viewing geometry and sun's position as well as the surfaces structural and optical properties [10]. This preprocessing is particularly important when the multiband images to be fused are acquired via different instruments, from different viewpoints, or at different times. After the preprocessing, the images should also be coregistered.

2.2. Linear Mixture Model

Under some mild assumptions, multiband images of natural scenes can be suitably described by a linear mixture model [8]. Specifically, the spectrum of each pixel can often be written as a linear mixture of a few archetypal spectral signatures known as endmembers. The number of endmembers, denoted by M, is usually much smaller than the spectral dimension of a hyperspectral image, i.e., $M \ll L$. Therefore, if we arrange M endmembers corresponding to $\mathbf{X}$ as columns of the matrix $\mathbf{E} \in \mathbb{R}^{L \times M}$, we can factorize $\mathbf{X}$ as

$$\mathbf{X} = \mathbf{EA} + \mathbf{P} \tag{3}$$

where $\mathbf{A} \in \mathbb{R}^{M \times N}$ is the matrix of endmember abundances and $\mathbf{P} \in \mathbb{R}^{L \times N}$ is a perturbation matrix that accounts for any possible inaccuracy or mismatch in the linear mixture mode. We assume that $\mathbf{P}$ is independent of $\mathbf{P}_k$, $k = 1, \ldots, K$, and has a matrix normal distribution as

$$\mathbf{P} \sim \mathcal{MN}_{L \times N}(\mathbf{0}_{L \times N}, \mathbf{\Sigma}, \mathbf{I}_N) \tag{4}$$

where $\mathbf{\Sigma} \in \mathbb{R}^{L \times L}$ is its row-covariance matrix. Every column of $\mathbf{A}$ contains the fractional abundances of the endmembers at a pixel. The fractional abundances are non-negative and often assumed to add up to one for each pixel.

The linear mixture model stated above has been widely used in various contexts and applications concerning multiband, particularly hyperspectral, images. Its popularity can mostly be attributed to its intuitiveness as well as relative simplicity and ease of implementation. However, remotely-sensed images of ground surface may suffer from strong nonlinear effects. These effects are generally due to ground characteristics such as non-planar surface and bidirectional reflectance, artefacts left by atmospheric removal procedures, and the presence of considerable atmospheric absorbance in the neighborhood of the bands of interest. The use of images of the same scene captured by different sensors, although coregistered, can also induce nonlinearly mainly owing to difference in observation geometry, lighting conditions, and miscalibration. Therefore, it should be taken into consideration that the mentioned nonlinear phenomena can impact the results of any procedure relying on the linear mixture model in any real-world application and the scale of the impact depends on the severity of the nonlinearities.

There are a few other caveats regarding the linear mixture model that should also be kept in mind. First, $\mathbf{X}$ in model (3) corresponds to a matrix of corrected (preprocessed) values, not raw ones that would typically be captured by a spectral imager of the same spatial and spectral resolutions. However, whether these values are radiance or reflectance has no impact on the validity of the model, though it certainly matters for further processing of the data. Second, the model (3) does not necessarily require each endmember to be the spectral signature of only one (pure) material. An endmember may be composed of the spectral signatures of multiple materials or may be seen as the spectral signature of a composite material made of several constituent materials. Additionally, depending on the application, the endmembers may be purposely defined in particular subjective ways. Third, in practice, an endmember may have slightly different spectral manifestations at different parts of a scene due to variable illumination, environmental, atmospheric, or temporal conditions. This so-called endmember variability [62] along with possible nonlinearities in the actual underlying mixing process [63] may introduce inaccuracies or inconsistencies in the linear mixture model and consequently in the endmember extraction or spectral unmixing techniques that rely on this model. Lastly, the sum-to-one assumption on the abundances of each pixel may not always hold, especially, when the linear mixture model is not able to account for every material in a pixel possibly because of the effects of endmember variability or nonlinear mixing.

2.3. Fusion Model

Substituting (3) into (1) gives

$$\mathbf{Y}_k = \mathbf{R}_k \mathbf{E} \mathbf{A} \mathbf{B}_k \mathbf{S}_k + \check{\mathbf{P}}_k \tag{5}$$

where the aggregate perturbation of the kth image is

$$\check{\mathbf{P}}_k = \mathbf{P}_k + \mathbf{R}_k \mathbf{P} \mathbf{B}_k \mathbf{S}_k. \tag{6}$$

Instead of estimating the target multiband image $\mathbf{X}$ directly, we consider estimating its abundance matrix $\mathbf{A}$ from the observations $\mathbf{Y}_k$, $k = 1, ..K$, given the endmember matrix $\mathbf{E}$. We can then obtain an estimate of the target image by multiplying the estimated abundance matrix by the endmember matrix. This way, we reduce the dimensionality of the fusion problem and consequently the associated computational burden. In addition, by estimating $\mathbf{A}$ first, we attain an unmixed fused image obviating the need to perform additional unmixing, if demanded by any application utilizing the fused image. However, this approach requires the prior knowledge of the endmember matrix $\mathbf{E}$. The columns of this matrix can be selected from a library of known spectral signatures, such as the U.S. Geological Survey digital spectral library [64], or extracted from the observed multiband images that have the appropriate spectral dimension.

3. Problem

3.1. Maximum-Likelihood Estimation

In order to facilitate our analysis, we define the following vectorized variables

$$\mathbf{y}_k = \text{vec}\{\mathbf{Y}_k\} \in \mathbb{R}^{L_k N_k \times 1} \tag{7a}$$

$$\mathbf{a} = \text{vec}\{\mathbf{A}\} \in \mathbb{R}^{MN \times 1} \tag{7b}$$

$$\mathbf{p}_k = \text{vec}\{\check{\mathbf{P}}_k\} \in \mathbb{R}^{L_k N_k \times 1} \tag{7c}$$

where $\text{vec}\{\cdot\}$ is the vectorization operator that stacks the columns of its matrix argument on top of each other. Applying $\text{vec}\{\cdot\}$ to both sides of (5) while using the property $\text{vec}\{\mathbf{ABC}\} = (\mathbf{C}^\top \otimes \mathbf{A})\text{vec}\{\mathbf{B}\}$ gives

$$\mathbf{y}_k = \left(\mathbf{S}_k^\top \mathbf{B}_k^\top \otimes \mathbf{R}_k \mathbf{E}\right)\mathbf{a} + \mathbf{p}_k \tag{8}$$

where $\otimes$ denotes the Kronecker product.

Since $\mathbf{P}_k$ and $\mathbf{P}$ have independent matrix normal distributions (see (2) and (4)), $\mathbf{p}_k$ has a multivariate normal distribution expressed as

$$\mathbf{p}_k \ \sim \ \mathcal{N}_{L_k N_k}\left(\mathbf{0}_{L_k N_k}, \mathbf{I}_{N_k} \otimes \Sigma_k + \mathbf{S}_k^{\top} \mathbf{B}_k^{\top} \mathbf{B}_k \mathbf{S}_k \otimes \mathbf{R}_k \Sigma \mathbf{R}_k^{\top}\right) \tag{9}$$

where $\mathbf{0}_{L_k N_k}$ stands for the $L_k N_k \times 1$ vector of zeroes. Using the approximation $\mathbf{S}_k^{\top} \mathbf{B}_k^{\top} \mathbf{B}_k \mathbf{S}_k \approx \xi_k \mathbf{I}_{N_k}$ with $\xi_k > 0$, we get

$$\mathbf{p}_k \ \sim \ \mathcal{N}_{L_k N_k}\left(\mathbf{0}_{L_k N_k}, \mathbf{I}_{N_k} \otimes \Lambda_k\right) \tag{10}$$

where

$$\Lambda_k = \Sigma_k + \xi_k \mathbf{R}_k \Sigma \mathbf{R}_k^{\top} \tag{11}$$

In view of (9) and (10), we have

$$\mathbf{y}_k \ \sim \ \mathcal{N}_{L_k N_k}\left(\left[\mathbf{S}_k^{\top} \mathbf{B}_k^{\top} \otimes \mathbf{R}_k \mathbf{E}\right]\mathbf{a}, \mathbf{I}_{N_k} \otimes \Lambda_k\right). \tag{12}$$

Hence, the probability density function of $\mathbf{y}_k$ parametrized over the unknown $\mathbf{a}$ can be written as

$$
\begin{aligned}
f_{\mathbf{y}_k}(\mathbf{y}_k; \mathbf{a}) \ &= \ \left|2\pi \mathbf{I}_{N_k} \otimes \Lambda_k\right|^{-\frac{1}{2}} \\
&\times \exp\left\{-\tfrac{1}{2}\left[\mathbf{y}_k - (\mathbf{S}_k^{\top} \mathbf{B}_k^{\top} \otimes \mathbf{R}_k \mathbf{E})\mathbf{a}\right]^{\top}\left(\mathbf{I}_{N_k} \otimes \Lambda_k\right)^{-1}\left[\mathbf{y}_k - (\mathbf{S}_k^{\top} \mathbf{B}_k^{\top} \otimes \mathbf{R}_k \mathbf{E})\mathbf{a}\right]\right\}.
\end{aligned}
\tag{13}
$$

Since the perturbations $\mathbf{p}_k$, $k = 1, \dots, K$, are independent of each other, the joint probability density function of the observations is written as

$$
\begin{aligned}
f_{\mathbf{y}_1, \dots, \mathbf{y}_K}(\mathbf{y}_1, \dots, \mathbf{y}_K; \mathbf{a}) \ &= \ \prod_{k=1}^{K} f_{\mathbf{y}_k}(\mathbf{y}_k; \mathbf{a}) \\
&= \ \prod_{k=1}^{K} \left|2\pi \mathbf{I}_{N_k} \otimes \Lambda_k\right|^{-\frac{1}{2}} \exp\left\{-\tfrac{1}{2}\sum_{k=1}^{K}\left\|\left(\mathbf{I}_{N_k} \otimes \Lambda_k^{-1/2}\right)\left[\mathbf{y}_k - (\mathbf{S}_k^{\top} \mathbf{B}_k^{\top} \otimes \mathbf{R}_k \mathbf{E})\mathbf{a}\right]\right\|^2\right\}
\end{aligned}
\tag{14}
$$

and the log-likelihood function of $\mathbf{a}$ given the observed data as

$$
\begin{aligned}
l(\mathbf{a}|\mathbf{y}_1, \dots, \mathbf{y}_K) \ &= \ \ln f_{\mathbf{y}_1, \dots, \mathbf{y}_K}(\mathbf{y}_1, \dots, \mathbf{y}_K; \mathbf{a}) \\
&= \ -\tfrac{1}{2}\ln\left(\prod_{k=1}^{K}\left|2\pi \mathbf{I}_{N_k} \otimes \Lambda_k\right|\right) - \tfrac{1}{2}\sum_{k=1}^{K}\left\|\left(\mathbf{I}_{N_k} \otimes \Lambda_k^{-1/2}\right)\left[\mathbf{y}_k - \left(\mathbf{S}_k^{\top} \mathbf{B}_k^{\top} \otimes \mathbf{R}_k \mathbf{E}\right)\mathbf{a}\right]\right\|^2.
\end{aligned}
\tag{15}
$$

Accordingly, the maximum-likelihood estimate of $\mathbf{a}$ is found by solving the following optimization problem

$$
\begin{aligned}
\hat{\mathbf{a}} \ &= \ \arg\max_{\mathbf{a}} l(\mathbf{a}|\mathbf{y}_1, \dots, \mathbf{y}_K) \\
&= \ \arg\min_{\mathbf{a}} \tfrac{1}{2}\sum_{k=1}^{K}\left\|\left(\mathbf{I}_{N_k} \otimes \Lambda_k^{-1/2}\right)\left[\mathbf{y}_k - (\mathbf{S}_k^{\top} \mathbf{B}_k^{\top} \otimes \mathbf{R}_k \mathbf{E})\mathbf{a}\right]\right\|^2.
\end{aligned}
\tag{16}
$$

This problem can be stated in terms of $\mathbf{A} = \mathrm{vec}^{-1}\{\mathbf{a}\}$ as

$$\hat{\mathbf{A}} = \arg\min_{\mathbf{A}} \frac{1}{2}\sum_{k=1}^{K}\left\|\Lambda_k^{-1/2}\left(\mathbf{Y}_k - \mathbf{R}_k \mathbf{E} \mathbf{A} \mathbf{B}_k \mathbf{S}_k\right)\right\|_F^2. \tag{17}$$

The Fisher information matrix (FIM) of the maximum-likelihood estimator $\hat{\mathbf{a}}$ in (16) is calculated as

$$\mathcal{F} = -\mathrm{E}[\mathcal{H}_l(\mathbf{a})] \tag{18}$$

where $\mathcal{H}_l(\mathbf{a})$ denotes the Hessian, i.e., the Jacobian of the gradient, of the log-likelihood function $l(\mathbf{a}|\mathbf{y}_1,\ldots,\mathbf{y}_K)$. The entry on the ith row and the jth column of $\mathcal{H}_l(\mathbf{a})$ is computed as

$$\frac{\partial^2}{\partial a_i \partial a_j} l(\mathbf{a}|\mathbf{y}_1,\ldots,\mathbf{y}_K) \tag{19}$$

where a_i and a_j denote the ith and jth entries of $\mathbf{a}$, respectively. Accordingly, we can show that

$$\mathcal{F} = \sum_{k=1}^{K}\left(\mathbf{B}_k\mathbf{S}_k\mathbf{S}_k^\top\mathbf{B}_k^\top \otimes \mathbf{E}^\top\mathbf{R}_k^\top\Lambda_k^{-1}\mathbf{R}_k\mathbf{E}\right). \tag{20}$$

If $\mathcal{F}$ is invertible, the optimization problem (16) has a unique solution given by

$$\hat{\mathbf{a}} = \left[\sum_{k=1}^{K}\left(\mathbf{B}_k\mathbf{S}_k\mathbf{S}_k^\top\mathbf{B}_k^\top \otimes \mathbf{E}^\top\mathbf{R}_k^\top\Lambda_k^{-1}\mathbf{R}_k\mathbf{E}\right)\right]^{-1}\sum_{k=1}^{K}\left(\mathbf{B}_k\mathbf{S}_k \otimes \mathbf{E}^\top\mathbf{R}_k^\top\Lambda_k^{-1}\right)\mathbf{y}_k \tag{21}$$

and the Cramer–Rao lower bound for the estimator $\hat{\mathbf{a}}$, which is a lower bound on the covariance of $\hat{\mathbf{a}}$, is the inverse of $\mathcal{F}$. The FIM $\mathcal{F}$ is guaranteed to be invertible when, for at least one image, the matrix $\mathbf{B}_k\mathbf{S}_k\mathbf{S}_k^\top\mathbf{B}_k^\top \otimes \mathbf{E}^\top\mathbf{R}_k^\top\Lambda_k^{-1}\mathbf{R}_k\mathbf{E}$ is full-rank.

The matrix $\mathbf{S}_k\mathbf{S}_k^\top$ has a rank of N_k hence for $D_k > 1$ is rank-deficient. The blurring matrix $\mathbf{B}_k$ does not change the rank of the matrix that it multiplies from the right. In addition, as Λ_k^{-1} is full-rank, $\mathbf{E}^\top\mathbf{R}_k^\top\Lambda_k^{-1}\mathbf{R}_k\mathbf{E}$ has a full rank of M when the rows of $\mathbf{R}_k\mathbf{E}$ are at least as many as its columns, i.e., $L_k \geq M$. Therefore, $\mathbf{A}$ and consequently $\mathbf{X}$ is guaranteed to be uniquely identifiable given $\mathbf{Y}_k$, $k = 1,\ldots,K$, only when at least one observed image, say the qth image, has full spatial resolution, i.e., $N_q = N$, with the number of its spectral bands being equal to or larger than the number of endmembers, i.e., $L_q \geq M$, so that, at least for the qth image, $\mathbf{B}_q\mathbf{S}_q\mathbf{S}_q^\top\mathbf{B}_q^\top \otimes \mathbf{E}^\top\mathbf{R}_q^\top\Lambda_q^{-1}\mathbf{R}_q\mathbf{E}$ is full-rank.

In practice, it is rarely possible to satisfy the abovementioned requirements as multiband images with high spectral resolution are generally spatially downsampled and the number of bands of the ones with full spatial resolution, such as panchromatic or multispectral images, is often less than the number of endmembers. Hence, the inverse problem of recovering $\mathbf{A}$ from $\mathbf{Y}_k$, $k = 1,\ldots,K$, is usually ill-posed or ill-conditioned. Thus, some prior knowledge need be injected into the estimation process to produce a unique and reliable estimate. The prior knowledge is intended to partially compensate for the information lost in spectral and spatial downsampling and usually stems from experimental evidence or common facts that may induce certain analytical properties or constraints. The prior information is commonly incorporated into the problem in the form of imposed constraints or additive regularization terms. Examples of prior knowledge about $\mathbf{A}$ that are regularly used in the literature are non-negativity and sum-to-one constraints, matrix normal distribution with known or estimated parameters [45], sparse representation with a learned or known dictionary or basis [41], and minimal total variation [38].

3.2. Regularization

To develop an algorithm for effective fusion of multiple multiband images with arbitrary spatial and spectral resolutions, we employ two mechanisms to regularize the maximum-likelihood cost function in (17).

As the first regularization mechanism, we impose a constraint on $\mathbf{A}$ such that its entries are non-negative and sum to one in all columns. We express this constraint as $\mathbf{A} \geq 0$ and $\mathbf{1}_M^\top\mathbf{A} = \mathbf{1}_N^\top$ where $\mathbf{A} \geq 0$ means all the entries of $\mathbf{A}$ are greater than or equal to zero. As the second regularization mechanism, we add an isotropic vector total-variation penalty term, denoted by $||\nabla\mathbf{A}||_{2,1}$, to the cost function. Here, $||\cdot||_{2,1}$ is the $l_{2,1}$-norm operator that returns the sum of l_2-norms of all the columns of its matrix argument. In addition, we define

$$\nabla \mathbf{A} = \begin{bmatrix} \mathbf{AD}_h \\ \mathbf{AD}_v \end{bmatrix} \in \mathbb{R}^{2M \times N} \tag{22}$$

where $\mathbf{D}_h$ and $\mathbf{D}_v$ are discrete differential matrix operators that, respectively, yield the horizontal and vertical first-order backward differences (gradients) of the row-vectorized image that they multiply from the right. Consequently, we formulate our regularized optimization problem for estimating $\mathbf{A}$ as

$$\min_{\mathbf{A}} \frac{1}{2} \sum_{k=1}^{K} ||\Lambda_k^{-1/2} \left(\mathbf{Y}_k - \mathbf{R}_k \mathbf{EAB}_k \mathbf{S}_k \right) ||_F^2 + \alpha ||\nabla \mathbf{A}||_{2,1} \tag{23}$$

$$\text{subject to : } \mathbf{A} \geq 0 \text{ and } \mathbf{1}_M^\top \mathbf{A} = \mathbf{1}_N^\top$$

where $\alpha \geq 0$ is the regularization parameter.

The non-negativity and sum-to-one constraints on $\mathbf{A}$, which force the columns of $\mathbf{A}$ to reside on the unit $(M-1)$-simplex, are naturally expected and help find a solution that is physically plausible. In addition, they implicitly induce sparseness in the solution. The total-variation penalty promotes solutions with a sparse gradient, a property that is known to be possessed by images of most natural scenes as they are usually made of piecewise homogeneous regions with few sudden changes at object boundaries or edges. Note that the subspace spanned by the endmembers is the one that the target image $\mathbf{X}$ lives in. Therefore, through the total-variation regularization of the abundance matrix $\mathbf{A}$, we regularize $\mathbf{X}$ indirectly.

4. Algorithm

Defining the set of values for $\mathbf{A}$ that satisfy the non-negativity and sum-to-one constraints as

$$\mathcal{S} = \left\{ \mathbf{A} \middle| \mathbf{A} \geq 0, \mathbf{1}_M^\top \mathbf{A} = \mathbf{1}_N^\top \right\} \tag{24}$$

and making use of the indicator function $\iota_{\mathcal{S}}(\mathbf{A})$ defined as

$$\iota_{\mathcal{S}}(\mathbf{A}) = \begin{cases} 0 & \mathbf{A} \in \mathcal{S} \\ +\infty & \mathbf{A} \notin \mathcal{S}, \end{cases} \tag{25}$$

we rewrite (23) as

$$\min_{\mathbf{A}} \frac{1}{2} \sum_{k=1}^{K} ||\Lambda_k^{-1/2} \left(\mathbf{Y}_k - \mathbf{R}_k \mathbf{EAB}_k \mathbf{S}_k \right) ||_F^2 + \alpha ||\nabla \mathbf{A}||_{2,1} + \iota_{\mathcal{S}}(\mathbf{A}). \tag{26}$$

4.1. Iterations

We use the alternating direction method of multipliers (ADMM), also known as the split-Bregman method, to solve the convex but nonsmooth optimization problem of (26). We split the problem to smaller and more manageable pieces by defining the auxiliary variables, $\mathbf{U}_k \in \mathbb{R}^{M \times N}$, $k = 1, \ldots, K$, $\mathbf{V} \in \mathbb{R}^{2M \times N}$, and $\mathbf{W} \in \mathbb{R}^{M \times N}$, and changing (26) into

$$\min_{\mathbf{A}, \mathbf{U}_1, \ldots, \mathbf{U}_K, \mathbf{V}, \mathbf{W}} \frac{1}{2} \sum_{k=1}^{K} ||\Lambda_k^{-1/2} \left(\mathbf{Y}_k - \mathbf{R}_k \mathbf{EU}_k \mathbf{S}_k \right) ||_F^2 + \alpha ||\mathbf{V}||_{2,1} + \iota_{\mathcal{S}}(\mathbf{W}) \tag{27}$$

$$\text{subject to : } \mathbf{U}_k = \mathbf{AB}_k, \ \mathbf{V} = \nabla \mathbf{A}, \ \mathbf{W} = \mathbf{A}.$$

Then, we write the augmented Lagrangian function associated with (27) as

$$\mathcal{L}(\mathbf{A}, \mathbf{U}_1, \ldots, \mathbf{U}_K, \mathbf{V}, \mathbf{W}, \mathbf{F}_1, \ldots, \mathbf{F}_K, \mathbf{G}, \mathbf{H}) = \frac{1}{2} \sum_{k=1}^{K} ||\mathbf{\Lambda}_k^{-1/2}(\mathbf{Y}_k - \mathbf{R}_k \mathbf{E} \mathbf{U}_k \mathbf{S}_k)||_{\mathrm{F}}^2 + \alpha ||\mathbf{V}||_{2,1} + \iota_S(\mathbf{W})$$
$$+ \frac{\mu}{2} \sum_{k=1}^{K} ||\mathbf{A}\mathbf{B}_k - \mathbf{U}_k - \mathbf{F}_k||_{\mathrm{F}}^2 + \frac{\mu}{2} ||\nabla \mathbf{A} - \mathbf{V} - \mathbf{G}||_{\mathrm{F}}^2 + \frac{\mu}{2} ||\mathbf{A} - \mathbf{W} - \mathbf{H}||_{\mathrm{F}}^2 \tag{28}$$

where $\mathbf{F}_k \in \mathbb{R}^{M \times N}$, $k = 1, \ldots, K$, $\mathbf{G} \in \mathbb{R}^{2M \times N}$, and $\mathbf{H} \in \mathbb{R}^{M \times N}$ are the scaled Lagrange multipliers and $\mu \geq 0$ is the penalty parameter.

Using the ADMM, we minimize the augmented Lagrangian function in an iterative fashion. At each iteration, we alternate the minimization with respect to the main unknown variable $\mathbf{A}$ and the auxiliary variables; then, we update the scaled Lagrange multipliers. Hence, we compute the iterates as

$$\mathbf{A}^{(n)} = \underset{\mathbf{A}}{\operatorname{argmin}} \, \mathcal{L}\left(\mathbf{A}, \mathbf{U}_1^{(n-1)}, \ldots, \mathbf{U}_K^{(n-1)}, \mathbf{V}^{(n-1)}, \mathbf{W}^{(n-1)}, \mathbf{F}_1^{(n-1)}, \ldots, \mathbf{F}_K^{(n-1)}, \mathbf{G}^{(n-1)}, \mathbf{H}^{(n-1)}\right) \tag{29}$$

$$\left\{\mathbf{U}_1^{(n)}, \ldots, \mathbf{U}_K^{(n)}, \mathbf{V}^{(n)}, \mathbf{W}^{(n)}\right\} = \underset{\mathbf{U}_1, \ldots, \mathbf{U}_K, \mathbf{V}, \mathbf{W}}{\operatorname{argmin}} \, \mathcal{L}\left(\mathbf{A}^{(n)}, \mathbf{U}_1, \ldots, \mathbf{U}_K, \mathbf{V}, \mathbf{W}, \mathbf{F}_1^{(n-1)}, \ldots, \mathbf{F}_K^{(n-1)}, \mathbf{G}^{(n-1)}, \mathbf{H}^{(n-1)}\right) \tag{30}$$

$$\mathbf{F}_k^{(n)} = \mathbf{F}_k^{(n-1)} - \left(\mathbf{A}^{(n)} \mathbf{B}_k - \mathbf{U}_k^{(n)}\right), \quad k = 1, \ldots, K \tag{31a}$$

$$\mathbf{G}^{(n)} = \mathbf{G}^{(n-1)} - \left(\nabla \mathbf{A}^{(n)} - \mathbf{V}^{(n)}\right) \tag{31b}$$

$$\mathbf{H}^{(n)} = \mathbf{H}^{(n-1)} - \left(\mathbf{A}^{(n)} - \mathbf{W}^{(n)}\right) \tag{31c}$$

where superscript (n) denotes the value of an iterate at iteration number $n \geq 0$. We repeat the iterations until convergence is reached up to a maximum allowed number of iterations.

Since we define the auxiliary variables independent of each other, the minimization of the augmented Lagrangian function (28) with respect to the auxiliary variables can be realized separately. Thus, (30) is equivalent to

$$\mathbf{U}_k^{(n)} = \underset{\mathbf{U}_k}{\operatorname{argmin}} \, \frac{1}{2} ||\mathbf{\Lambda}_k^{-1/2}(\mathbf{Y}_k - \mathbf{R}_k \mathbf{E} \mathbf{U}_k \mathbf{S}_k)||_{\mathrm{F}}^2 + \frac{\mu}{2} ||\mathbf{A}^{(n)} \mathbf{B}_k - \mathbf{U}_k - \mathbf{F}_k^{(n-1)}||_{\mathrm{F}}^2, \quad k = 1, \ldots, K \tag{32}$$

$$\mathbf{V}^{(n)} = \underset{\mathbf{V}}{\operatorname{argmin}} \, \alpha ||\mathbf{V}||_{2,1} + \frac{\mu}{2} ||\nabla \mathbf{A}^{(n)} - \mathbf{V} - \mathbf{G}^{(n-1)}||_{\mathrm{F}}^2 \tag{33}$$

$$\mathbf{W}^{(n)} = \underset{\mathbf{W}}{\operatorname{argmin}} \, \iota_S(\mathbf{W}) + \frac{\mu}{2} ||\mathbf{A}^{(n)} - \mathbf{W} - \mathbf{H}^{(n-1)}||_{\mathrm{F}}^2. \tag{34}$$

4.2. Solutions of Subproblems

Considering (28), (29) can be written as

$$\mathbf{A}^{(n)} = \underset{\mathbf{A}}{\operatorname{argmin}} \, \sum_{k=1}^{K} ||\mathbf{A}\mathbf{B}_k - \mathbf{U}_k^{(n-1)} - \mathbf{F}_k^{(n-1)}||_{\mathrm{F}}^2 + ||\nabla \mathbf{A} - \mathbf{V}^{(n-1)} - \mathbf{G}^{(n-1)}||_{\mathrm{F}}^2 + ||\mathbf{A} - \mathbf{W}^{(n-1)} - \mathbf{H}^{(n-1)}||_{\mathrm{F}}^2. \tag{35}$$

Calculating the gradient of the cost function in (35) with respect to $\mathbf{A}$ and setting it to zero gives

$$\mathbf{A}^{(n)} = \left[\sum_{k=1}^{K} \left(\mathbf{U}_k^{(n-1)} + \mathbf{F}_k^{(n-1)}\right) \mathbf{B}_k^\top + \mathbf{Q}_1^{(n-1)} \mathbf{D}_h^\top + \mathbf{Q}_2^{(n-1)} \mathbf{D}_v^\top + \mathbf{W}^{(n-1)} + \mathbf{H}^{(n-1)}\right]$$
$$\times \left(\sum_{k=1}^{K} \mathbf{B}_k \mathbf{B}_k^\top + \mathbf{D}_h \mathbf{D}_h^\top + \mathbf{D}_v \mathbf{D}_v^\top + \mathbf{I}_N\right)^{-1} \tag{36}$$

where, for the convenience of presentation, we define $Q_1^{(n-1)}$ and $Q_2^{(n-1)}$ as

$$\begin{bmatrix} Q_1^{(n-1)} \\ Q_2^{(n-1)} \end{bmatrix} = V^{(n-1)} + G^{(n-1)}. \tag{37}$$

To make the computation of $A^{(n)}$ in (36) more efficient, we assume that the two-dimensional convolutions represented by B_k, $k = 1, \ldots, K$, are cyclic. In addition, we assume that the differential matrix operators D_h and D_v apply with periodic boundaries. Consequently, multiplications by $B_k^\top$, $D_h^\top$, and $D_v^\top$ as well as by $\left(\sum_{k=1}^{K} B_k B_k^\top + D_h D_h^\top + D_v D_v^\top + I_N \right)^{-1}$ can be performed through the use of the fast Fourier transform (FFT) algorithm and the circular convolution theorem. This theorem states that the Fourier transform of a circular convolution is the pointwise product of the Fourier transforms, i.e., a circular convolution can be expressed as the inverse Fourier transform of the product of the individual spectra [65].

Equating the gradient of the cost function in (32) with respect to U_k to zero results in

$$E^\top R_k^\top \Lambda_k^{-1} R_k E U_k^{(n)} S_k S_k^\top + \mu U_k^{(n)} = E^\top R_k^\top \Lambda_k^{-1} Y_k S_k^\top + \mu \left(A^{(n)} B_k - F_k^{(n-1)} \right). \tag{38}$$

Multiplying both sides of (38) from the right by the masking matrix $M_k = S_k S_k^\top$ and its complement $I_N - M_k$ yields

$$U_k^{(n)} M_k = \left(E^\top R_k^\top \Lambda_k^{-1} R_k E + \mu I_N \right)^{-1} \left[E^\top R_k^\top \Lambda_k^{-1} Y_k S_k^\top + \mu \left(A^{(n)} B_k - F_k^{(n-1)} \right) M_k \right] \tag{39}$$

and

$$U_k^{(n)} (I_N - M_k) = \left(A^{(n)} B_k - F_k^{(n-1)} \right) (I_N - M_k), \tag{40}$$

respectively. Note that we have $S_k^\top S_k = I_N$ and M_k is idempotent, i.e., $M_k M_k = M_k$. Summing both sides of (39) and (40) gives the solution of (32) for $k = 1, \ldots, K$ as

$$\begin{aligned} U_k^{(n)} &= U_k^{(n)} M_k + U_k^{(n)} (I_N - M_k) \\ &= \left(E^\top R_k^\top \Lambda_k^{-1} R_k E + \mu I_N \right)^{-1} \left[E^\top R_k^\top \Lambda_k^{-1} Y_k S_k^\top + \mu \left(A^{(n)} B_k - F_k^{(n-1)} \right) M_k \right] + \left(A^{(n)} B_k - F_k^{(n-1)} \right) (I_N - M_k). \end{aligned} \tag{41}$$

The terms $\left(E^\top R_k^\top \Lambda_k^{-1} R_k E + \mu I_N \right)^{-1}$ and $E^\top R_k^\top \Lambda_k^{-1} Y_k S_k^\top$ do not change during the iterations and can be precomputed.

The subproblem (33) can be decomposed pixelwise and its solution is linked to the so-called Moreau proximity operator of the $\ell_{2,1}$-norm given by column-wise vector-soft-thresholding [66,67]. If we define

$$Z^{(n)} = \nabla A^{(n)} - G^{(n-1)}, \tag{42}$$

the jth column of $V^{(n)}$, denoted by $v_j^{(n)}$, is given in terms of the jth column of $Z^{(n)}$, denoted by $z_j^{(n)}$, as

$$v_j^{(n)} = \frac{\max \left\{ ||z_j^{(n)}||_2 - \frac{\alpha}{\mu}, 0 \right\}}{||z_j^{(n)}||_2} z_j^{(n)}. \tag{43}$$

The solution of (34) is the value of the proximity operator of the indicator function $\iota_S(W)$ at the point $A^{(n)} - H^{(n-1)}$, which is the projection of $A^{(n)} - H^{(n-1)}$ onto the set S defined by (24). Therefore, we have

$$\begin{aligned} W^{(n)} &= \underset{W \in S}{\operatorname{argmin}} || A^{(n)} - H^{(n-1)} - W ||_F^2 \\ &= \Pi_S \left\{ A^{(n)} - H^{(n-1)} \right\} \end{aligned} \tag{44}$$

where $\Pi_S \{ \cdot \}$ denotes the projection onto S. We implement this projection onto the unit $(M - 1)$-simplex employing the algorithm proposed in [68].

Algorithm 1 presents a summary of the proposed algorithm.

Algorithm 1 The proposed algorithm

1: initialize

2:     $\mathbf{E} \leftarrow \mathrm{VCA}(\mathbf{Y}_l)$ % if $\mathbf{E}$ is not known and $\mathbf{Y}_l$ has full spectral resolution

3:     $\mathbf{A}^{(0)} \leftarrow$ upscale and interpolate the output of $\mathrm{SUnSAL}(\mathbf{Y}_l, \mathbf{E})$

4:     for $k = 1, \ldots, K$

5:         $\mathbf{U}_k^{(0)} = \mathbf{A}^{(0)}$

6:         $\mathbf{F}_k^{(n)} = \mathbf{0}_{M \times N}$

7:         $\mathbf{V}^{(0)} = \mathbf{A}^{(0)}, \mathbf{W}^{(0)} = \mathbf{A}^{(0)}$

8:         $\mathbf{G}^{(0)} = \mathbf{0}_{M \times N}, \mathbf{H}^{(0)} = \mathbf{0}_{M \times N}$

9: for $n = 1, 2, \ldots$ % until a convergence criterion is met or a given maximum number of iterations is reached

10:         $\begin{bmatrix} \mathbf{Q}_1^{(n-1)} \\ \mathbf{Q}_2^{(n-1)} \end{bmatrix} = \mathbf{V}^{(n-1)} + \mathbf{G}^{(n-1)}$

11:         $\mathbf{A}^{(n)} = \left[\sum_{k=1}^{K} \left(\mathbf{U}_k^{(n-1)} + \mathbf{F}_k^{(n-1)} \right) \mathbf{B}_k^\top + \mathbf{Q}_1^{(n-1)} \mathbf{D}_h^\top + \mathbf{Q}_2^{(n-1)} \mathbf{D}_v^\top + \mathbf{W}^{(n-1)} + \mathbf{H}^{(n-1)} \right]$
$\times \left(\sum_{k=1}^{K} \mathbf{B}_k \mathbf{B}_k^\top + \mathbf{D}_h \mathbf{D}_h^\top + \mathbf{D}_v \mathbf{D}_v^\top + \mathbf{I}_N \right)^{-1}$

12:         for $k = 1, \ldots, K$

13:             $\mathbf{U}_k^{(n)} = \left(\mathbf{E}^\top \mathbf{R}_k^\top \mathbf{\Lambda}_k^{-1} \mathbf{R}_k \mathbf{E} + \mu \mathbf{I}_N \right)^{-1} \left[\mathbf{E}^\top \mathbf{R}_k^\top \mathbf{\Lambda}_k^{-1} \mathbf{Y}_k \mathbf{S}_k^\top + \mu \left(\mathbf{A}^{(n)} \mathbf{B}_k - \mathbf{F}_k^{(n-1)} \right) \mathbf{M}_k \right]$
$+ \left(\mathbf{A}^{(n)} \mathbf{B}_k - \mathbf{F}_k^{(n-1)} \right) (\mathbf{I}_N - \mathbf{M}_k)$

14:         $\mathbf{Z}^{(n)} = \nabla \mathbf{A}^{(n)} - \mathbf{G}^{(n-1)}$

15:         for $j = 1, \ldots, N$

16:             $\mathbf{v}_j^{(n)} = \dfrac{\max\left\{ ||\mathbf{z}_j^{(n)}||_2 - \frac{\alpha}{\mu}, 0 \right\}}{||\mathbf{z}_j^{(n)}||_2} \mathbf{z}_j^{(n)}$

17:         $\mathbf{W}^{(n)} = \Pi_\mathcal{S} \left\{ \mathbf{A}^{(n)} - \mathbf{H}^{(n-1)} \right\}$

18:         for $k = 1, \ldots, K$

19:             $\mathbf{F}_k^{(n)} = \mathbf{F}_k^{(n-1)} - \left(\mathbf{A}^{(n)} \mathbf{B}_k - \mathbf{U}_k^{(n)} \right)$

20:         $\mathbf{G}^{(n)} = \mathbf{G}^{(n-1)} - \left(\nabla \mathbf{A}^{(n)} - \mathbf{V}^{(n)} \right)$

21:         $\mathbf{H}^{(n)} = \mathbf{H}^{(n-1)} - \left(\mathbf{A}^{(n)} - \mathbf{W}^{(n)} \right)$

22: calculate the fused image

23:         $\hat{\mathbf{X}} = \mathbf{E} \mathbf{A}^{(n)}$

4.3. Convergence

By defining

$$\mathcal{U} = [\mathbf{U}_1, \cdots, \mathbf{U}_K, \mathbf{V}, \mathbf{W}]^\top \tag{45}$$

and

$$\mathcal{C} = [\mathbf{B}_1, \cdots, \mathbf{B}_K, \mathbf{D}_h, \mathbf{D}_v, \mathbf{I}_N]^\top, \tag{46}$$

(27) can be expressed as

$$\min_{\mathbf{A}} f(\mathcal{U}) \text{ subject to } \mathcal{U} = \mathcal{C} \mathbf{A}^\top \tag{47}$$

where

$$f(\mathcal{U}) = \frac{1}{2} \sum_{k=1}^{K} ||\mathbf{\Lambda}_k^{-1/2} (\mathbf{Y}_k - \mathbf{R}_k \mathbf{E} \mathbf{U}_k \mathbf{S}_k)||_\mathrm{F}^2 + \alpha ||\mathbf{V}||_{2,1} + \iota_\mathcal{S}(\mathbf{W}). \tag{48}$$

The function $f(\mathcal{U})$ is closed, proper, and convex as it is a sum of closed, proper, and convex functions and $\mathcal{C}$ has full column rank. Therefore, according to Theorem 8 of [56], if (47) has a solution, the proposed algorithm converges to this solution, regardless of the initial values as long as the penalty parameter μ is positive. If no solution exists, at least one of $\mathbf{A}^{(n)}$ and $\mathcal{U}^{(n)}$ will diverge.

5. Simulations

To examine the performance of the proposed algorithm in comparison with the state-of-the-art, we simulate the fusion of three multiband images, viz. a panchromatic image, a multispectral image, and a hyperspectral image. To this end, we adopt the popular practice known as the Wald's protocol [69], which is to use a reference image with high spatial and spectral resolutions to generate the lower-resolution images that are fused and evaluate the fusion performance by comparing the fused image with the reference image.

We obtain the reference images of our experiments by cropping five publicly available hyperspectral images to the spatial resolutions given in Table 1. These images are called Botswana [70], Indian Pines [71], Washington DC Mall [71], Moffett Field [17], and Kennedy Space Center [70]. The Botswana image has been captured by the Hyperion sensor aboard the Earth Observing 1 (EO-1) satellite, the Washington DC Mall image by the airborne-mounted Hyperspectral Digital Imagery Collection Experiment (HYDICE), and the Indian Pines, Moffett Filed, and Kennedy Space Center images by the NASA Airborne Visible/Infrared Imaging Spectrometer (AVIRIS) instrument. All images cover the visible near-infrared (VNIR) and short-wavelength-infrared (SWIR) ranges with uncalibrated, excessively noisy, and water-absorbed bands removed. The spectral resolution of each image is also given in Table 1. The data as well as the MATLAB code used to produce the results of this paper can be found at [72].

Table 1. The spatial and spectral dimensions of the considered hyperspectral datasets (reference images) and the value of the regularization parameter used in the proposed algorithm with each dataset.

Image	No. of Rows	No. of Columns	No. of Bands	α
Botswana	400	240	145	5
Indian Pines	400	400	200	7
Washington DC Mall	400	300	191	5
Moffett Field	480	320	176	22
Kennedy Space Center	500	400	176	28

We generate three multiband images (panchromatic, multispectral, and hyperspectral) using each reference image. We obtain the hyperspectral images by applying a rotationally-symmetric 2D Gaussian blur filter with a kernel size of 13 × 13 and a standard deviation of 2.12 to each reference image followed by downsampling with a ratio of 4 in both spatial dimensions for all bands. For the multispectral images, we use a Gaussian blur filter with a kernel size of 7 × 7 and a standard deviation of 1.06 and downsampling with a ratio of 2 in both spatial dimensions for all bands of each reference image. Afterwards, we downgrade the resultant images spectrally by applying the spectral responses of the Landsat 8 multispectral sensor. This sensor has eight multispectral bands and one panchromatic band. Figure 1 depicts the spectral responses of all the bands of this sensor [73]. We create the panchromatic images from the reference images using the panchromatic band of the Landsat 8 sensor without applying any spatial blurring or downsampling. We add zero-mean Gaussian white noise to each band of the produced multiband images such that the band-specific signal-to-noise ratio (SNR) is 30 dB for the multispectral and hyperspectral images and 40 dB for the panchromatic image. In practice, SNR may vary along the bands of a multiband sensor and the noise may be non-zero-mean or non-Gaussian. Our use of the same SNR for all bands and zero-mean Gaussian noise is a simplification adopted for the purpose of evaluating the proposed algorithm and comparing it with the considered benchmarks.

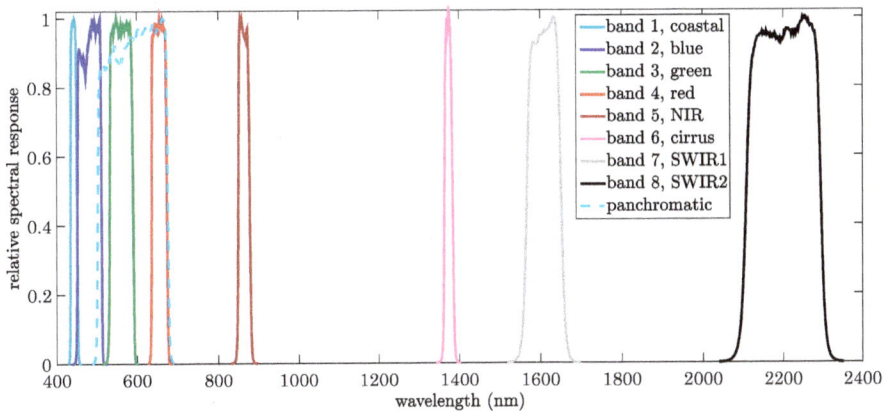

Figure 1. The spectral responses of the Landsat 8 multispectral and panchromatic sensors.

Note that we have selected the standard deviations of the abovementioned 2D Gaussian blur filters such that the normalized magnitude of the modulation transfer function (MTF) of both filters is approximately 0.25 at the Nyquist frequency in both spatial dimensions [74] as shown in Figure 2. We have also selected the filter kernel sizes in accordance with the downsampling ratios and the selected standard deviations. In our simulations, we use symmetric 2D Gaussian blur filters for simplicity and ease of computations. Gaussian blur filters are known to well approximate the real acquisition MTFs, which may be affected by a number of physical processes. With a pushbroom acquisition, the MTF is different in the across-track and along-track directions. This is because, in the along-track direction, the blurring is due to the effects of the spatial resolution and the apparent motion of the scene. On the other hand, in the across-track direction, the blurring is mainly attributable to the resolution of the instrument, fixed by both detector and optics. There is also the contribution to MTF by the propagation trough a scattering atmosphere.

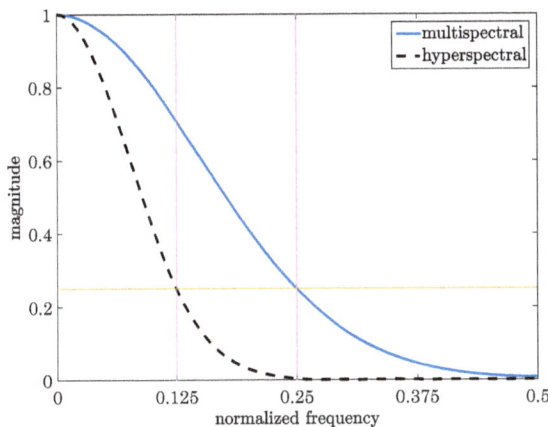

Figure 2. The modulation transfer function (normalized spatial-frequency response) of the used 2D Gaussian blur filters in both spatial dimensions. The solid curve corresponds to the filter used to generate the multispectral images and the dashed curve corresponds to the filter used to generate the hyperspectral images.

The current multiband image fusion algorithms published in the literature are designed to fuse two images at a time. In order to compare the performance of the proposed algorithm with the state-of-the-art, we consider fusing the abovementioned three multiband images in three different ways, which we refer to as Pan + HS, Pan + (MS + HS), and (Pan + MS) + HS, using the existing algorithms for pansharpening, hyperspectral pansharpening, and hyperspectral-multispectral fusion. In Pan + HS, we only fuse the panchromatic and hyperspectral images. In Pan + (MS + HS), and (Pan + MS) + HS, we fuse the given images in two cascading stages. In Pan + (MS + HS), first, we fuse the multispectral and hyperspectral images. Then, we fuse the resultant hyperspectral image with the panchromatic image. We use the same algorithm at both stages, albeit with different parameter values. In (Pan + MS) + HS, we first fuse the panchromatic image with the multispectral one. Then, we fuse the pansharpened multispectral image with the hyperspectral image. We use two different algorithms at each of the two stages resulting in four combined solutions.

For pansharpening, which is the fusion of a panchromatic image with a multispectral one, we use two algorithms called the band-dependent spatial detail (BDSD) [75] and the modulation-transfer-function generalized Laplacian pyramid with high-pass modulation (MTF-GLP-HPM) [76–78]. The BDSD algorithm belongs to the class of component substitution methods and the MTF-GLP-HPM algorithm falls into the category of multiresolution analysis. In [18], where several pansharpening algorithms are studied, it is shown that the BDSD and MTF-GLP-HPM algorithms exhibit the best performance among all the considered ones.

For fusing a panchromatic or multispectral image with a hyperspectral image, we use two algorithms proposed in [38,79,80], which are called HySure and R-FUSE-TV, respectively. These algorithms are based on total-variation regularization and are among the best performing and most efficient hyperspectral pansharpening and multispectral–hyperspectral fusion algorithms currently available [21,81].

We use three performance metrics for assessing the quality of a fused image with respect to its reference image. The metrics are the relative dimensionless global error in synthesis (ERGAS) [82], spectral angle mapper (SAM) [83], and $Q2^n$ [84]. The metric $Q2^n$ is a generalization of the universal image quality index (UIQI) proposed in [85] and an extension of the Q4 index [86] to hyperspectral images based on hypercomplex numbers.

We extract the endmembers (columns of $\mathbf{E}$) from each hyperspectral image using the vertex component analysis (VCA) algorithm [87]. The VCA is a fast unsupervised unmixing algorithm that assumes the endmembers as the vertices of a simplex encompassing the hyperspectral data cloud. We utilize the SUnSAL algorithm [88] together with the extracted endmembers to unmix each hyperspectral image and obtain its abundance matrix. Then, we upscale the resulting matrix by a factor of four and apply two-dimensional spline interpolation on each of its rows (abundance bands) to generate the initial estimate for the abundance matrix $\mathbf{A}^{(0)}$. We initialize the proposed algorithm as well as the HySure and R-FUSE-TV algorithms by this matrix.

To make our comparisons fair, we tune the values of the parameters in the HySure and R-FUSE-TV algorithms to yield the best possible performance in all experiments. In addition, in order to use the BDSD and MTF-GLP-HPM algorithms to their best potential, we provide these algorithms with the true point-spread function, i.e., the blurring kernel, used to generate the multispectral images.

Apart from the number of endmembers, which can be estimated using, for example, the HySime algorithm [31], the proposed algorithm has two tunable parameters, the total-variation regularization parameter α and the ADMM penalty parameter μ. The automatic tuning of the values of these parameters is an interesting and challenging subject. There are a number of strategies that can be employed such as those proposed in [66,89]. We found through experimentations that although the value of μ impacts the convergence speed of the proposed algorithm, as long as it is within an appropriate range, it has little influence on the accuracy of the proposed algorithm. Therefore, we set it to $\mu = 1.5 \times 10^3$ in all experiments. The value of α affects the performance of the proposed algorithm in subtle ways as shown in Figure 3 where we plot the performance metrics, ERGAS, SAM, and $Q2^n$, against α for the Botswana and Washington DC Mall images. The results in Figure 3 suggest that, for different values of α, there is a trade-off between the performance metrics, specifically, ERGAS and $Q2^n$ on one side and SAM on the other. Therefore, we tune the value of α for each experiment only roughly to obtain a reasonable set of values for all three performance metrics. We give the values of α used in the proposed algorithm in Table 1.

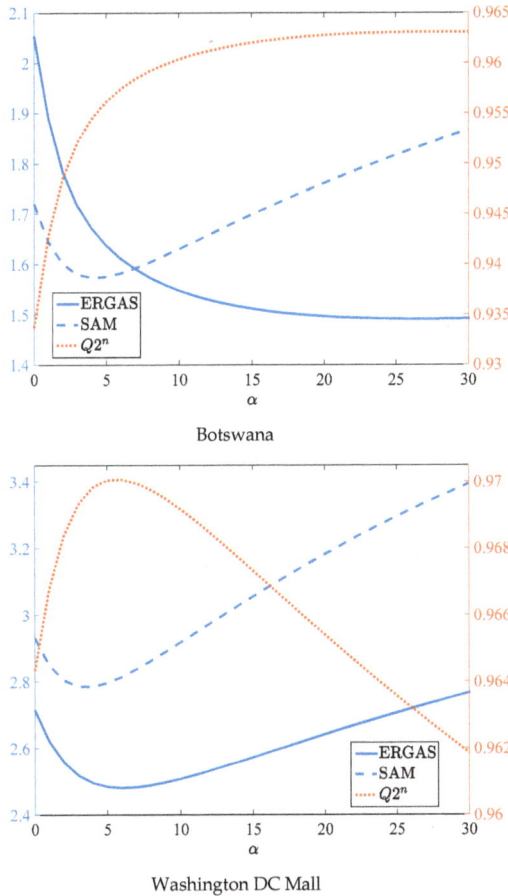

Botswana

Washington DC Mall

Figure 3. The values of the performance metrics versus the regularization parameter α for the experiments with Botswana and Washington DC Mall images. The left y-axis corresponds to ERGAS and SAM and the right y-axis to $Q2^n$.

In Table 2, we give the values of the performance metrics to assess the quality of the images fused using the proposed algorithm and the considered benchmarks. We provide the performance metrics for the case of considering only the bands within the spectrum of the panchromatic image as well as the case of considering all bands, i.e., the entire spectrum of the reference image. We also give the time taken by each algorithm to produce the fused images. We used MATLAB (The MathWorks, Natick, MA, USA) with a 2.9-GHz Core-i7 CPU and 24 GB of DDR3 RAM and ran each of the proposed, HySure, and R-FUSE-TV algorithms for 200 iterations as they always converged sufficiently after this number of iterations. According to the results in Table 2, the proposed algorithm significantly outperforms the considered benchmarks. It is also evident from the required processing times that the computational (time) complexity of the proposed algorithm is lower than those of its contenders.

Table 2. The values of the performance metrics for assessing the fusion quality as well as the runtimes of the considered algorithms for different datasets.

Botswana

Fusion	Algorithm(s)	Spectrum of Pan			Entire Spectrum			Time (s)
		ERGAS	SAM (°)	$Q2^n$	ERGAS	SAM (°)	$Q2^n$	
Pan + MS + HS	proposed	0.900	1.355	0.980	1.637	1.575	0.956	47.01
Pan + HS	HySure	1.273	1.975	0.967	1.839	2.435	0.946	61.20
	R-FUSE-TV	1.272	1.974	0.967	1.840	2.436	0.946	61.17
Pan + (MS + HS)	HySure	1.256	1.721	0.962	1.992	2.101	0.937	78.28
	R-FUSE-TV	1.265	1.734	0.961	2.002	2.113	0.937	79.44
(Pan + MS) + HS	BDSD & HySure	1.393	1.971	0.955	2.458	2.359	0.912	62.58
	BDSD & R-FUSE-TV	1.392	1.977	0.956	2.461	2.365	0.912	62.10
	MTF-GLP-HPM & HySure	1.441	2.120	0.957	2.181	2.442	0.931	62.78
	MTF-GLP-HPM & R-FUSE-TV	1.440	2.124	0.957	2.185	2.446	0.931	62.20

Indian Pines

Fusion	Algorithm(s)	Spectrum of Pan			Entire Spectrum			Time (s)
		ERGAS	SAM (°)	$Q2^n$	ERGAS	SAM (°)	$Q2^n$	
Pan + MS + HS	proposed	0.304	0.293	0.990	0.500	0.761	0.969	80.21
Pan + HS	HySure	0.420	0.547	0.986	0.813	1.108	0.632	106.75
	R-FUSE-TV	0.425	0.555	0.986	0.813	1.113	0.632	106.47
Pan + (MS + HS)	HySure	0.656	0.641	0.961	0.834	1.117	0.594	134.79
	R-FUSE-TV	0.695	0.642	0.953	0.875	1.120	0.573	134.32
(Pan + MS) + HS	BDSD & HySure	0.538	0.517	0.972	0.803	1.183	0.670	108.33
	BDSD & R-FUSE-TV	0.539	0.520	0.972	0.794	1.182	0.674	107.34
	MTF-GLP-HPM & HySure	0.566	0.563	0.972	0.959	1.268	0.626	108.48
	MTF-GLP-HPM & R-FUSE-TV	0.567	0.567	0.972	0.947	1.270	0.628	107.51

Washington DC Mall

Fusion	Algorithm(s)	Spectrum of Pan			Entire Spectrum			Time (s)
		ERGAS	SAM (°)	$Q2^n$	ERGAS	SAM (°)	$Q2^n$	
Pan + MS + HS	proposed	0.731	1.116	0.997	2.484	2.795	0.970	59.52
Pan + HS	HySure	1.171	2.047	0.992	3.822	4.539	0.930	79.02
	R-FUSE-TV	1.171	2.042	0.992	3.832	4.537	0.930	78.38
Pan + (MS + HS)	HySure	0.937	1.718	0.994	3.233	3.592	0.949	99.74
	R-FUSE-TV	1.204	1.738	0.991	3.270	3.664	0.947	100.53
(Pan + MS) + HS	BDSD & HySure	1.114	2.039	0.992	4.174	5.048	0.918	79.68
	BDSD & R-FUSE-TV	1.104	2.060	0.992	4.251	5.033	0.916	78.41
	MTF-GLP-HPM & HySure	1.308	1.870	0.991	4.380	5.147	0.911	79.28
	MTF-GLP-HPM & R-FUSE-TV	1.298	1.884	0.991	4.440	5.114	0.910	78.13

Moffett Field

Fusion	Algorithm(s)	Spectrum of Pan			Entire Spectrum			Time (s)
		ERGAS	SAM (°)	$Q2^n$	ERGAS	SAM (°)	$Q2^n$	
Pan + MS + HS	proposed	0.572	0.786	0.992	4.232	3.148	0.885	77.37
Pan + HS	HySure	0.902	1.151	0.985	6.507	4.233	0.823	107.73
	R-FUSE-TV	0.914	1.152	0.984	6.416	4.210	0.827	106.20
Pan + (MS + HS)	HySure	0.826	1.004	0.986	5.078	3.603	0.868	134.78
	R-FUSE-TV	0.964	1.014	0.977	5.100	3.670	0.845	135.20
(Pan + MS) + HS	BDSD & HySure	1.061	1.135	0.980	5.325	4.065	0.829	108.91
	BDSD & R-FUSE-TV	1.058	1.134	0.980	5.244	4.039	0.834	106.12
	MTF-GLP-HPM & HySure	1.396	1.122	0.968	5.924	4.384	0.824	108.98
	MTF-GLP-HPM & R-FUSE-TV	1.396	1.123	0.969	5.835	4.360	0.830	106.28

Kennedy Space Center

Fusion	Algorithm(s)	Spectrum of Pan			Entire Spectrum			Time (s)
		ERGAS	SAM (°)	$Q2^n$	ERGAS	SAM (°)	$Q2^n$	
Pan + MS + HS	proposed	1.024	1.628	0.984	2.468	3.211	0.909	99.94
Pan + HS	HySure	1.451	2.426	0.979	3.544	3.995	0.890	138.16
	R-FUSE-TV	1.518	2.496	0.974	3.680	3.795	0.886	134.97
Pan + (MS + HS)	HySure	1.462	2.203	0.967	2.851	3.546	0.909	172.18
	R-FUSE-TV	1.875	2.343	0.939	2.986	4.155	0.878	172.25
(Pan + MS) + HS	BDSD & HySure	1.738	2.594	0.949	3.727	4.824	0.850	138.66
	BDSD & R-FUSE-TV	1.691	2.547	0.953	3.534	4.584	0.865	135.74
	MTF-GLP-HPM & HySure	6.801	3.250	0.912	9.532	5.183	0.805	138.60
	MTF-GLP-HPM & R-FUSE-TV	8.143	3.264	0.914	11.130	5.197	0.816	135.58

In Figures 4 and 5, we plot the sorted per-pixel normalized root mean-square error (NRMSE) values of the proposed algorithm and the best performing algorithms from each of the Pan + HS, Pan + (MS + HS), and (Pan + MS) + HS categories. Figure 4 corresponds to the case of considering only the spectrum of the panchromatic image and Figure 5 to the case of considering the entire spectrum. We define the per-pixel NRMSE as $||\mathbf{x}_j - \hat{\mathbf{x}}_j||_2 / ||\mathbf{x}_j||_2$ where $\mathbf{x}_j$ and $\hat{\mathbf{x}}_j$ are the jth column of the reference image $\mathbf{X}$ and the fused image $\hat{\mathbf{X}}$, respectively. We sort the NRMSE values in the ascending order.

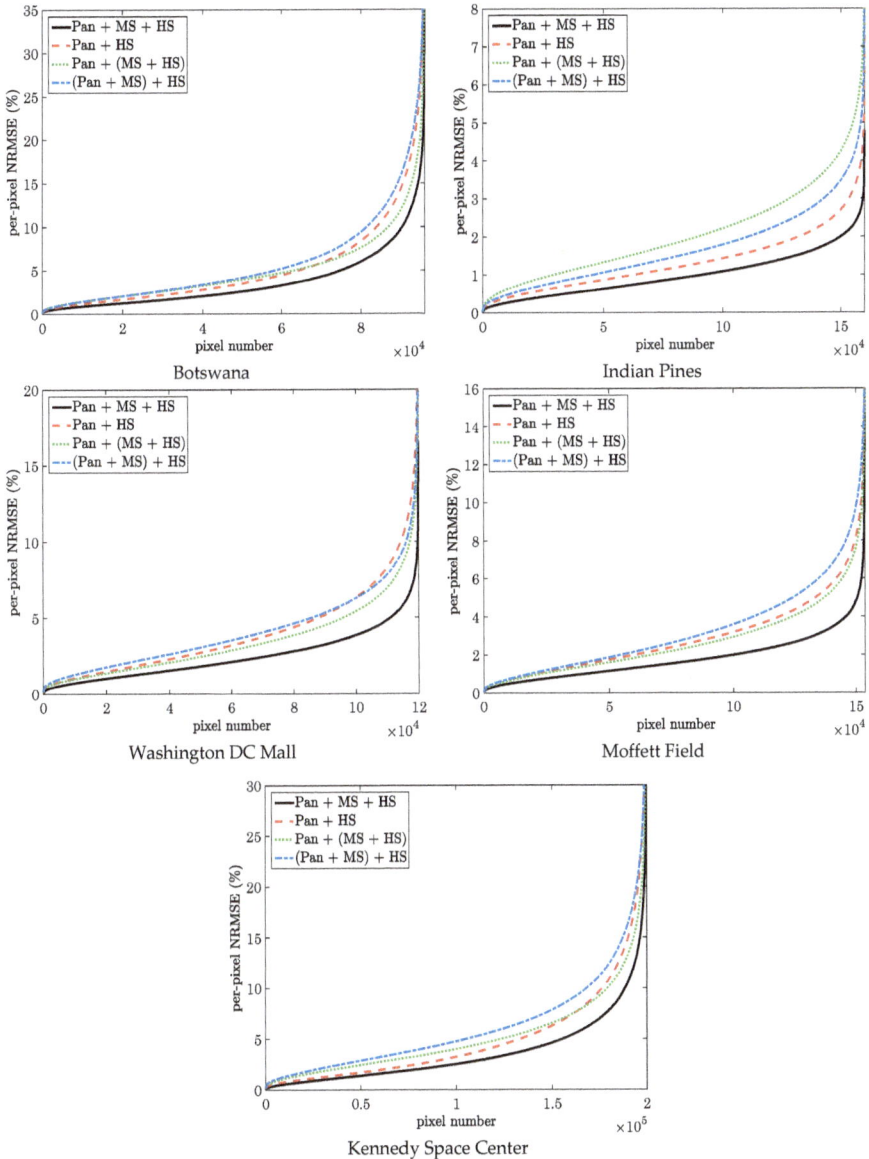

Figure 4. The sorted per-pixel NRMSE of different algorithms measured only on the spectrum of the panchromatic image in experiments with different images.

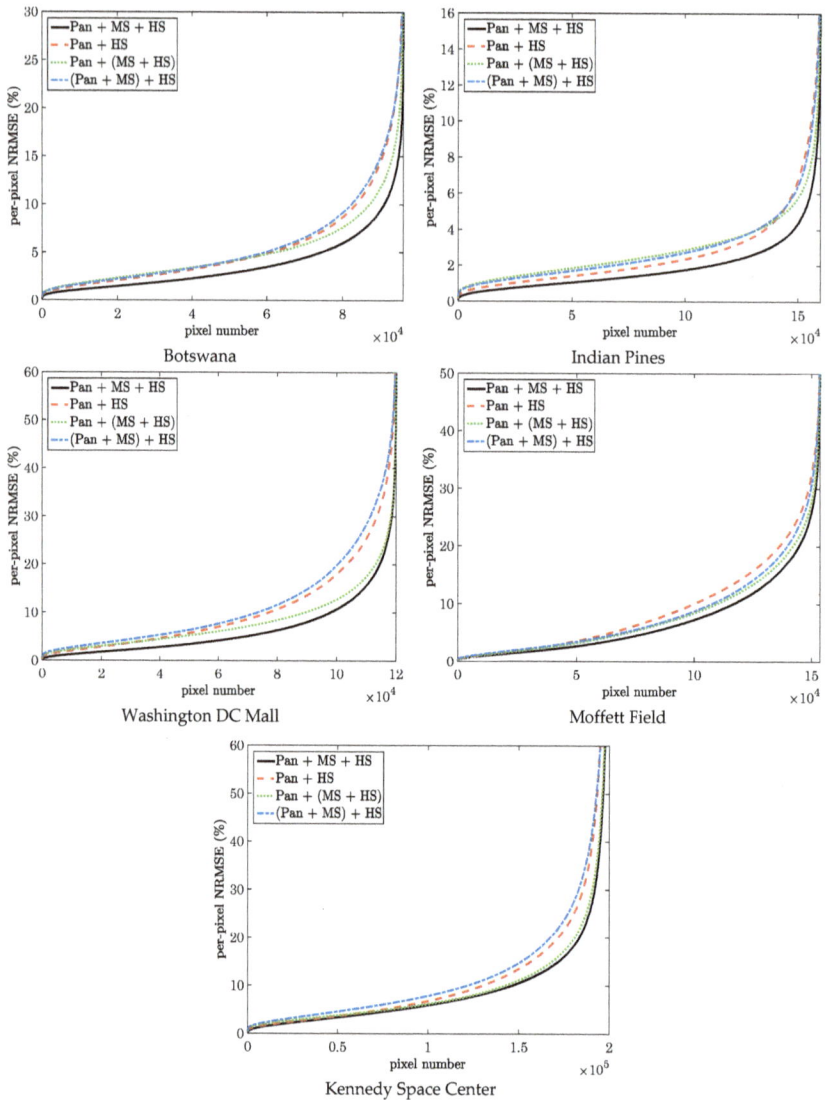

Figure 5. The sorted per-pixel NRMSE of different algorithms measured on the entire spectrum in experiments with different images.

In Figure 6, we show RGB renderings of the reference images together with the panchromatic, multispectral, and hyperspectral images generated from them and used for the fusion. We also show the fused images yielded by the proposed algorithm and Pan + (MS + HS) fusion using the HySure algorithm, which generally performs better than the other considered benchmarks. The multispectral images are depicted using their red, green, and blue bands. The RGB representations of the hyperspectral images are rendered through transforming the spectral data to the CIE XYZ color space and then transforming the XYZ values to the sRGB color space. From visual inspection of the reference and fused images shown in Figure 6, it is observed that the images fused by the proposed algorithm match their corresponding reference images better than the ones produced by the Pan + (MS + HS) fusion using the HySure algorithm do.

Panchromatic	Multi-spectral	Hyper-spectral	Reference	Proposed	Pan + (MS + HS) by HySure

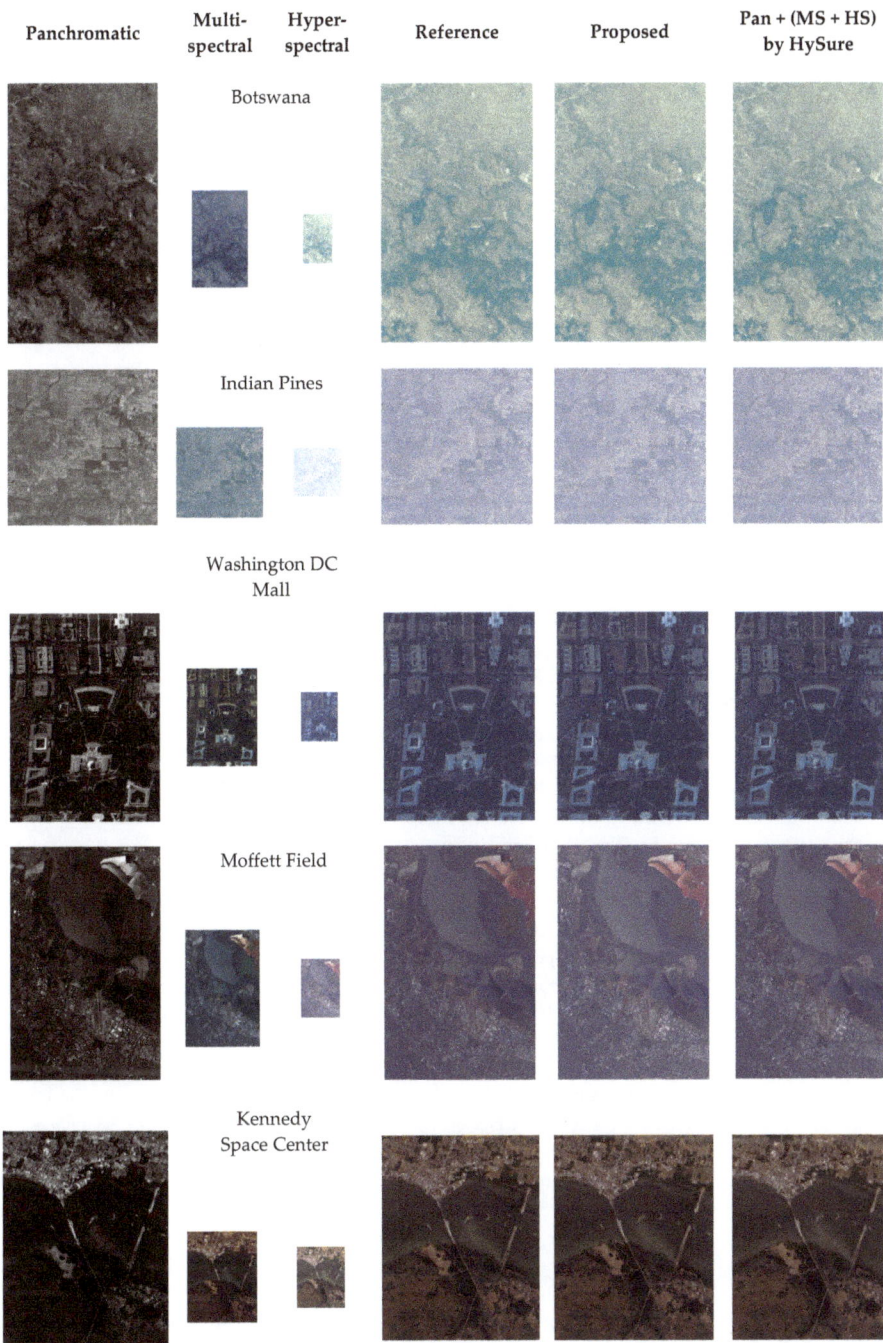

Botswana

Indian Pines

Washington DC Mall

Moffett Field

Kennedy Space Center

Figure 6. The panchromatic, multispectral, and hyperspectral images that are fused together, the reference hyperspectral image, and the fused images produced by the proposed algorithm and the Pan + (MS + HS) method using the HySure algorithm.

J. Imaging **2018**, *4*, 118

6. Conclusions

We proposed a new image fusion algorithm that can simultaneously fuse multiple multiband images. We utilized the well-known forward observation model together with the linear mixture model to cast the fusion problem as a reduced-dimension linear inverse problem. We used a vector total-variation penalty as well as non-negativity and sum-to-one constraints on the endmember abundances to regularize the associated maximum-likelihood estimation problem. The regularization encourages the estimated fused image to have low rank with a sparse representation in the spectral domain while preserving the edges and discontinuities in the spatial domain. We solved the regularized problem using the alternating direction method of multipliers. We demonstrated the advantages of the proposed algorithm in comparison with the state-of-the-art via experiments with five real hyperspectral images that were done following the Wald's protocol.

Author Contributions: The author was responsible for all aspects of this work.

Funding: This work received no external funding.

Conflicts of Interest: The author declares no conflict of interest.

References

1. Arablouei, R. Fusion of multiple multiband images with complementary spatial and spectral resolutions. In Proceedings of the 2018 IEEE International Conference on Acoustics, Speech and Signal Processing (ICASSP), Calgary, AB, Canada, 15–20 April 2018.
2. Courbot, J.-B.; Mazet, V.; Monfrini, E.; Collet, C. Extended faint source detection in astronomical hyperspectral images. *Signal Process.* **2017**, *135*, 274–283. [CrossRef]
3. Du, B.; Zhang, Y.; Zhang, L.; Zhang, L. A hypothesis independent subpixel target detector for hyperspectral Images. *Signal Process.* **2015**, *110*, 244–249. [CrossRef]
4. Mohammadzadeh, A.; Tavakoli, A.; Zoej, M.J.V. Road extraction based on fuzzy logic and mathematical morphology from pansharpened IKONOS images. *Photogramm. Rec.* **2006**, *21*, 44–60. [CrossRef]
5. Souza, C., Jr.; Firestone, L.; Silva, L.M.; Roberts, D. Mapping forest degradation in the Eastern amazon from SPOT 4 through spectral mixture models. *Remote Sens. Environ.* **2003**, *87*, 494–506. [CrossRef]
6. Du, B.; Zhao, R.; Zhang, L.; Zhang, L. A spectral-spatial based local summation anomaly detection method for hyperspectral images. *Signal Process.* **2016**, *124*, 115–131. [CrossRef]
7. Licciardi, G.A.; Villa, A.; Khan, M.M.; Chanussot, J. Image fusion and spectral unmixing of hyperspectral images for spatial improvement of classification maps. In Proceedings of the 2012 IEEE International Geoscience and Remote Sensing Symposium, Munich, Germany, 22–27 July 2012.
8. Bioucas-Dias, J.M.; Plaza, A.; Dobigeon, N.; Parente, M.; Du, Q.; Gader, P.; Chanussot, J. Hyperspectral unmixing overview: Geometrical, statistical, and sparse regression-based approaches. *IEEE J. Sel. Top. Appl. Earth Obs. Remote Sens.* **2012**, *5*, 354–379. [CrossRef]
9. Caiafa, C.F.; Salerno, E.; Proto, A.N.; Fiumi, L. Blind spectral unmixing by local maximization of non-Gaussianity. *Signal Process.* **2008**, *88*, 50–68. [CrossRef]
10. Bioucas-Dias, J.M.; Plaza, A.; Camps-Valls, G.; Scheunders, P.; Nasrabadi, N.; Chanussot, J. Hyperspectral remote sensing data analysis and future challenges. *IEEE Geosci. Remote Sens. Mag.* **2013**, *1*, 6–36. [CrossRef]
11. Shaw, G.A.; Burke, H.-H.K. Spectral imaging for remote sensing. *Lincoln Lab. J.* **2003**, *14*, 3–28.
12. Greer, J. Sparse demixing of hyperspectral images. *IEEE Trans. Image Process.* **2012**, *21*, 219–228. [CrossRef] [PubMed]
13. Arablouei, R.; de Hoog, F. Hyperspectral image recovery via hybrid regularization. *IEEE Trans. Image Process.* **2016**, *25*, 5649–5663. [CrossRef] [PubMed]
14. Arablouei, R. Spectral unmixing with perturbed endmembers. *IEEE Trans. Geosci. Remote Sens* **2018**, in press. [CrossRef]
15. Arablouei, R.; Goan, E.; Gensemer, S.; Kusy, B. Fast and robust push-broom hyperspectral imaging via DMD-based scanning. In *Novel Optical Systems Design and Optimization XIX, Proceedings of the Optical Engineering + Applications 2016—Part of SPIE Optics + Photonics, San Diego, CA, USA, 6–10 August 2016*; SPIE: Bellingham, WA, USA, 2016; Volume 9948.

16. Satellite Sensors. Available online: https://www.satimagingcorp.com/satellite-sensors/ (accessed on 10 October 2018).

17. NASA's airborne visible/infrared imaging spectrometer (AVIRIS). Available online: https://aviris.jpl.nasa.gov/data/free_data.html (accessed on 10 October 2018).

18. Vivone, G.; Alparone, L.; Chanussot, J.; Mura, M.D.; Garzelli, A.; Licciardi, G.A.; Restaino, R.; Wald, L. A critical comparison among pansharpening algorithms. *IEEE Trans. Geosci. Remote Sens.* **2015**, *53*, 2565–2586. [CrossRef]

19. Alparone, L.; Wald, L.; Chanussot, J.; Thomas, C.; Gamba, P.; Bruce, L.M. Comparison of pansharpening algorithms: Outcome of the 2006 GRS-S data-fusion contest. *IEEE Trans. Geosci. Remote Sens.* **2007**, *45*, 3012–3021. [CrossRef]

20. Aiazzi, B.; Alparone, L.; Baronti, S.; Garzelli, A.; Selva, M. 25 years of pansharpening: A critical review and new developments. In *Signal and Image Processing for Remote Sensing*, 2nd ed.; Chen, C.H., Ed.; CRC Press: Boca Raton, FL, USA, 2011.

21. Loncan, L.; Almeida, L.B.; Bioucas-Dias, J.M.; Briottet, X.; Chanussot, J.; Dobigeon, N.; Fabre, S.; Liao, W.; Licciardi, G.A.; Simões, M.; et al. Hyperspectral pansharpening: A review. *IEEE Geosci. Remote Sens. Mag.* **2015**, *3*, 27–46. [CrossRef]

22. Yin, H. Sparse representation based pansharpening with details injection model. *Signal Process.* **2015**, *113*, 218–227. [CrossRef]

23. Pajares, G.; de la Cruz, J.M. A wavelet-based image fusion tutorial. *Pattern Recognit.* **2004**, *37*, 1855–1872. [CrossRef]

24. Selva, M.; Aiazzi, B.; Butera, F.; Chiarantini, L.; Baronti, S. Hyper-sharpening: A first approach on SIM-GA data. *IEEE J. Sel. Top. Appl. Earth Obs. Remote Sens.* **2015**, *8*, 3008–3024. [CrossRef]

25. Eismann, M.T.; Hardie, R.C. Application of the stochastic mixing model to hyperspectral resolution enhancement. *IEEE Trans. Geosci. Remote Sens.* **2004**, *42*, 1924–1933. [CrossRef]

26. Huang, B.; Song, H.; Cui, H.; Peng, J.; Xu, Z. Spatial and spectral image fusion using sparse matrix factorization. *IEEE Trans. Geosci. Remote Sens.* **2014**, *52*, 1693–1704. [CrossRef]

27. Zhang, Y.; de Backer, S.; Scheunders, P. Noise-resistant wavelet-based Bayesian fusion of multispectral and hyperspectral images. *IEEE Trans. Geosci. Remote Sens.* **2009**, *47*, 3834–3843. [CrossRef]

28. Kawakami, R.; Wright, J.; Tai, Y.-W.; Matsushita, Y.; Ben-Ezra, M.; Ikeuchi, K. High-resolution hyperspectral imaging via matrix factorization. In Proceedings of the IEEE Conference on Computer Vision and Pattern Recognition(CVPR 2011), Colorado Springs, CO, USA, 20–25 June 2011; pp. 2329–2336.

29. Wycoff, E.; Chan, T.-H.; Jia, K.; Ma, W.-K.; Ma, Y. A non-negative sparse promoting algorithm for high resolution hyperspectral imaging. In Proceedings of the 2013 IEEE International Conference on Acoustics, Speech and Signal Processing, Vancouver, BC, Canada, 26–31 May 2013; pp. 1409–1413.

30. Cawse-Nicholson, K.; Damelin, S.; Robin, A.; Sears, M. Determining the intrinsic dimension of a hyperspectral image using random matrix theory. *IEEE Trans. Image Process.* **2013**, *22*, 1301–1310. [CrossRef] [PubMed]

31. Bioucas-Dias, J.; Nascimento, J. Hyperspectral subspace identification. *IEEE Trans. Geosci. Remote Sens.* **2008**, *46*, 2435–2445. [CrossRef]

32. Landgrebe, D. Hyperspectral image data analysis. *IEEE Signal Process. Mag.* **2002**, *19*, 17–28. [CrossRef]

33. Yuan, Y.; Fu, M.; Lu, X. Low-rank representation for 3D hyperspectral images analysis from map perspective. *Signal Process.* **2015**, *112*, 27–33. [CrossRef]

34. Zurita-Milla, R.; Clevers, J.G.P.W.; Schaepman, M.E. Unmixing-based landsat TM and MERIS FR data fusion. *IEEE Geosci. Remote Sens. Lett.* **2008**, *5*, 453–457. [CrossRef]

35. Rudin, L.; Osher, S.; Fatemi, E. Nonlinear total variation based noise removal algorithms. *Phys. D Nonlinear Phenom.* **1992**, *60*, 259–268. [CrossRef]

36. Beck, A.; Teboulle, M. Fast gradient-based algorithms for constrained total variation image denoising and deblurring problems. *IEEE Trans. Image Process.* **2009**, *18*, 2419–2434. [CrossRef] [PubMed]

37. Bresson, X.; Chan, T. Fast dual minimization of the vectorial total variation norm and applications to color image processing. *Inverse Probl. Imaging* **2008**, *2*, 455–484. [CrossRef]

38. Simões, M.; Bioucas-Dias, J.; Almeida, L.B.; Chanussot, J. A convex formulation for hyperspectral image superresolution via subspace-based regularization. *IEEE Trans. Geosci. Remote Sens.* **2015**, *53*, 3373–3388. [CrossRef]

39. He, X.; Condat, L.; Bioucas-Dias, J.; Chanussot, J.; Xia, J. A new pansharpening method based on spatial and spectral sparsity priors. *IEEE Trans. Image Process.* **2014**, *23*, 4160–4174. [CrossRef] [PubMed]

40. Palsson, F.; Sveinsson, J.R.; Ulfarsson, M.O. A new pansharpening algorithm based on total variation. *IEEE Geosci. Remote Sens. Lett.* **2014**, *11*, 318–322. [CrossRef]

41. Wei, Q.; Bioucas-Dias, J.; Dobigeon, N.; Tourneret, J. Hyperspectral and multispectral image fusion based on a sparse representation. *IEEE Trans. Geosci. Remote Sens.* **2015**, *53*, 3658–3668. [CrossRef]

42. Dong, W.; Fu, F.; Shi, G.; Cao, X.; Wu, J.; Li, G.; Li, X. Hyperspectral image super-resolution via non-negative structured sparse representation. *IEEE Trans. Image Process.* **2016**, *25*, 2337–2352. [CrossRef] [PubMed]

43. Grohnfeldt, C.; Zhu, X.X.; Bamler, R. Jointly sparse fusion of hyperspectral and multispectral imagery. In Proceedings of the 2013 IEEE International Geoscience and Remote Sensing Symposium (IGARSS), Melbourne, Australia, 21–26 July 2013; pp. 4090–4093.

44. Akhtar, N.; Shafait, F.; Mian, A. Bayesian sparse representation for hyperspectral image super resolution. In Proceedings of the 2015 IEEE Conference on Computer Vision and Pattern Recognition (CVPR), Boston, MA, USA, 7–12 June 2015; pp. 3631–3640.

45. Wei, Q.; Dobigeon, N.; Tourneret, J.-Y. Bayesian fusion of multiband images. *IEEE J. Sel. Top. Signal Process.* **2015**, *9*, 1117–1127. [CrossRef]

46. Hardie, R.C.; Eismann, M.T.; Wilson, G.L. MAP estimation for hyperspectral image resolution enhancement using an auxiliary sensor. *IEEE Trans. Image Process.* **2004**, *13*, 1174–1184. [CrossRef] [PubMed]

47. Zhang, Y.; Duijster, A.; Scheunders, P. A Bayesian restoration approach for hyperspectral images. *IEEE Trans. Geosci. Remote Sens.* **2012**, *50*, 3453–3462. [CrossRef]

48. Veganzones, M.A.; Simões, M.; Licciardi, G.; Yokoya, N.; Bioucas-Dias, J.M.; Chanussot, J. Hyperspectral super-resolution of locally low rank images from complementary multisource data. *IEEE Trans. Image Process.* **2016**, *25*, 274–288. [CrossRef] [PubMed]

49. Berne, O.; Helens, A.; Pilleri, P.; Joblin, C. Non-negative matrix factorization pansharpening of hyperspectral data: An application to mid-infrared astronomy. In Proceedings of the 2010 2nd Workshop on Hyperspectral Image and Signal Processing: Evolution in Remote Sensing, Reykjavik, Iceland, 14–16 June 2010; pp. 1–4.

50. Yokoya, N.; Yairi, T.; Iwasaki, A. Coupled nonnegative matrix factorization unmixing for hyperspectral and multispectral data fusion. *IEEE Trans. Geosci. Remote Sens.* **2012**, *50*, 528–537. [CrossRef]

51. Lanaras, C.; Baltsavias, E.; Schindler, K. Hyperspectral superresolution by coupled spectral unmixing. In Proceedings of the IEEE ICCV, Santiago, Chile, 7–13 December 2015; pp. 3586–3594.

52. Wei, Q.; Bioucas-Dias, J.; Dobigeon, N.; Tourneret, J.-Y.; Chen, M.; Godsill, S. Multiband image fusion based on spectral unmixing. *IEEE Trans. Geosci. Remote Sens.* **2016**, *54*, 7236–7249. [CrossRef]

53. Kwan, C.; Budavari, B.; Bovik, A.C.; Marchisio, G. Blind quality assessment of fused WorldView-3 images by using the combinations of pansharpening and hypersharpening paradigms. *IEEE Geosci. Remote Sens. Lett.* **2017**, *14*, 1835–1839. [CrossRef]

54. Yokoya, N.; Yairi, T.; Iwasaki, A. Hyperspectral, multispectral, and panchromatic data fusion based on coupled non-negative matrix factorization. In Proceedings of the 2011 3rd Workshop on Hyperspectral Image and Signal Processing: Evolution in Remote Sensing, Lisbon, Portugal, 6–9 June 2011; pp. 1–4.

55. Afonso, M.; Bioucas-Dias, J.M.; Figueiredo, M. An augmented Lagrangian approach to the constrained optimization formulation of imaging inverse problems. *IEEE Trans. Image Process.* **2011**, *20*, 681–695. [CrossRef] [PubMed]

56. Eckstein, J.; Bertsekas, D.P. On the Douglas–Rachford splitting method and the proximal point algorithm for maximal monotone operators. *Math. Program.* **1992**, *55*, 293–318. [CrossRef]

57. Gabay, D.; Mercier, B. A dual algorithm for the solution of nonlinear variational problems via finite-element approximation. *Comput. Math. Appl.* **1976**, *2*, 17–40. [CrossRef]

58. Glowinski, R.; Marroco, A. Sur l'approximation, par éléments finis d'ordre un, et la résolution, par pénalisation-dualité d'une classe de problèmes de Dirichlet non linéaires, Revue française d'automatique, informatique, recherche opérationnelle. *Analyse Numérique* **1975**, *9*, 41–76. [CrossRef]

59. Boyd, S.; Parikh, N.; Chu, E.; Peleato, B.; Eckstein, J. Distributed optimization and statistical learning via the alternating direction method of multipliers. *Found. Trends Mach. Learn.* **2011**, *3*, 1–122. [CrossRef]

60. Esser, E. *Applications of Lagrangian-Based Alternating Direction Methods and Connections to Split-Bregman*; CAM Reports 09-31; Center for Computational Applied Mathematics, University of California: Los Angeles, CA, USA, 2009.

61. Gao, B.-C.; Montes, M.J.; Davis, C.O.; Goetz, A.F. Atmospheric correction algorithms for hyperspectral remote sensing data of land and ocean. *Remote Sens. Environ.* **2009**, *113*, S17–S24. [CrossRef]

62. Zare, A.; Ho, K.C. Endmember variability in hyperspectral analysis. *IEEE Signal Process. Mag.* **2014**, *31*, 95–104. [CrossRef]

63. Dobigeon, N.; Tourneret, J.-Y.; Richard, C.; Bermudez, J.C.M.; McLaughlin, S.; Hero, A.O. Nonlinear unmixing of hyperspectral images: Models and algorithms. *IEEE Signal Process. Mag.* **2014**, *31*, 89–94. [CrossRef]

64. USGS Digital Spectral Library 06. Available online: https://speclab.cr.usgs.gov/spectral.lib06/ (accessed on 10 October 2018).

65. Stockham, T.G., Jr. High-speed convolution and correlation. In Proceedings of the ACM Spring Joint Computer Conference, New York, NY, USA, 26–28 April 1966; pp. 229–233.

66. Donoho, D.; Johnstone, I. Adapting to unknown smoothness via wavelet shrinkage. *J. Am. Stat. Assoc.* **1995**, *90*, 1200–1224. [CrossRef]

67. Combettes, P.; Pesquet, J.-C. Proximal splitting methods in signal processing. In *Fixed-Point Algorithms for Inverse Problems in Science and Engineering*; Springer: New York, NY, USA, 2011; pp. 185–212.

68. Condat, L. Fast projection onto the simplex and the l_1 ball. *Math. Program.* **2016**, *158*, 575–585. [CrossRef]

69. Wald, L.; Ranchin, T.; Mangolini, M. Fusion of satellite images of different spatial resolutions: Assessing the quality of resulting image. *IEEE Trans. Geosci. Remote Sens.* **2005**, *43*, 1391–1402.

70. Hyperspectral Remote Sensing Scenes. Available online: www.ehu.eus/ccwintco/?title=Hyperspectral_Remote_Sensing_Scenes (accessed on 10 October 2018).

71. Baumgardner, M.; Biehl, L.; Landgrebe, D. *220 band AVIRIS Hyperspectral Image Data set: June 12, 1992 Indian Pine Test Site 3*; Purdue University Research Repository; Purdue University: West Lafayette, IN, USA, 2015.

72. Fusing multiple multiband images. Available online: https://github.com/Reza219/Multiple-multiband-image-fusion (accessed on 10 October 2018).

73. Landsat 8. Available online: https://landsat.gsfc.nasa.gov/landsat-8/ (accessed on 10 October 2018).

74. Alparone, L.; Baronti, S.; Aiazzi, B.; Garzelli, A. Spatial methods for multispectral pansharpening: Multiresolution analysis demystified. *IEEE Trans. Geosci. Remote Sens.* **2016**, *54*, 2563–2576. [CrossRef]

75. Garzelli, A.; Nencini, F.; Capobianco, L. Optimal MMSE pan sharpening of very high resolution multispectral images. *IEEE Trans. Geosci. Remote Sens.* **2008**, *46*, 228–236. [CrossRef]

76. Aiazzi, B.; Alparone, L.; Baronti, S.; Garzelli, A.; Selva, M. MTF-tailored multiscale fusion of high-resolution MS and Pan imagery. *Photogramm. Eng. Remote Sens.* **2006**, *72*, 591–596. [CrossRef]

77. Lee, J.; Lee, C. Fast and efficient panchromatic sharpening. *IEEE Trans. Geosci. Remote Sens.* **2010**, *48*, 155–163.

78. Aiazzi, B.; Alparone, L.; Baronti, S.; Garzelli, A.; Selva, M. An MTF-based spectral distortion minimizing model for pan-sharpening of very high resolution multispectral images of urban areas. In Proceedings of the 2nd GRSS/ISPRS Joint Workshop Remote Sensing Data Fusion URBAN Areas, Berlin, Germany, 22–23 May 2003; pp. 90–94.

79. Wei, Q.; Dobigeon, N.; Tourneret, J.-Y.; Bioucas-Dias, J.M.; Godsill, S. R-FUSE: Robust fast fusion of multiband images based on solving a Sylvester equation. *IEEE Signal Process. Lett.* **2016**, *23*, 1632–1636. [CrossRef]

80. Wei, Q.; Dobigeon, N.; Tourneret, J.-Y. Fast fusion of multi-band images based on solving a Sylvester equation. *IEEE Trans. Image Process.* **2015**, *24*, 4109–4121. [CrossRef] [PubMed]

81. Yokoya, N.; Grohnfeldt, C.; Chanussot, J. Hyperspectral and multispectral data fusion: A comparative review of the recent literature. *IEEE Geosci. Remote Sens. Mag.* **2017**, *5*, 29–56. [CrossRef]

82. Wald, L. Quality of high resolution synthesised images: Is there a simple criterion? In Proceedings of the Fusion of Earth data: Merging point measurements, raster maps and remotely sensed images, Nice, France, 26–28 January 2000; pp. 99–103.

83. Kruse, F.A.; Lefkoff, A.B.; Boardman, J.W.; Heidebrecht, K.B.; Shapiro, A.T.; Barloon, P.J.; Goetz, A.F.H. The spectral image processing system (SIPS): Interactive visualization and analysis of imaging spectrometer data. *Remote Sens. Environ.* **1993**, *44*, 145–163. [CrossRef]

84. Garzelli, A.; Nencini, F. Hypercomplex quality assessment of multi/hyperspectral images. *IEEE Geosci. Remote Sens. Lett.* **2009**, *6*, 662–665. [CrossRef]

85. Wang, Z.; Bovik, A.C. A universal image quality index. *IEEE Signal Process. Lett.* **2002**, *9*, 81–84. [CrossRef]

86. Alparone, L.; Baronti, S.; Garzelli, A.; Nencini, F. A global quality measurement of pan-sharpened multispectral imagery. *IEEE Geosci. Remote Sens. Lett.* **2004**, *1*, 313–317. [CrossRef]

87. Nascimento, J.; Bioucas-Dias, J. Vertex component analysis: A fast algorithm to unmix hyperspectral data. *IEEE Trans. Geosci. Remote Sens.* **2005**, *43*, 898–910. [CrossRef]

88. Bioucas-Dias, J.; Figueiredo, M. Alternating direction algorithms for constrained sparse regression: Application to hyperspectral unmixing. In Proceedings of the 2010 2nd Workshop on Hyperspectral Image and Signal Processing: Evolution in Remote Sensing, Reykjavik, Iceland, 14–16 June 2010; pp. 1–4.

89. Golub, G.; Heath, M.; Wahba, G. Generalized cross-validation as a method for choosing a good ridge parameter. *Technometrics* **1979**, *21*, 215–223. [CrossRef]

Journal of
Imaging

MDPI

Article

Efficient Lossless Compression of Multitemporal Hyperspectral Image Data

Hongda Shen [1,†], Zhuocheng Jiang [2] and W. David Pan [2,*]

[1] Bank of America Corporation, New York, NY 10020, USA; hongdadeeplearning@gmail.com
[2] Department of Electrical and Computer Engineering, University of Alabama in Huntsville, Huntsville, AL 35899, USA; zj0004@uah.edu
* Correspondence: pand@uah.edu
† Note: This paper is continuation of the first author's PhD work, which is not associated with his current affiliation.

Received: 14 November 2018; Accepted: 27 November 2018; Published: 2 December 2018

Abstract: Hyperspectral imaging (HSI) technology has been used for various remote sensing applications due to its excellent capability of monitoring regions-of-interest over a period of time. However, the large data volume of four-dimensional multitemporal hyperspectral imagery demands massive data compression techniques. While conventional 3D hyperspectral data compression methods exploit only spatial and spectral correlations, we propose a simple yet effective predictive lossless compression algorithm that can achieve significant gains on compression efficiency, by also taking into account temporal correlations inherent in the multitemporal data. We present an information theoretic analysis to estimate potential compression performance gain with varying configurations of context vectors. Extensive simulation results demonstrate the effectiveness of the proposed algorithm. We also provide in-depth discussions on how to construct the context vectors in the prediction model for both multitemporal HSI and conventional 3D HSI data.

Keywords: lossless compression; multitemporal hyperspectral images; information theoretic analysis; predictive coding

1. Introduction

Hyperspectral imaging (HSI) technologies have been widely used in many applications of remote sensing (RS) owing to the high spatial and spectral resolutions of hyperspectral images [1]. In some applications (e.g., hyperspectral imaging change detection [2–5]), we need to collect a sequence of hyperspectral images over the same spatial area at different times. The set of hyperspectral images collected over one location at varying time points is called multitemporal hyperspectral images [6,7]. From these multitemporal images, changes of observed locations over time can be detected and analyzed. Figure 1 illustrates a typical multitemporal hyperspectral image dataset. Each stack represents one 3D HSI. A sequence of 3D HSI stacks are captured by the HSI sensor over time.

Hyperspectral datasets tend to be of very large sizes. In the case of 4D multitemporal HSI datasets, the accumulated data volume increases very rapidly (to the Gigabyte or even Terabyte level), thereby making data acquisition, storage and transmission very challenging, especially when network bandwidth is severely constrained. As the number of hyperspectral images grows, it is clear that data compression techniques play a crucial role in the development of hyperspectral imaging techniques [8,9]. Lossy compression can significantly improves the compression efficiency, albeit at the cost of selective information loss. However, the fact that human visual systems are not sensitive to certain types and levels of distortions caused by information loss makes lossy compression useful. While lossy compression methods typically provide much larger data reduction than lossless methods, they might not be suitable for many accuracy-demanding hyperspectral imaging applications, where the images are intended to be analyzed automatically by computers. Since lossless compression methods can strictly guarantee no loss in the reconstructed data, lossless compression would be more desirable in these applications.

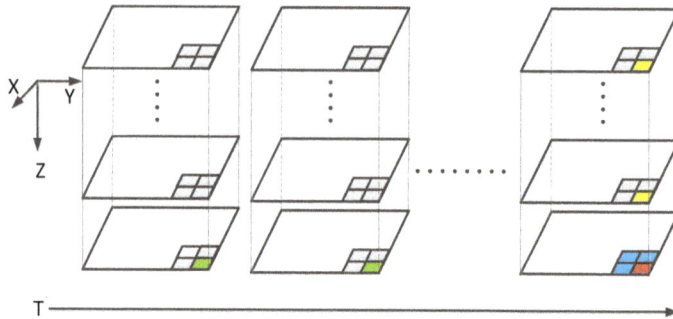

Figure 1. A multitemporal hyperspectral image dataset, where X and Y are the spatial directions, Z is the spectral direction, and T is the temporal direction.

Many efforts have been made to develop efficient lossless compression algorithms for 3D HSI data. LOCO-I [10] and 2D-CALIC [11] utilize spatial redundancy to reduce the entropy of prediction residuals. To take advantage of strong spectral correlations in HSI data, 3D compression methods have been proposed, which includes 3D-CALIC [12], M-CALIC [13], LUT [14] and its variants, SLSQ [15] and CCAP [16]. Also, some transform-based methods, such as SPIHT [17], SPECK [18], etc., can be easily extended to lossless compression even though they were designed for lossy compression.

Recently, clustering techniques have been introduced into 3D HSI data lossless compression and produced state-of-the-art performance over publicly available datasets. In [19], B. Aiazzi et al. proposed a predictive method leveraging crisp or fuzzy clustering to produce state-of-the-art results. Later, authors in both [20,21] again utilized the K-means clustering algorithm to improve the compression efficiency. Although these methods can yield higher compression, their computational costs are significantly higher than regular linear predictive methods. Plus, it is very difficult to parallel the process to leverage hardware acceleration if clustering technique is required as a preprocessing step in those approaches. In addition to the goal of reducing the entropy of either prediction residuals or transform coefficients, low computational complexity is another influential factor because many sensing platforms have very limited computing resources. Therefore, a low-complexity method called the "Fast Lossless" (FL) method, proposed by the NASA Jet Propulsion Lab (JPL) in [22], was selected as the core predictor in the Consultative Committee for Space Data Systems (CCSDS) new standard for multispectral and hyperspectral data compression [23], to provide efficient compression on 3D HSI data. This low-complexity merit also enables efficient multitemporal HSI data compression.

Multitemporal HSI data has an additional temporal dimension compared to 3D HSI data. Therefore, we can take advantage of temporal correlations to improve the overall compression efficiency of 4D HSI data. Nonetheless, there is very sparse work on lossless compression of multitemporal HSI data in the literature. Mamun et al. proposed a 4D lossless compression algorithm in [24], albeit lacking details on the prediction algorithms. In [25], a combination of Karhunen-Loève Transform (KLT), Discrete Wavelet Transform (DWT), and JPEG 2000 was applied to reduce the spectral and temporal redundancy of 4D remote sensing image data. However, the method can only achieve lossy compression. Additionally, Zhu et al. proposed another lossy compression approach for multitemporal HSI data in [7], based on a combination of linear prediction and a spectral concatenation of images. For the first time, we addressed lossless compression of multitemporal HSI data in [6], by introducing a *correntropy* based Least Mean Square filter for the Fast Lossless (FL) predictor. While the benefit of exploiting temporal correlations in compression has been demonstrated by some papers such as [26,27], in this work, we conduct an in-depth information-theoretic analysis on the amount of compression achievable on multitemporal HSI data, by taking into account both the spectral and temporal correlations. On the other hand, this additional temporal decorrelation definitely poses a greater challenge to data processing speed especially for those powerful but computationally expensive algorithms, e.g., [19–21]. Therefore, we propose a low-complexity linear prediction algorithm, which extends the well-known FL method into a 4D version to achieve higher data compression, by better adapting to the underlying statistics of multitemporal HSI data. Note that most existing 3D HSI compression methods can be extended into 4D versions with proper modifications. However, this is beyond the scope of this paper.

The remainder of this paper is organized as follows. First, in Section 2, we give an overview of the multitemporal HSI datasets used in the study, which include three publicly available datasets, as well as two multitemporal HSI datasets we generated by using hyperspectral cameras. In Section 3, we present the information-theoretic analysis, followed by the introduction of a new algorithm for multitemporal HSI data lossless compression in Section 4. Finally, we present simulation results in Section 5 and make some concluding remarks in Section 6.

2. Datasets

Since there is little prior work on multi-temporal hyperspectral image compression, publicly available multi-temporal HSI datasets are very rare. Currently, the time-lapse hyperspectral radiance images of natural scenes [28] are the only available datasets to our best knowledge. Therefore, we created another two datasets capturing two scenes of Alabama A&M University campus using the portable Surface Optics Corporation (SOC) 700 hyperspectral camera [29] to enrich the relevant resources and facilitate further research. Hence, we introduce both data sources especially our datasets in detail before the actual analysis and algorithm development.

2.1. Time-Lapse Hyperspectral Imagery

Time-lapse hyperspectral imaging technology has been used for various remote sensing applications due to its excellent capability of monitoring regions-of-interest over a period of time. Time-lapse hyperspectral imagery is a sequence of 3D HSIs captured over the same scene but at different time stamps (often at a fixed time interval). Therefore, time-lapse hyperspectral imagery can be considered as a 4D dataset, whose size increases significantly with the total number of time stamps.

In [28], the authors made public several sequences of hyperspectral radiance images of scenes undergoing natural illumination changes. In each scene, hyperspectral images were acquired at about one-hour intervals. We randomly selected three 4D time-lapse HSI datasets, Levada, Gualtar and Nogueiro. Basic information of these three datasets are listed in the Table 1. Detailed information of these datasets can be found in [30]. Each single HSI has the same spatial size, 1024×1344, with 33 spectral bands. Both Gualtar and Nogueiro have nine time stamps while Levada has seven. Note that the original data for these datasets was linearly mapped into $[0, 1]$ and stored using "double" floating point format (64 bits) [28]. In order to evaluate the prediction-based lossless compression performance of algorithms, we pre-process the datasets by re-mapping the data samples back to their original precision of 12 bits. The resulting sizes of the datasets range from 454.78 MB (for seven frames) to 584.71 MB (for nine frames).

Figure 2 shows the Levada, Noguerio and Gualtar sequences from top to bottom. Detailed information about the Levada sequence can be found in [28]. Note that only 2D color-rendered RGB (Red, Green and Blue) images are shown in Figure 2 instead of the actual HSI data for display purpose. Since time-lapse HSIs are captured over the same scene at different time instants with gradually changing natural illumination, we can see that images at different time instants are very similar in Figure 2. These temporal correlations can be exploited to improve the overall compression efficiency.

Table 1. Multitemporal hyperspectral image datasets.

Dataset	Size	The Number of Time Frames	Precision (bits)
Levada	$1024 \times 1344 \times 33$	7	12
Gualtar	$1024 \times 1344 \times 33$	9	12
Noguerio	$1024 \times 1344 \times 33$	9	12
Scene-1	$640 \times 640 \times 120$	21	12
Scene-2	$640 \times 640 \times 120$	16	12

Figure 2. Some sample images at different time instants from the time-lapse hyperspectral image datasets (from top to bottom: Levada, Nogueiro and Gualtar).

2.2. AAMU Datasets

Due to very few 4D HSI datasets available in the public domain, we created some new datasets to increase the data diversity of our study. To this end, we used a SOC 700 hyperspectral camera (manufactured by Surface Optics Corporstion, CA, USA) and produced 4D datasets for two scenes on the campus of Alabama A&M University (AAMU). The SOC 700 camera can record and process hyperspectral imagery at a rate of 15 megabytes of data every second (120-band elements per second at 12-bit resolution, 640 pixels per row, 100 rows per second). The imaging system's spectral response covers the visible and near-infrared spectral range (from 0.43 to 0.9 microns), and can be used in normal to low lighting conditions with variable exposure times and display gains. More detailed about the SOC 700 system can be found at [29].

We placed the camera at two distinct locations of the AAMU campus and generated two datasets, which we call Scene-1 and Scene-2. 3D HSI cubes in Scene-1 and Scene-2 are of the same size: $640 \times 640 \times 120$ with 21 and 16 time frames, respectively. The overall dataset sizes of Scene-1 and Scene-2 are roughly 1.70 GB and 850 MB, respectively. Compared to three time-lapse datasets discussed earlier, these two AAMU datasets are much larger, making themselves more suitable for evaluating compression efficiencies. In contrast to the time-lapse datasets, the images of the AAMU datasets were acquired at time-varying rates of approximately one per five minutes or one per minute, thereby introducing time-varying temporal correlations through the entire dataset. This special feature will allow us to investigate the relationship between prediction accuracy and correlations at different levels.

Figure 3 shows the 2D color-rendered RGB images for a few time instants for the AAMU multitemporal HSI datasets. While changing illumination conditions over time can be observed in both datasets, temporal similarity in both pixel intensity and image structure is also obvious, similar to the three time-lapse datasets shown in Figure 2. In order to quantify the potential gain on compression achievable by exploiting the temporal correlations in 4D HSI datasets, we conducted an information-theoretic analysis as detailed in the next section.

Figure 3. Sample images at different time instants from the AAMU hyperspectral image datasets (top: Scene-1, and bottom: Scene-2).

3. Problem Analysis

While the actual amount of compression achieved depends on the choice of specific compression algorithms [31], information theoretic analysis can provide us an upper bound on the amount of compression achievable. Here we focus on analyzing how temporal correlation can help improve the compression of 4D hyperspectral image datasets, as opposed to the baseline 3D-compression case where only spatial and spectral correlations are considered.

Let X_j^t be a 4D hyperspectral image at the tth time instant and jth spectral band where X represents a two-dimensional image with K distinct pixel values v_i ($i \in \{1, \cdots, K\}$) within each band. Then the entropy of this source can be obtained based on the probabilities $p(v_i)$ of these values by

$$H(X_j^t) = -\sum_{i=1}^{K} p(v_i) \cdot \log_2 \left[p(v_i) \right]. \tag{1}$$

If we assume that there are no dependencies between pixels of X_j^t, at least $H(X_j^t)$ bits must be spent on average for each pixel of this image. However, for typical 4D hyperspectral images, this assumption does not hold given the existence of spatial, spectral and temporal correlations. The value of a particular pixel might be similar to some other pixels from its spatial, spectral or temporal neighborhoods (contexts). Considering these correlations can lead to reduced information (fewer bits to code each pixel on average) than the entropy $H(X_j^t)$. The conditional entropy of the image captures the correlations as follows:

$$H(X_j^t | C_j^t) = -\sum_{i=1}^{K} p(v_i | C_j^t) \cdot \log_2 \left[p(v_i | C_j^t) \right]. \tag{2}$$

where C_j^t denoted as *context*, which represents a group of correlated pixels. In general, conditioning reduces entropy, $H(X_j^t | C_j^t) \leq H(X_j^t)$.

The choice of context largely determines how much compression we can achieve by using prediction-based lossless compression schemes. One should include highly-correlated pixels into the context. Spectral and temporal correlations are typically much stronger than spatial correlations in multitemporal hyperspectral images. For example, ref. [20] claims that explicit spatial de-correlation is not always necessary to achieve good compression [31]. Ref. [31] shows that a linear prediction scheme was adequate for spectral and/or temporal prediction because of high degree of correlations, in contrast to non-linearity nature of spatial de-correlation. Therefore, we construct the context vector C_j^t using only pixels from previous bands at the same spatial location, as well as pixels from the same spectral band at the same location but at previous time points. Specifically, we denote pixels from previous N_b bands at the same spatial location as X_{j-m}^t, $m \in \{1, 2, ..., N_b\}$ (yellow pixels in Figure 1), and pixels from the same spectral band at the same location but from previous N_t temporal positions X_j^{t-n}, $n \in \{1, 2, ..., N_t\}$ (green pixels in Figure 1), respectively. Then, the conditional entropy in Equation (2) becomes

$$H(X_j^t | X_{j-m}^t, X_j^{t-n}) = -\sum_{i=1}^{K} p(v_i | X_{j-m}^t, X_j^{t-n}) \cdot \log_2 \left[p(v_i | X_{j-m}^t, X_j^{t-n}) \right]. \tag{3}$$

By using the relation between joint entropy and conditional entropy, we can further rewrite Equation (3) as

$$H(X_j^t | X_{j-m}^t, X_j^{t-n}) = H(X_j^t, X_{j-m}^t, X_j^{t-n}) - H(X_{j-m}^t, X_j^{t-n}), \tag{4}$$

which enables a simple algorithm for estimation of the conditional entropies. It suffices to estimate the above two joint entropies by counting the occurrence frequency of each $(N_t + N_b + 1)$-tuple in the set $(X_j^t, X_{j-m}^t, X_j^{t-n})$, and $(N_t + N_b)$-tuple in the set (X_{j-m}^t, X_j^{t-n}), respectively. However, as pointed out in [31], the entropy estimates become very inaccurate when two or more previous bands are used for prediction in practice. The reason is that, as the entropy is conditioned upon multiple bands, the set $(X_j^t, X_{j-m}^t, X_j^{t-n})$ takes on values from the alphabet $\chi^{(N_t+N_b+1)}$, whose size can become extremely large, e.g., $(2^{12})^{N_t+N_b+1}$ for our datasets. As a consequence, a band might not contain enough pixels to provide statistically meaningful estimates of the probabilities. Similar to the "data source transform" trick proposed in [31], we consider each bit-plane of X_j^t as one separate binary source. Although binary sources greatly reduce the alphabet size, which makes it possible to obtain accurate entropy estimates, results obtained for the binary source are not very representative of the actual bit rates obtained by a practical coder since statistical dependencies between those bit-planes cannot be neglected. However, using bit-plane sources would be useful for our study since our main goal is to evaluate the relative instead of the absolute performance gain achievable by using different contexts based on various combinations of spectral and temporal bands. Therefore, we will compute the conditional entropy in Equation 3, for all the bit-planes separately, and then take their average to be the overall performance gain for a specific prediction context. In this sense, we extend the algorithm in [31] by incorporating previous temporal bands into the context vector, which allows us to estimate also the temporal correlations.

We applied this estimation algorithm on five multitemporal HSI datasets to estimate the potential compression performance of multitemporal hyperspectral image with a combination of various spectral and temporal bands. $H(p, q)$ is the entropy conditioned to p previous bands at the current time point and q bands at current spectral band but from previous time points for prediction. Using the binary-source based estimation method, we summed up the conditional entropies of all the bit-planes (a total of 12 bit-planes for all our datasets) as the estimation of $H(p, q)$ for each band of the dataset. Then the averages of $H(p, q)$ over all bands are reported in Figure 4 for all five datasets. Due to limited space, we only show results for parameters p and q chosen between 0 and 5. More detailed results for other datasets can be found in Tables A1–A5 in Appendix A.

From Figure 4, we can observe that as either p or q increases, the general trend is that the conditional entropy decreases; however, as p or q further increases (e.g, from 4 to 5), the reduction of entropy becomes smaller than the case of either p or q going from 0 to 1. This means that including a few previous bands either spectrally or temporally in the context can be very useful to improving the performance of the prediction-based compression algorithms, but the return of adding more bands from distant past will diminish as the correlations get weaker, let alone the increased computational cost associated with involving excessive number of bands for prediction. In addition, the conditional entropy tends to decrease faster with an increased p than with an increased q. This is indicative of stronger spectral correlations than temporal correlations. For example, the fourth image in the first row of Figure 3 represents a dramatic change of illumination conditions during image capturing, thereby weakening the temporal correlations. However, there still exist significant temporal correlations which, if exploited properly, can lead to improved compression by considering only spectral correlations. To this end, we propose a compression algorithm, which exploits temporal correlations in multitemporal HSI data to enhance the overall compression performance.

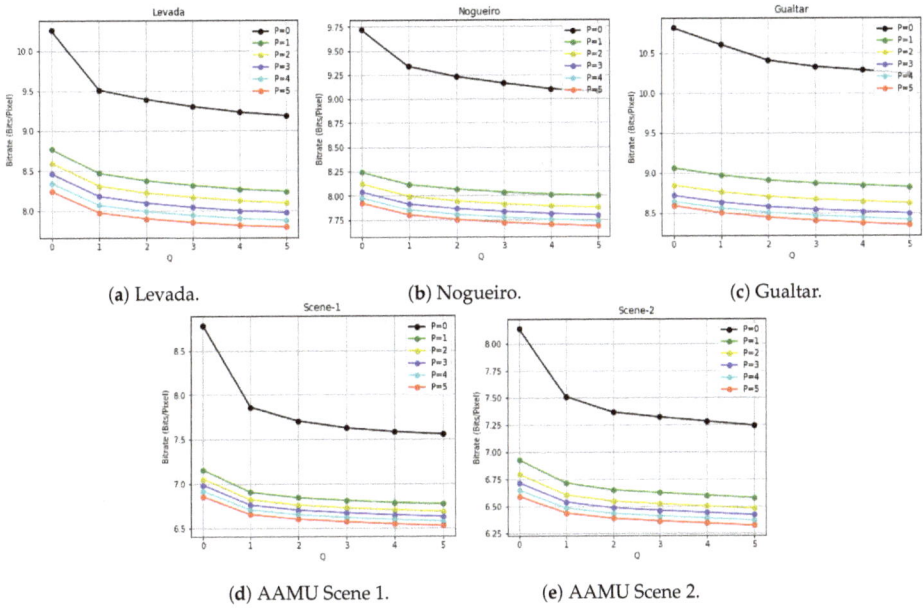

(a) Levada.

(b) Nogueiro.

(c) Gualtar.

(d) AAMU Scene 1.

(e) AAMU Scene 2.

Figure 4. Conditional entropies over five datasets for different P and Q combination.

4. Proposed Algorithm

Our lossless compression algorithm is based on predicting the pixels to be coded, by using a linear combination of those pixels already coded (in a neighboring causal context). Prediction residuals are obtained by subtracting the actual pixel values from their estimates. The residuals are then encoded using entropy coders.

For multitemporal hyperspectral image, the estimate of a pixel value can be obtained by

$$\hat{x}_{p,q}^{j,t} = \mathbf{w}_{j,t}^T \mathbf{y}_{p,q}^{j,t}. \tag{5}$$

where $\hat{x}_{p,q}^{j,t}$ represents an estimate of a pixel, $x_{p,q}^{j,t}$, at spatial location (p, q), the jth band and the tth time point, while $\mathbf{w}_{j,t}$ denotes the weights for linearly combining the pixel values $\mathbf{y}_{p,q}^{j,t}$. These pixels are drawn from a causal context of several previously coded bands either at the same time point or at previous time points. More specifically, $\mathbf{y}_{p,q}^{j,t} = \left[\mathbf{x}_{p,q}^{j-m,t}, \mathbf{x}_{p,q}^{j,t-n} \right]$, $m \in \{1, 2, ..., N_b\}$, $n \in \{1, 2, ..., N_t\}$.

For accurate prediction, weights should be able to adapt to locally changing statistics of pixel values in the multitemporal HSI data. For this sake, learning algorithms were introduced for lossless compression of 3D HSI data [22,32]. Adaptive learning was also used in the so-called Fast Lossless (FL) method. Due to its low-complexity and effectiveness, the FL method has been selected as a new compression standard for multispectral and hyperspectral data by CCSDS (Consultative Committee for Space Data Systems) [23]. The core learning algorithm of the FL method is the *sign* algorithm, which is a variant of least mean square (LMS). In prior work, we proposed another LMS variant, called *correntropy* based LMS (CLMS) algorithm, which uses the Maximum Correntropy Criterion [6,8] for lossless compression of 3D and multi-temporal HSI data. By replacing the cost function for LMS based learning with correntropy [33], the CLMS method introduces a new term in the weight update function, which allows the learning rate to change, in order to improve on the conventional LMS based method with a constant learning rate. However, good performance of the CLMS method depends heavily on proper tuning of the kernel variance, which is an optimization parameter used by the "kernel trick" associated with the correntropy. To avoid the excessive need to tune the kernel variance for various types of images in the multitemporal HSI datasets, we adopted the sign algorithm used by the FL predictor with an expanded context vector.

In order to exploit spatial correlations also found in hyperspectral datasets, we follow the simple approach in [22], where local-mean-removed pixel values are used as input to our linear predictor. Specifically, for an arbitrary pixel $x_{p,q}^{j,t}$ in a multi-temporal HSI, the spatial local mean is

$$\mu_{p,q}^{j,t} = [x_{p-1,q-1}^{j,t} + x_{p-1,q}^{j,t} + x_{p,q-1}^{j,t}]/3. \tag{6}$$

After mean subtraction, the causal context becomes

$$\mathbf{U}_{p,q}^{j,t} = \left[\mathbf{x}_{p,q}^{j-m,t} - \mu_{p,q}^{j-m,t}, \mathbf{x}_{p,q}^{j,t-n} - \mu_{p,q}^{j,t-n} \right], \tag{7}$$

where $m \in \{1, 2, ..., N_b\}$, and $n \in \{1, 2, ..., N_t\}$. To simplify the notation, we represent the spatial location (p, q) with a single index, k, in that $k = (p-1) * N_y + q$, where N_y refers to number of pixels in each row within one band. In other words, we line up the pixels in a 1D vector, where the pixels will be processed sequentially in the iterative optimization process of the sign algorithm. Now the predicted value for an arbitrary pixel in each band of multi-temporal HSI dataset is given by

$$\hat{x}_k^{j,t} = \mu_k^{j,t} + \mathbf{w}_{j,t}^T \mathbf{U}_k^{j,t}, \tag{8}$$

where $\mathbf{w}_{j,t}^T$ are the weights to be adapted sequentially for each band. If follows that the prediction residual can be obtained as

$$e_k^{j,t} = x_k^{j,t} - \hat{x}_k^{j,t} = x_k^{j,t} - \left(\mu_k^{j,t} + \mathbf{w}_{j,t}^T \mathbf{U}_k^{j,t} \right). \tag{9}$$

We apply the sign algorithm to iteratively update the weights as

$$\mathbf{w}_{j,t}^{k+1} = \mathbf{w}_{j,t}^k + \rho(k) \cdot \text{sign}(e_k^{j,t}) \cdot \mathbf{U}_k^{j,t}, \tag{10}$$

where $\rho(k)$ is an adaptive learning rate proposed in [23] to achieve fast convergence to solutions close to the global optimum. Our study found that using this adaptive learning rate can provide good results on multitemporal datasets. Note that we need to reset the weights and learning rate for each new band in the dataset to account for potentially varying statistics.

After prediction, all the residuals are mapped to non-negative values [23] and then coded into bitstream losslessly by using the Golomb-Rice Codes (GRC) [34]. Although GRC is selected as the entropy coder because of its computational efficiency [35], we observed that using arithmetic coding can offer slightly lower bitrates, albeit at a much higher computational cost. Pseudo Algorithm 1 of this 4D extension of Fast-Lossless is given to better show its structure and workflow.

Algorithm 1 Fast-Lossless-4D Predictor

Input:
1) 4D HSI data **X**.
2) T (Total # of time frames of **X**).
3) B (Total # of spectral bands for each time frame of **X**)
4) P (# of previous spectral bands).
5) Q (# of bands from previous time frames).
6) μ (learning rate).
for t = 1:T **do**
    for b = 1:B **do**
        Local mean subtracted data **U** using Equations (6) and (7).
        initialize: $\mathbf{w}_{j,t} = \mathbf{0}$.
        for each pixel in this band in raster scan order **do**
            Output residual $\mathbf{e}^{j,t}$ using Equation (9).
            Updating $\mathbf{w}_{j,t}$ using Equation (10).
        end for
    end for
end for

---

5. Simulation Results

We tested the proposed algorithm on all five multitemporal HSI datasets (Levada, Gualtar, Nogueiro, AAMU scene_1 and AAMU scene_2). To show the performance of our algorithm, we present the bitrates after compression in Figure 5 (Detailed results can be found in Tables A6–A10). Similar to the conditional entropy estimation results in Section 3, the bitrates were obtained by using various combinations of p (spectral) and q (temporal) number of bands for causal contexts.

We can see that for the case of $p = 0$ and $q = 0$, where we simply use mean subtraction (for spatial decorrelation) without spectral and temporal decorrelation, we can already achieve significant amount of compression on the input data by lowering the original bitrate from 12 bits/pixel to about 6 bits/pixel. If we consider either spectral or temporal correlations, or both, we can achieve additional compression gains on multitemporal HSI data. For example, the bit rate can be reduced by approximately 1 bit/pixel or 0.2 bit/pixel by including in the prediction context one more previous band spectrally or temporally. Generally, the bitrates decrease with more bands being included in the context, which agrees well with the results on condition entropy estimation in Section 3. Furthermore, if we fix the p value and increase the q value, and vice versa, we can achieve better compression. However, the return on including more bands will diminish gradually as p and q further increase. In some cases, we can even have less compression if the context includes some remote bands that might be weakly correlated with the pixels to be predicted. For example, in Table A7, when $p = 5$ and $q = 4$, the corresponding bit rate is 4.4953, which is 0.0026 higher than 4.4927 (when $p = 5$ and $q = 3$). Similar examples can be found in Table A9 (when $p = 5$ and $q = 4$) and in Table A10 (when $p = 5$ and $q = 1$). Including weakly correlated or totally uncorrelated pixels might lower the quality of the context, leading to degraded compression performance. In this same spirit, we can see that spectral decorrelation turns out to be more effective in reducing the bitrates than temporal decorrelation. This means that spectral correlations are stronger than temporal correlations in the datasets we tested. The reason can be that each hyperspectral image cube in these multitemporal HSI datasets was captured at time intervals of at least a few minutes, during which significant change of pixel values (e.g, caused by illumination condition changes) might have taken place. If the image capturing time interval is reduced, then we expect the stronger temporal correlations.

On the other hand, prediction using only one previous spectral band, and/or the same spectral band but from previous time instance can offer a low-complexity compressor with sufficiently good compression performance. The bitrate results show the wide range of tradeoffs for us to explore in order to balance compression performance with computational complexity.

We also compared the proposed algorithm (based on Fast-Lossless algorithm) with our previous work (based on correntropy LMS learning) in [6], namely CLMS, which seems to be the only existing work on lossless compression of multitemporal HSI data. For fair comparison, we use the same parameter setting in [6]. Although it would be straightforward to show the bitrates for both algorithms in multiple tables, we choose to visualize bitrates changes as number of previous bands or number of previous time frames increases for both algorithms. To reduce the complexity of this visualization, we only present $P = 3$ case since it is the default setting in the FL method. Figure 6 shows the bitrate changes for the five datasets. Note that when $Q = 0$, our method is essentially equivalent to 3D FL method. Therefore, we use green dashed line to mark FL method performance in Figure 6. While blue, green and red curves represent CLMS, FL method and ours respectively, it is clear that our methods produce the lowest bitrates consistently. Although bitrates of our method was only slightly lower than applying FL method directly to each one of time framed HSI, it outperformed CLMS method by a significant margin. The improvements on time-lapse datasets are more significant than on AAMU datasets in general. Consistent with results shown previously, we have higher compression gains in the spectral dimension than the temporal dimension. However, the results show that our algorithm can take advantage of the temporal correlations available to bring additional improvements on the overall compression performance.

(**a**) Levada.　　　　　　(**b**) Nogueiro.　　　　　　(**c**) Gualtar.

(**d**) AAMU Scene 1.　　　　　(**e**) AAMU Scene 2.

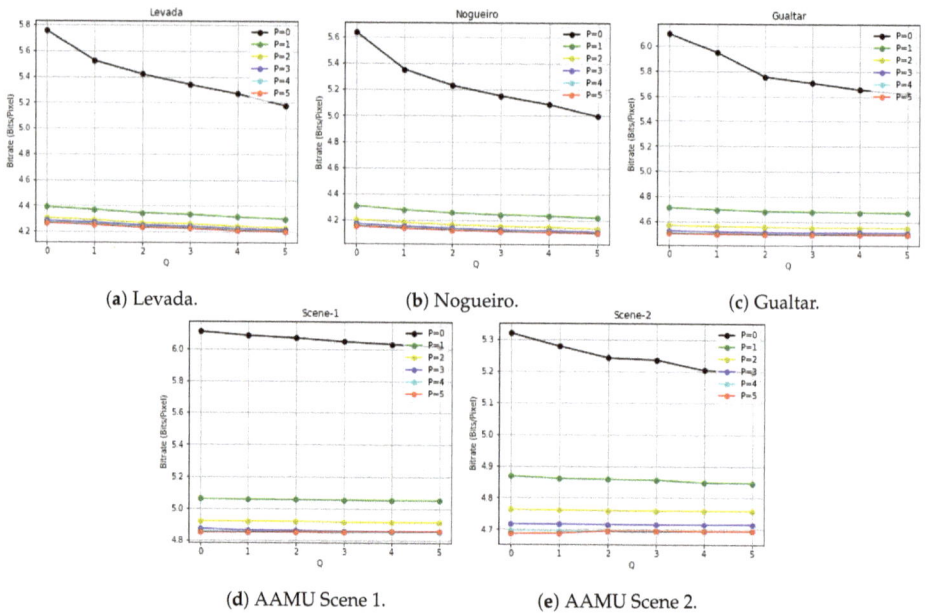

Figure 5. Bitrates over five datasets for different P and Q combination.

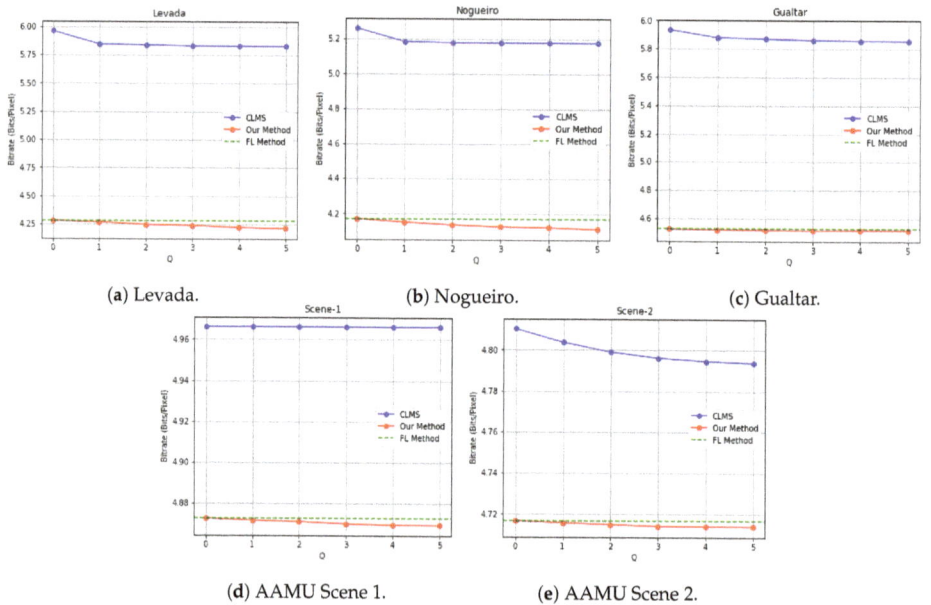

(**a**) Levada.　　　　　　(**b**) Nogueiro.　　　　　　(**c**) Gualtar.

(**d**) AAMU Scene 1.　　　　　(**e**) AAMU Scene 2.

Figure 6. Bitrates comparison with CLMS and FL methods over five datasets for different Q when $P = 3$.

6. Conclusions

We have proposed a new predictive lossless compression algorithm for multitemporal time-lapse hyperspectral image data using a low-complexity sign algorithm with an expanded prediction context. Simulation results have demonstrated the outstanding capability of this algorithm to compress multitemporal HSI data through spectral and temporal decorrelation. The actual compression results are congruent with the information theoretic analysis and estimation based on conditional entropy. We show that increasing the number of previous bands for prediction can yield better compression performance, by exploiting the spectral and temporal correlations in the datasets.

As future work, we intend to study how to adaptively select bands to build an optimal context vector for prediction. Also, we will investigate how to fully integrate the proposed algorithm and the analytic framework to achieve real-time compression on streaming hyperspectral data. Furthermore, the proposed algorithm can be extended to lossless compression of regions-of-interest in hyperspectral images, which can offer much higher compression than compressing the entire hyperspectral image dataset.

Author Contributions: Conceptualization, H.S. and W.D.P.; Methodology, H.S. and W.D.P.; Software, H.S.; Validation, H.S., Z.J., and W.D.P.; Formal Analysis, H.S.; Investigation, H.S.; Resources, H.S. and W.D.P.; Data Curation, H.S.; Writing-Original Draft Preparation, H.S.; Writing-Review & Editing, H.S., Z.J. and W.D.P.; Visualization, H.S.; Supervision, W.D.P.; Project Administration, W.D.P.; Funding Acquisition, W.D.P.

Funding: This research received no external funding.

Acknowledgments: We would like to thank Joel Fu of the Computer Science Program, Alabama A&M University, Normal, AL, for providing facilities including a SOC 700 hyperspectral camera for data collection in this research.

Conflicts of Interest: The authors declare no conflict of interest.

Appendix A. Condition Entropy Estimation Empirical Experimental Results

Table A1. Conditional entropies $H(p,q)$ (bits/pixels), computed for various values of p and q, over "Levada".

p	$q = 0$	$q = 1$	$q = 2$	$q = 3$	$q = 4$	$q = 5$
0	10.2558	9.5120	9.3905	9.3022	9.2279	9.1827
1	8.7675	8.4673	8.3744	8.3141	8.2663	8.2366
2	8.5951	8.3063	8.2192	8.1633	8.1179	8.0895
3	8.4613	8.1782	8.0949	8.0410	7.9979	7.9719
4	8.3440	8.0707	7.9908	7.9394	7.8984	7.8742
5	8.2414	7.9733	7.8977	7.8497	7.8113	7.7888

Table A2. Conditional entropies $H(p,q)$ (bits/pixels), computed for various values of p and q, over "Gualtar".

p	$q = 0$	$q = 1$	$q = 2$	$q = 3$	$q = 4$	$q = 5$
0	10.8137	10.6057	10.4057	10.3266	10.2818	10.2387
1	9.0550	8.9657	8.9007	8.8638	8.8393	8.8176
2	8.8388	8.7553	8.6971	8.6623	8.6374	8.6155
3	8.7107	8.6283	8.5715	8.5362	8.5100	8.4870
4	8.6375	8.5534	8.4964	8.4592	8.4310	8.4064
5	8.5837	8.4956	8.4371	8.3975	8.3670	8.3406

Table A3. Conditional entropies $H(p,q)$ (bits/pixels), computed for various values of p and q, over "Nogueiro".

p	$q = 0$	$q = 1$	$q = 2$	$q = 3$	$q = 4$	$q = 5$
0	9.7167	9.3381	9.2314	9.1641	9.1023	9.0740
1	8.2400	8.1105	8.0619	8.0304	8.0064	7.9918
2	8.1165	7.9890	7.9395	7.9080	7.8841	7.8695
3	8.0343	7.9097	7.8603	7.8292	7.8054	7.7904
4	7.9722	7.8497	7.8005	7.7695	7.7453	7.7299
5	7.9200	7.7979	7.7492	7.7180	7.6936	7.6775

Table A4. Conditional entropies $H(p,q)$ (bits/pixels), computed for various values of p and q, over "Scene-1".

p	$q=0$	$q=1$	$q=2$	$q=3$	$q=4$	$q=5$
0	8.7785	7.8624	7.7055	7.6271	7.5834	7.5548
1	7.1538	6.9037	6.8398	6.8070	6.7836	6.7671
2	7.0491	6.8166	6.7555	6.7233	6.6999	6.6829
3	6.9839	6.7599	6.7003	6.6683	6.6448	6.6272
4	6.9190	6.7052	6.6475	6.6160	6.5923	6.5736
5	6.8533	6.6513	6.5956	6.5646	6.5402	6.5193

Table A5. Conditional entropies $H(p,q)$ (bits/pixels), computed for various values of p and q, over "Scene-2".

p	$q=0$	$q=1$	$q=2$	$q=3$	$q=4$	$q=5$
0	8.1367	7.5082	7.3654	7.3194	7.2787	7.2397
1	6.9248	6.7109	6.6466	6.6202	6.5955	6.5719
2	6.7897	6.6004	6.5421	6.5169	6.4945	6.4731
3	6.7126	6.5362	6.4812	6.4568	6.4354	6.4148
4	6.6461	6.4810	6.4289	6.4052	6.3843	6.3637
5	6.5892	6.4334	6.3836	6.3601	6.3390	6.3173

Table A6. Bit rates (bits/pixels), obtained for various values of p and q, on "Levada".

p	$q=0$	$q=1$	$q=2$	$q=3$	$q=4$	$q=5$
0	5.7538	5.5248	5.4193	5.3400	5.2689	5.1784
1	4.3876	4.3674	4.3422	4.3313	4.3122	4.2987
2	4.3029	4.2878	4.2660	4.2570	4.2395	4.2293
3	4.2813	4.2679	4.2471	4.2389	4.2218	4.2125
4	4.2704	4.2578	4.2377	4.2298	4.2131	4.2043
5	4.2631	4.2508	4.2312	4.2235	4.2070	4.1985

Table A7. Bit rates (bits/pixels) obtained for various values of p and q, on "Gualtar".

p	$q=0$	$q=1$	$q=2$	$q=3$	$q=4$	$q=5$
0	6.0941	5.9463	5.7547	5.7057	5.6541	5.6218
1	4.7064	4.6928	4.6802	4.6754	4.6732	4.6695
2	4.5655	4.5577	4.5518	4.5495	4.5505	4.5487
3	4.5214	4.5155	4.5117	4.5102	4.5121	4.5113
4	4.5057	4.5006	4.4977	4.4966	4.4990	4.4984
5	4.5004	4.4960	4.4936	4.4927	4.4953	4.4948

Table A8. Bit rates (bits/pixels), obtained for various values of p and q, on "Nogueiro".

p	$q=0$	$q=1$	$q=2$	$q=3$	$q=4$	$q=5$
0	5.6329	5.3524	5.2312	5.1521	5.0874	4.9953
1	4.3070	4.2769	4.2545	4.2407	4.2324	4.2186
2	4.2013	4.1800	4.1631	4.1520	4.1463	4.1355
3	4.1701	4.1515	4.1364	4.1260	4.1212	4.1112
4	4.1586	4.1413	4.1272	4.1171	4.1127	4.1031
5	4.1525	4.1363	4.1228	4.1131	4.1088	4.0994

Table A9. Bit rates (bits/pixels), obtained for various values of p and q, on "AAMU Scene-1".

p	$q = 0$	$q = 1$	$q = 2$	$q = 3$	$q = 4$	$q = 5$
0	6.1097	6.0844	6.0691	6.0478	6.0322	6.0156
1	5.0580	5.0557	5.0534	5.0518	5.0492	5.0491
2	4.9186	4.9181	4.9168	4.9133	4.9126	4.9122
3	4.8727	4.8717	4.8711	4.8701	4.8695	4.8694
4	4.8555	4.8537	4.8517	4.8507	4.8502	4.8501
5	4.8491	4.8495	4.8506	4.8538	4.8556	4.8556

Table A10. Bit rates (bits/pixels), obtained for various values of p and q, on "AAMU Scene-2".

p	$q = 0$	$q = 1$	$q = 2$	$q = 3$	$q = 4$	$q = 5$
0	5.3202	5.2791	5.2429	5.2363	5.2045	5.1977
1	4.8675	4.8597	4.8569	4.8554	4.8475	4.8461
2	4.7615	4.7599	4.7586	4.7572	4.7569	4.7569
3	4.7167	4.7156	4.7148	4.7140	4.7138	4.7137
4	4.6953	4.6939	4.6928	4.6920	4.6919	4.6920
5	4.6842	4.6862	4.6936	4.6940	4.6925	4.6925

References

1. Shen, H.; Pan, W.D.; Wu, D. Predictive Lossless Compression of regions-of-interest in Hyperspectral Images With No-Data Regions. *IEEE Trans. Geosci. Remote Sens.* **2017**, *55*, 173–182. [CrossRef]
2. Thouvenin, P.A.; Dobigeon, N.; Tourneret, J.Y. A Hierarchical Bayesian Model Accounting for Endmember Variability and Abrupt Spectral Changes to Unmix Multitemporal Hyperspectral Images. *IEEE Trans. Comput. Imag.* **2018**, *4*, 32–45. [CrossRef]
3. Marinelli, D.; Bovolo, F.; Bruzzone, L. A novel change detection method for multitemporal hyperspectral images based on a discrete representation of the change information. In Proceedings of the 2017 IEEE International Geoscience and Remote Sensing Symposium (IGARSS), Fort Worth, TX, USA, 23–28 July 2017; pp. 161–164.
4. Liu, S.; Bruzzone, L.; Bovolo, F.; Du, P. Unsupervised Multitemporal Spectral Unmixing for Detecting Multiple Changes in Hyperspectral Images. *IEEE Trans. Geosci. Remote Sens.* **2016**, *54*, 2733–2748. [CrossRef]
5. Ertürk, A.; Iordache, M.D.; Plaza, A. Sparse Unmixing-Based Change Detection for Multitemporal Hyperspectral Images. *IEEE J. Sel. Top. Appl. Earth Obs.* **2016**, *9*, 708–719. [CrossRef]
6. Shen, H.; Pan, W.D.; Dong, Y. Efficient Lossless Compression of 4D Hyperspectral Image Data. In Proceedings of the 3rd International Conference on Advances in Big Data Analytics, Las Vegas, NV, USA, 25–28 July 2016.
7. Zhu, W.; Du, Q.; Fowler, J.E. Multitemporal Hyperspectral Image Compression. *IEEE Geosci. Remote Sens. Lett.* **2011**, *8*, 416–420. [CrossRef]
8. Shen, H.; Pan, W.D. Predictive lossless compression of regions-of-interest in hyperspectral image via Maximum Correntropy Criterion based Least Mean Square learning. In Proceedings of the 2016 IEEE International Conference on Image Processing (ICIP), Phoenix, AZ, USA, 25–28 September 2016; pp. 2182–2186.
9. Liaghati, A.; Pan, W.D.; Jiang, Z. Biased Run-Length Coding of Bi-Level Classification Label Maps of Hyperspectral. *IEEE J. Sel. Top. Appl. Earth Obs. Remote Sens.* **2017**, *10*, 4580–4588. [CrossRef]
10. Weinberger, M.J.; Seroussi, G.; Sapiro, G. The LOCO-I lossless image compression algorithm: Principles and standardization into JPEG-LS. *IEEE Trans. Image Process.* **2000**, *9*, 1309–1324. [CrossRef]
11. Wu, X.; Memon, N. Context-based lossless interband compression-extending CALIC. *IEEE Trans. Image Process.* **2000**, *9*, 994–1001.

12.	Magli, E.; Olmo, G.; Quacchio, E. Optimized onboard lossless and near-lossless compression of hyperspectral data using CALIC. *IEEE Trans. Geosci. Remote Sens.* **2004**, *1*, 21–25. [CrossRef]

13.	Wu, X.; Memon, N. Context-based, adaptive, lossless image coding. *IEEE Trans. Commun.* **1997**, *45*, 437–444. [CrossRef]

14.	Mielikainen, J. Lossless compression of hyperspectral images using lookup tables. *IEEE Signal Process. Lett.* **2006**, *13*, 157–160. [CrossRef]

15.	Rizzo, F.; Carpentieri, B.; Motta, G.; Storer, J.A. Low-complexity lossless compression of hyperspectral imagery via linear prediction. *IEEE Signal Process. Lett.* **2005**, *12*, 138–141. [CrossRef]

16.	Wang, H.; Babacan, S.D.; Sayood, K. Lossless Hyperspectral-Image Compression Using Context-Based Conditional Average. *IEEE Trans. Geosci. Remote Sens.* **2007**, *45*, 4187–4193. [CrossRef]

17.	Said, A.; Pearlman, W.A. A new, fast, and efficient image codec based on set partitioning in hierarchical trees. *IEEE Trans. Circuits Syst. Video Technol.* **1996**, *6*, 243–250. [CrossRef]

18.	Pearlman, W.A.; Islam, A.; Nagaraj, N.; Said, A. Efficient, low-complexity image coding with a set-partitioning embedded block coder. *IEEE Trans. Circuits Syst. Video Technol.* **2004**, *14*, 1219–1235. [CrossRef]

19.	Aiazzi, B.; Alparone, L.; Baronti, S.; Lastri, C. Crisp and Fuzzy Adaptive Spectral Predictions for Lossless and Near-Lossless Compression of Hyperspectral Imagery. *IEEE Geosci. Remote Sens. Lett.* **2007**, *4*, 532–536. [CrossRef]

20.	Mielikainen, J.; Huang, B. Lossless Compression of Hyperspectral Images Using Clustered Linear Prediction with Adaptive Prediction Length. *IEEE Geosci. Remote Sens. Lett.* **2012**, *9*, 1118–1121. [CrossRef]

21.	Wu, J.; Kong, W.; Mielikainen, J.; Huang, B. Lossless Compression of Hyperspectral Imagery via Clustered Differential Pulse Code Modulation with Removal of Local Spectral Outliers. *IEEE Signal Process. Lett.* **2015**, *22*, 2194–2198. [CrossRef]

22.	Klimesh, M. *Low-Complexity Lossless Compression of Hyperspectral Imagery via Adaptive Filtering*; The Interplanetary Network Progress Report; NASA Jet Propulsion Laboratory (JPL): Pasadena, CA, USA, 2005; pp. 1–10.

23.	Lossless Multispectral & Hyperspectral Image Compression CCSDS 123.0-B-1, Blue Book, May 2012. 2015. Available online: https://public.ccsds.org/Pubs/123x0b1ec1.pdf (accessed on 10 August 2018).

24.	Mamun, M.A.; Jia, X.; Ryan, M. Sequential multispectral images compression for efficient lossless data transmission. In Proceedings of the 2010 Second IITA International Conference on Geoscience and Remote Sensing, Qingdao, China, 28–31 August 2010; Volume 2, pp. 615–618.

25.	Muñoz-Gomez, J.; Bartrina-Rapesta, J.; Blanes, I.; Jimenez-Rodriguez, L.; Aulí-Llinàs, F.; Serra-Sagristà, J. 4D remote sensing image coding with JPEG2000. *Proc. SPIE* **2010**, *7810*, 1–9.

26.	Ricci, M.; Magli, E. On-board lossless compression of solar corona images. In Proceedings of the 2015 IEEE International Geoscience and Remote Sensing Symposium (IGARSS), Milan, Italy, 26–31 July 2015; pp. 2091–2094.

27.	Mamun, M.; Jia, X.; Ryan, M.J. Nonlinear Elastic Model for Flexible Prediction of Remotely Sensed Multitemporal Images. *IEEE Geosci. Remote Sens. Lett.* **2014**, *11*, 1005–1009. [CrossRef]

28.	Foster, D.H.; Amano, K.; Nascimento, S.M. Time-lapse ratios of cone excitations in natural scenes. *Vision Res.* **2016**, *120*, 45–60. [CrossRef] [PubMed]

29.	SOC700 Series Hyperspectral Imaging Systems. 2018. Available online: https://surfaceoptics.com/products/hyperspectral-imaging/soc710-portable-hyperspectral-camera/ (accessed on 10 August 2018).

30.	Time-Lapse Hyperspectral Radiance Images of Natural Scenes 2015. 2015. Available online: http://personalpages.manchester.ac.uk/staff/david.foster/Time-Lapse_HSIs/Time-Lapse_HSIs_2015.html (accessed on 1 March 2015).

31.	Magli, E. Multiband Lossless Compression of Hyperspectral Images. *IEEE Trans. Geosci. Remote Sens.* **2009**, *47*, 1168–1178. [CrossRef]

32.	Shen, H.; Pan, W.D.; Wang, Y. A Novel Method for Lossless Compression of Arbitrarily Shaped Regions of Interest in Hyperspectral Imagery. In Proceedings of the 2015 IEEE SoutheastCon, Fort Lauderdale, FL, USA, 9–12 April 2015; pp. 1–6.

33.	Liu, W.; Pokharel, P.P.; Principe, J.C. Correntropy: Properties and Applications in Non-Gaussian Signal Processing. *IEEE Trans. Signal Process.* **2007**, *55*, 5286–5298. [CrossRef]

34.   Golomb, S. Run-length encodings (Corresp.). *IEEE Trans. Inf. Theory* **1966**, *12*, 399–401. [CrossRef]
35.   Shen, H.; Pan, W.D.; Dong, Y.; Jiang, Z. Golomb-Rice Coding Parameter Learning Using Deep Belief Network for Hyperspectral Image Compression. In Proceedings of the IEEE International Geoscience and Remote Sensing Symposium (IGARSS), Fort Worth, TX, USA, 23–28 July 2017; pp. 2239–2242.

Journal of
Imaging

MDPI

Review

Compressive Sensing Hyperspectral Imaging by Spectral Multiplexing with Liquid Crystal

Yaniv Oiknine, Isaac August, Vladimir Farber, Daniel Gedalin and Adrian Stern *

Department of Electro-Optical Engineering, Ben-Gurion University of the Negev, Beer-Sheva 84105, Israel;
oiknine@post.bgu.ac.il (Y.O.); augusty@post.bgu.ac.il (I.A.); farberv@post.bgu.ac.il (V.F.);
danielge@post.bgu.ac.il (D.G.)
* Correspondence: stern@bgu.ac.il

Received: 28 October 2018; Accepted: 18 December 2018; Published: 22 December 2018

Abstract: Hyperspectral (HS) imaging involves the sensing of a scene's spectral properties, which are often redundant in nature. The redundancy of the information motivates our quest to implement Compressive Sensing (CS) theory for HS imaging. This article provides a review of the Compressive Sensing Miniature Ultra-Spectral Imaging (CS-MUSI) camera, its evolution, and its different applications. The CS-MUSI camera was designed within the CS framework and uses a liquid crystal (LC) phase retarder in order to modulate the spectral domain. The outstanding advantage of the CS-MUSI camera is that the entire HS image is captured from an order of magnitude fewer measurements of the sensor array, compared to conventional HS imaging methods.

Keywords: compressive sensing; hyperspectral imaging; multiplexing system; liquid crystal; three-dimensional imaging; integral imaging; remote sensing; point target detection; CS-MUSI

1. Introduction

Hyperspectral (HS) imaging has gained increasing interest in many fields and applications. These techniques can be found in airborne and remote sensing applications [1–4], biomedical and medical studies [5–7], food and agricultural monitoring [8–10], forensic applications [11,12], and many more. The HS images captured for these applications are usually arranged in three-dimensional (3D) datacubes, which include two dimensions (2D) for the spatial information and one additional dimension (1D) for the spectral information. With a 2D spatial domain of megapixel size and with the third (spectral) dimension typically containing hundreds of spectral bands, the HS data is usually huge. Consequently, its scanning, storage and digital processing is challenging.

Studies have shown that the huge HS datacubes are often highly redundant [13–17] and, therefore, very compressible or sparse. This gives the incentive to implement Compressive Sensing (CS) theory in HS systems. CS is a sampling framework that facilitates efficient acquisition of sparse signals. Numerous techniques that employ the CS framework for spectral imaging [18–26] have been proposed in order to reduce the scanning efforts. Most of these techniques involve spatial–spectral multiplexing, which is suitable for CS; however, this multiplexing inevitably impairs both the spatial and spectral domains. In References [27,28], we introduced a novel CS HS camera dubbed the Compressive Sensing Miniature Ultra-Spectral imaging (CS-MUSI) camera. The CS-MUSI camera overcomes the impairment in the spatial domain by performing only spectral multiplexing without any spatial multiplexing. Figure 1 provides a schematic description of HS datacubes that undergo only spectral multiplexing. The figure presents three examples of a HS datacube modulated at three different exposures. Different spectral multiplexing is obtained by applying different conditions on the modulator. The spectrally multiplexed data is ultimately integrated by a focal plane array (FPA).

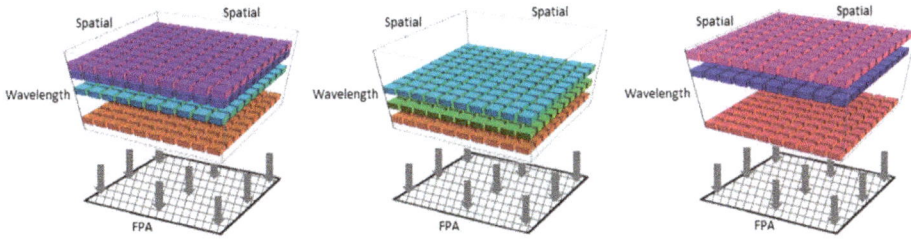

Figure 1. Spectral multiplexing. The figure represents three different examples of spectral multiplexing. Each sub-figure illustrates multiplexing of a few spectral bands onto a FPA.

Within the framework of CS, the CS-MUSI camera can reconstruct HS images with hundreds of spectral bands from only spectrally multiplexed shots, numbering an order of magnitude less than would be required using conventional systems. Furthermore, the CS-MUSI camera benefits from high optical throughput, small volume, light weight, and reduced acquisition time. In this paper, we review the evolution of the CS-MUSI camera and its different applications.

In this article, we first review the innovative concept behind our CS-MUSI camera. We describe spectral multiplexing within the CS framework using a liquid crystal (LC) phase retarder, and expand on the sensing and reconstruction processes. In the following, we outline the optical setup of the CS-MUSI camera and its realization. Lastly, we describe different possible applications for the CS-MUSI camera, including HS staring imaging, HS scanning imaging, 4D imaging, and target detection tasks.

2. Spectral Modulation for CS with LC

The core of the CS-MUSI camera is a single LC phase retarder (Figure 2), which we designed to work as a spectral modulator that is compliant with the CS framework [27]. By using the LC phase retarder, the signal multiplexing is accomplished entirely in the spectral domain, and no spectral-to-spatial transformations are required. The LC phase retarder is built by placing a LC cell between two polarizers. The spectral transmission is controlled by the voltage applied on the LC cell, causing variations in the cell's birefringence, which, in turn, cause refractive index changes. For the case where the optical axis of the LC cell is at 45° to two perpendicular polarizers, the spectral transmission response of the LC phase retarder can be described by [29]:

$$\phi_{LC}(\lambda, V_i) = \frac{1}{2} - \frac{1}{2}\cos\left(\frac{2\pi\Delta n(V_i)d}{\lambda}\right), \tag{1}$$

where $\Delta n(V_i)$ is the birefringence produced by voltage V_i, d is the cell thickness and λ is the wavelength. The voltage applied to the LC cell is an AC voltage, usually in the form of sine or square wave and with a typical frequency in the order of kHz.

We designed the LC cell to have a relatively thick cavity (tens of microns) to facilitate modulation over a broad spectrum with oscillatory behavior, as can been seen in the different spectral responses presented in Figure 3. The figure presents 15 plots of 15 different measured spectral responses for different AC voltages (2 kHz square wave) applied on the LC cell. It can be noticed that as the voltage decreases [higher birefringence, $\Delta n(V_i)$], the number of peaks in the spectral transmission graphs rises. Theoretically, these spectral responses should follow expression (1), spanning the entire range from 0 to 1 for all the wavelength range. However, in practice, we can see that the modulation depths in Figure 3 are not equal for all the peaks. This is due to the quality of the polarizers.

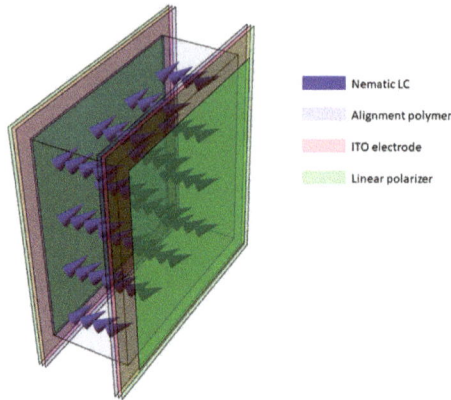

Figure 2. LC cell phase retarder. The LC phase retarder is made of a Nematic LC layer (blue arrow) sandwiched between two glass plates and two linear polarizers (green layers). The glass plates are coated with Indium Tin Oxide (ITO, pink layers) and a polymer alignment layer (purple layers).

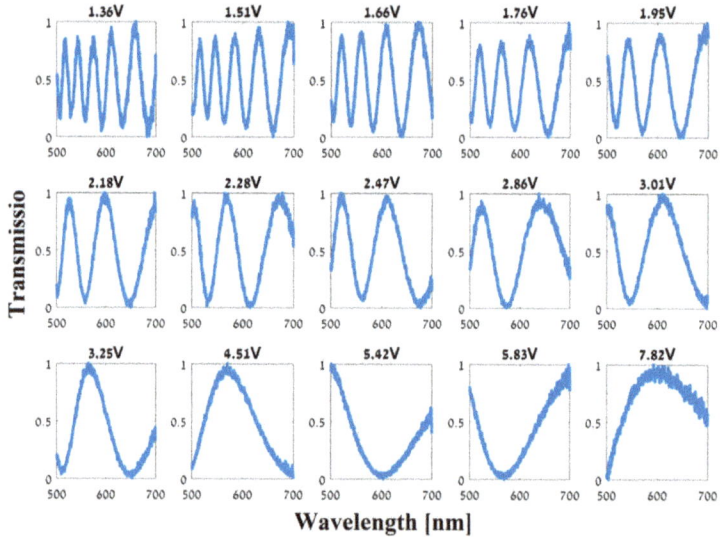

Figure 3. Measured spectral responses (intensity transmission vs. wavelength in nm) of the fabricated LC phase retarder. Each graph represents the spectral modulation with a different voltage applied on the LC cell (15 different voltages).

The acquisition process and the optical scheme of the CS-MUSI camera are shown in Figure 4. The spatial–spectral power distribution of the HS object, $F(x, y, \lambda)$, is modulated by the LC phase retarder transmittance function, $\phi_{LC}(\lambda, V_i)$ (the graph from Figure 3). The i'th modulated spectral signal is spectrally multiplexed and integrated at each pixel in the 2D sensor array, which gives the encoded measurements:

$$G_i(x, y) = \int \phi_{LC}(\lambda, V_i)F(x, y, \lambda)d\lambda. \tag{2}$$

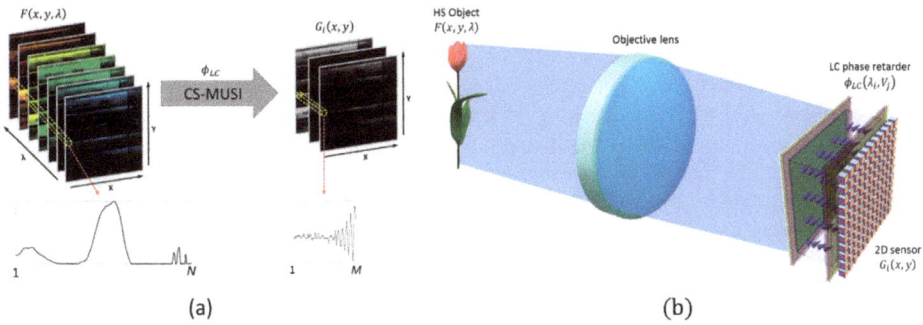

Figure 4. (**a**) CS-MUSI acquisition process. (**b**) CS-MUSI optical scheme diagram. The HS object $F(x, y, \lambda)$ is modulated according to $\phi_{LC}(\lambda, V_i)$, yielding the multiplexed measurement $G_i(x, y)$.

As the sensor samples discrete values, and for CS analysis and reconstruction purposes, it is more convenient to present the sensing process from Equation (2) in a matrix-vector format. Let us denote the spectral signal with N spectral bands by $\mathbf{f} \in \Re^{N \times 1}$ and the multiplexed measured spectral signal with M entries by $\mathbf{g} \in \Re^{M \times 1}$. By using these vectors, the measurement process can be described by:

$$\mathbf{g} = \mathbf{\Phi} \mathbf{f}, \tag{3}$$

where $\mathbf{\Phi} \in \Re^{M \times N}$ represents the CS sensing matrix. Compressive sensing [22,30–32] provides a framework to capture and to recover signals from fewer measurements than required by the Shannon-Nyquist sampling theorem (i.e., $M < N$). The CS framework relies on three main ingredients. First, the sparsity of the signal, namely, the spectral signal with N spectral bands can be expressed by $\mathbf{f} = \mathbf{\Psi} \alpha$, where α is a K-sparse vector (containing $K \ll N$ non-zero elements) and $\mathbf{\Psi}$ is the sparsifying operator. In accordance with the CS framework, Equation (3) can be written as:

$$\mathbf{g} = \mathbf{\Phi} \mathbf{\Psi} \alpha = \Omega \alpha, \tag{4}$$

where $\Omega = \mathbf{\Phi} \mathbf{\Psi}$. Second, the CS systems needs an appropriate sensing design, which is represented by the system sensing matrix, $\mathbf{\Phi}$. Third, the CS framework relies on the existence of an appropriate reconstruction algorithm, upon which we will expand in the next section.

3. Reconstruction Process

The CS-MUSI camera was designed in accordance with the CS framework. Consequently, the captured data is a compressed version of the scene's HS datacube. Therefore, a reconstruction process that solves Equation (4) should be performed. Over the years, several CS algorithms have been developed [32–37] in order to recover the original signal. A common class of these reconstruction algorithms solve the $\ell_2 - \ell_1$ minimization problem:

$$\tilde{\alpha} = \underset{\alpha}{\operatorname{argmin}} \left\{ \frac{1}{2} \|\mathbf{g} - \mathbf{\Phi} \mathbf{\Psi} \alpha\|_2^2 + \tau \|\alpha\|_1 \right\}. \tag{5}$$

where $\tilde{\alpha} \in \Re^{N \times 1}$ is the estimated K-sparse signal, τ is a regularization parameter and $\| \cdot \|_p$ is the l_p norm. The algorithms recover the original signal by using the known system sensing matrix, $\mathbf{\Phi}$, and the signal sparsifying operator, $\mathbf{\Psi}$ [22,37–39]. The sparsifying operator can be a mathematical transform (DCT, Wavelet, Curvelets etc.) or a learned dictionary; the latter has shown promising results [40] and will be described in the next subsection. Common algorithms developed to solve Equation (5) are TwIST [33], GPSR [34], SpaRSA [35], TVAL3 [36], etc.

Dictionary for Sparse Representation

The first ingredient that the CS framework relies on is the sparsity of the signal. The sparser the representation of the signal is, the better the CS reconstruction algorithms perform. We found [40] that using a learned dictionary [38] as the sparsifying operator can significantly improve the reconstruction accuracy in comparison to using a mathematical basis. In addition, using a dictionary can reduce both the time and the number of measurements required in order to reconstruct the original signal. A dictionary is a learned sparsifier from exemplars. In order to be able to use a dictionary in CS algorithms, a preprocessing stage has to be performed, and it is done only once.

First, a large database of spectral signals, $S \in \Re^{N \times N_S}$, is collected. This database contains N_S spectra with N spectral bands:

$$S = [s_1, s_2, \ldots, s_{N_S}], \tag{6}$$

where s_i is the i'th spectrum in the database. Then, an over-complete spectral dictionary with N_d atoms to train, $\Psi_d \in \Re^{N \times N_d}$, is created by applying a dictionary-learning algorithm, such as the K-SVD algorithm [38], to the spectral database S:

$$\Psi_d = \text{K-SVD}\{S\}. \tag{7}$$

Ψ_d is a dictionary that relates the spectral data f to its sparse representation, $f = \Psi_d \alpha$. Each column of Ψ_d is referred to as an atom of the dictionary. Therefore, the spectrum f can be viewed as a linear combination of atoms in Ψ_d according to weights in α. Based on Equation (7), a corresponding system dictionary $\Omega_d \in \Re^{M \times N_d}$ is created by the inner products of the spectral dictionary with the CS-MUSI sensing matrix, Φ:

$$\Omega_d = \Phi \Psi_d. \tag{8}$$

After the dictionary has been prepared, it can be used as the sparsifying operator to reconstruct the original spectral signal by finding the estimated atom weights vector, $\tilde{\alpha}_d$. These weights could be found, for example, by the $l_2 - l_1$ minimization problem from Equation (5) that becomes:

$$\tilde{\alpha}_d = \underset{\alpha}{\text{argmin}} \left\{ \frac{1}{2} \|g - \Omega_d \alpha\|_2^2 + \tau \|\alpha\|_1 \right\}. \tag{9}$$

Once the atom weights are found the original spectral signal is estimated by applying the atom weights on the spectral dictionary we created, Ψ_d (Equation (7)):

$$\tilde{f} = \Psi_d \tilde{\alpha}_d. \tag{10}$$

For more details on the utilization of dictionaries for CS-MUSI data reconstruction and its advantages over other sparsifiers, the reader is referred to [40].

4. Compressive Hyperspectral and Ultra-Spectral Imaging

The CS-MUSI camera we built is shown in Figure 5a. It is slightly different from the optical setup that was designed and presented in Figure 4, but is optically equivalent. It has a 1:1 optical relay that projects the LC plane onto a 2D sensor array, thus avoiding the need to attach the LC cell to the 2D sensor array. A LC cell is placed in the image plane of a zoom lens. The light transmitted through the LC cell is conjugated to a sensor array using a 1:1 relay lens. The optical sensor of the camera is a uEye CMOS UI-3240CP-C-HQ with 1280 × 1024 pixels, with a pixel size of 5.3 μm × 5.3 μm and 8-bit grayscale level radiometric sampling. The camera LC cell from Figure 5b was manufactured in-house and has a cell gap of approximately 50 μm and a clear aperture of about 8 mm × 8 mm. The LC cell was fabricated from two flat glass plates coated with Indium Tin Oxide (ITO) and a polymer alignment layer. The cavity is filled with LC material E44 (Merck, Darmstadt, Germany). Together with two linear polarizers on both sides of the LC cell, the LC phase retarder is created (Figure 2).

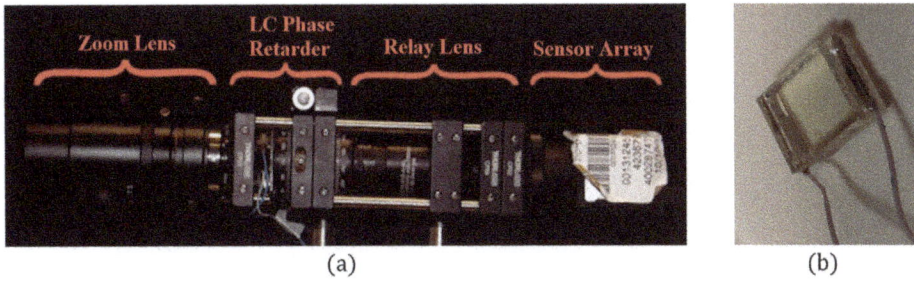

Figure 5. (a) Realization of the CS-MUSI camera. (b) In-house manufactured LC cell.

4.1. Camera Calibration

The theoretical expression in Equation (1) cannot be used directly in order to obtain the system-sensing matrix, because the dependence of the birefringence on the voltage and the material dispersion are unknown. Consequently, a calibration process in which the spectral responses of the camera were precisely measured, was performed once. Using a point light source as the object and by replacing the sensor of the camera with a commercial high-precision grating spectrometer, the spectral responses of the camera were measured. The calibration process was performed by using a halogen light source with a pre-measured spectrum and by applying voltages from 0 V to 10 V on the LC cell with steps of 2 mV. Figure 6, presents the CS-MUSI system's spectral response map that was measured in the calibration process.

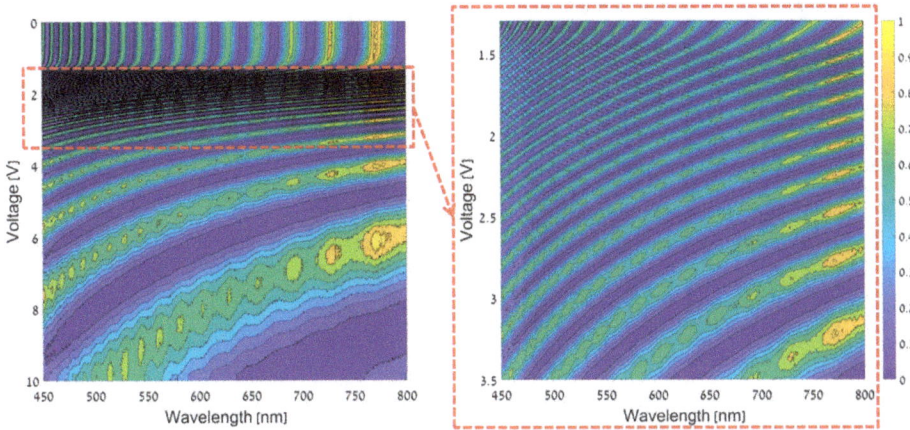

Figure 6. CS-MUSI spectral response map for voltages from 0 V to 10 V (**left map**) and a zoom in on the area where the voltages are from 1.3 V to 3.5 V (**right map**).

The system sensing matrix, Φ, is obtained by selecting M rows from the CS-MUSI system's spectral response map (Figure 6, left). This is done by using a sequential forward floating selection method [28,41] that aims to achieve a highly incoherent set of measurements.

4.2. Staring Mode

The basic imaging mode of the CS-MUSI camera is the staring mode, so that the camera and scene do not move. Scanning mode acquisition is described in the next subsection. In staring mode, each spectrally multiplexed shot covers the same field-of-view (FOV) and, by taking M shots, a compressed HS datacube is captured. Figures 7–9 demonstrate the reconstruction of spectral images (HS and ultra-spectral images) attained by the CS-MUSI camera in staring mode. Figure 7 presents the results of an experiment where the emission spectra of three arrays of red, green and blue light sources (Thorlabs LIU001, LIU002 and LIU003 LED arrays) were imaged using the CS-MUSI camera. Figure 7a shows the image of the light sources captured by a standard RGB color camera. The imaging experiment was performed by capturing 32 spectrally multiplexed images containing 1024 × 1280 pixels. Figure 7b–e shows four captured images that represent four single grayscale frames from the spectrally-compressed measurements. The images show the total optical intensity that has passed through the LC phase retarder and was collected by the sensor array at a given shot with a given LC voltage. From the captured data, a window of 700 × 700 pixels was used in the reconstruction process. Using the TwIST solver [33] and orthogonal Daubechies-5 wavelet as the sparsifying operator, a HS datacube with 391 spectral bands (410–800 nm) was reconstructed, yielding a compression ratio of about 12:1. Figure 7f presents a pseudo-color image obtained by projecting the reconstructed HS datacube onto the RGB space. Figure 7g–i displays three images from the reconstructed datacube at different wavelengths (460 nm, 520 nm and 650 nm). Figure 7k–m demonstrate spectrum reconstruction for three points in the HS datacube and a comparison to the measured spectra of the three respective LEDs with a commercial grating-based spectrometer. The reconstruction PSNR is 32.4 dB, 34.8 dB and 27.9 dB for the blue, green and red LED points, respectively.

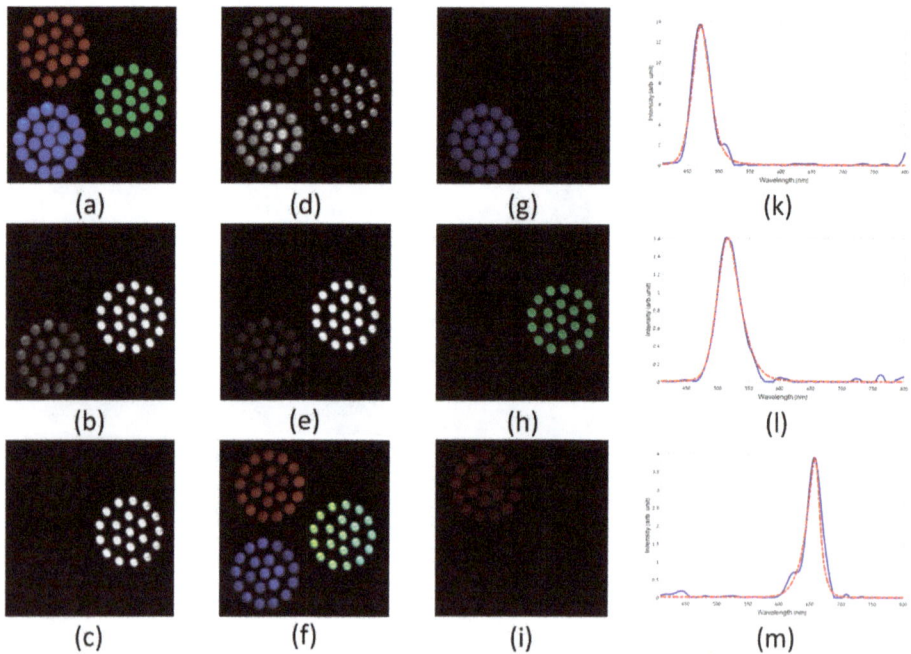

Figure 7. Staring mode reconstruction result of three LED arrays. (**a**) RGB color image of three LED arrays that were used as objects to be imaged with CS-MUSI. (**b–e**) Representative single exposure images for LC cell voltage of 0 V, 5.8373 V, 7.6301 V and 8.6552 V, respectively. (**f**) RGB representation of the reconstructed HS image (700 × 700 pixels× 391 bands). (**g–i**) Reconstructed images at 460 nm, 520 nm and 650 nm, respectively. (**k–m**) Spectrum reconstruction for three points in the HS datacube and comparison to the measured spectra of the three respective LEDs with a commercial grating-based spectrometer.

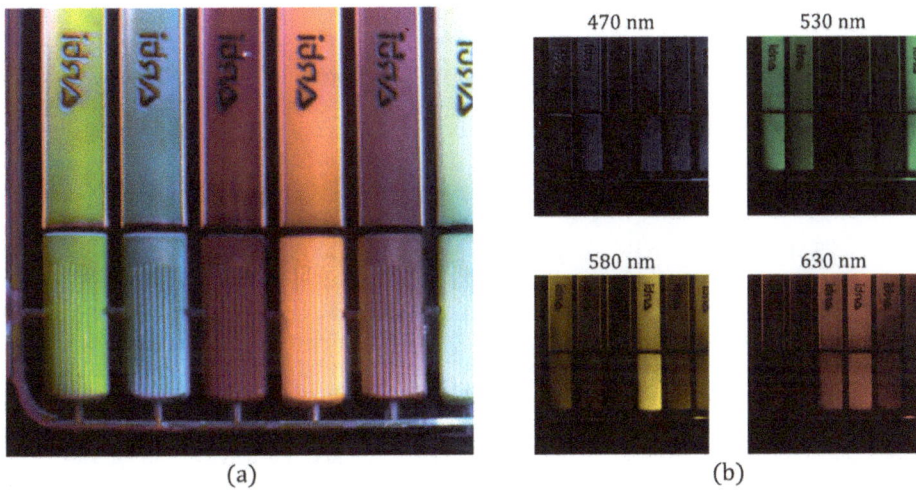

Figure 8. Staring mode reconstruction result of six different markers. (**a**) RGB representation of the reconstructed HS image (800 × 900 pixels × 1171 bands). (**b**) Four reconstructed images at four different wavelengths (470 nm, 530 nm, 580 nm, and 630 nm).

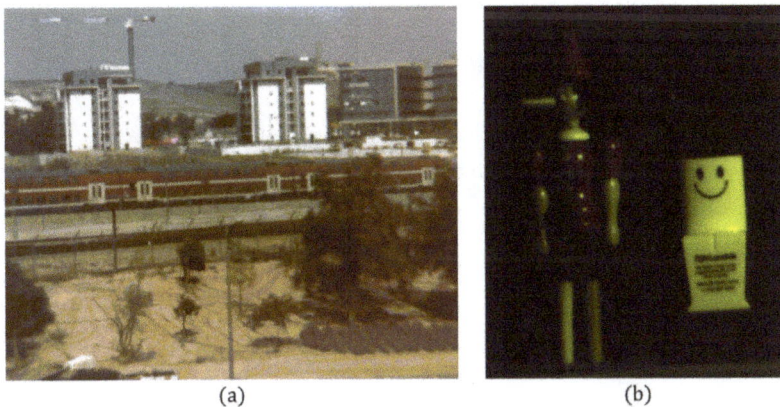

Figure 9. Staring mode reconstruction results with the dictionary of (**a**) outdoor and (**b**) indoor HS images taken with CS-MUSI camera. The figures show RGB representation of the reconstructed HS datacube.

Figure 8 presents an indoor scene where six different markers are imaged using the CS-MUSI camera. The imaging experiment was performed by capturing 100 spectrally multiplexed images containing 1024 × 1280 pixels. From the captured data, a window of 800 × 900 pixels was used in the reconstruction process. Using the TwIST solver [33] and orthogonal Daubechies-5 wavelet as the sparsifying operator, a HS datacube with 1171 spectral bands (410–800 nm) was restored, yielding a compression ratio of almost 12:1. Figure 8a presents a pseudo-color image obtained by projecting the reconstructed HS datacube onto the RGB space. Figure 8b displays four images from the reconstructed datacube at different wavelengths (470 nm, 530 nm, 580 nm and 630 nm).

The quality and time expenditure of reconstructed HS images depends significantly on the sparsifying operator. We found that by using a learned dictionary as the sparsifying operator [40] the quality and time improves. Figure 9 illustrates the reconstruction of HS images attained by the CS-MUSI camera using a spectral dictionary as the sparsifying operator (Section 3). The size of this dictionary was $N_d = 1000$ atoms and was computed by using $N_S > 100{,}000$ spectrum exemplars taken form a large database of HS images [42] and from a library of different spectra [40,43]. The HS images reconstructed from the CS-MUSI camera are in the range of 500 nm to 700 nm with 579 spectral bands, and were reconstructed from only 32 measurements, thus yielding a compression ratio of approximately 18:1. Figure 9a,b shows an RGB representation of the reconstructed HS images of outdoor and indoor scenes, respectively.

4.3. Scanning Mode

The CS-MUSI camera can also be applied in a mode where the camera, the scene, or objects in the scene, are not stationary [44]. Such scenarios include microscope applications with moving cells or scanning platforms, and airborne and remote sensing systems. By capturing a sequence of spectrally multiplexed shots and tracking the object, it is possible to reconstruct HS data by an appropriate registration. It is required that the object appears in M shots. For example, in the case of along-track scanning [44] (Figure 10) the CS-MUSI camera needs $2M$ measurements in order to capture a scene of the size of the camera FOV. A second requirement is image registration, since the FOV of each shot is slightly different. As a result, before solving Equation (4) it is necessary to register all the measured images along a common spatial grid. This can be done with one of the many available algorithms [45–47].

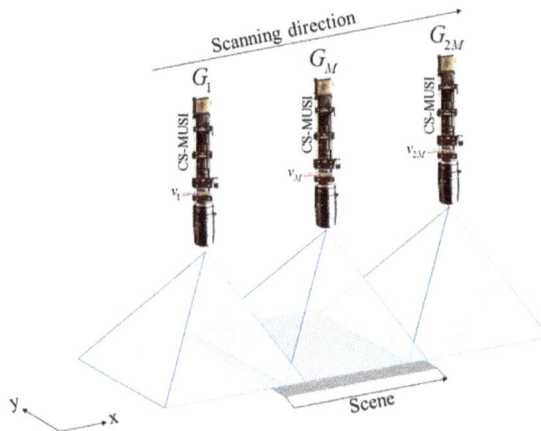

Figure 10. CS-MUSI camera along-track scanning. Each shot of the CS-MUSI camera, G_i, captures a shifted scene with a different LC spectral transmission (which depends on the voltage v_i).

Figure 11 shows experimental results that demonstrate the ability to reconstruct HS images in a scanning mode. In this experiment, the scanning is along-track [44], which is similar to the push broom scanning technique [3]. The spectral multiplexed imaging acquisition process was conducted while the CS-MUSI camera was moving in front of three arrays of LED light sources (Thorlabs LIU001, LIU002, and LIU003) (Figure 11a). While the CS-MUSI camera was moving, a set of 100 voltages between 0 V and 10 V was applied repetitively. Figure 11b–e presents four representative spectrally multiplexed intensity measurements (shot #30, #100, #150 and #300, respectively) from a total of 300 measurements (Supplementary Materials). These images illustrated the along-track scanning of the camera from right to left, where in the first shot only part of the blue and red LEDs appears in the spectral multiplexed image, and in the last shot only the green LED appears.

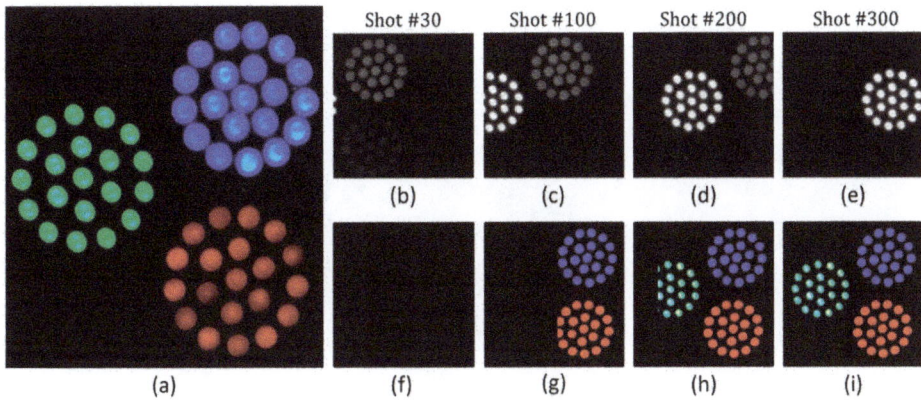

Figure 11. Scanning mode (Figure 10) reconstruction result. (**a**) RGB color image of three LED arrays. (**b**–**e**) representative single exposure images (frame #30, #90, #150 and #300, respectively) and (**f**–**i**) the RGB representation of the reconstructed HS image up to the appropriate column.

Since the reconstruction was performed column by column [48], the reconstruction process can start before the scanning process is completed. Once a column is measured in $M = 100$ shots it can be reconstructed. Figure 11f–i shows an RGB representation of the reconstructed HS images up to different shot numbers. It can be noticed that at 30 shots (Figure 11f), no image column can be reconstructed as the total number of shots is smaller than M and no object column has enough measurements in order to be reconstructed. However, after $M = 100$ shots, some of the column images can be reconstructed after they have been measured M times. In this example, the HS image was reconstructed using the SpaRSA solver [35] and orthogonal Daubechies-4 wavelet as the sparsifying operator. From the 100 shots of each column, a HS image with 1171 spectral bands (410–800 nm) was restored, yielding a compression ratio of almost 12:1.

5. 4D Imaging

Integrating the CS-MUSI camera with an appropriate 3D imaging technique enables achieving a four-dimensional (4D) camera that can efficiently capture 3D spatial images together with their spectral information [49–52]. Joint spectral and volumetric data can be very useful for object shape detection [51,52] and material classification in various engineering and medical applications. The CS approach facilitated the acquisition effort associated with the huge dimensionality of the 4D spectral-volumetric data.

For 3D imaging we used Integral Imaging (InIm) [53–56], since its implementation is relatively simple. The first step of InIm is the acquisition of an actual 3D scene. In this step, multiple 2D images from slightly different perspectives are captured. Each of these images is called an elemental image (EI). Generally, the acquisition process can be implemented by a lenslet array (or pinhole array) or by synthetic aperture InIm. Synthetic aperture InIm (Figure 12) can be realized by an array of cameras distributed on the image plane, or by a single moving camera which moves perpendicularly to the system's optical axis. Replacing the moving camera with our CS-MUSI camera enables capturing 3D spatial images together with their spectral information. By using the captured InIm data it is possible to synthesize depth maps, virtual perspectives and refocused images.

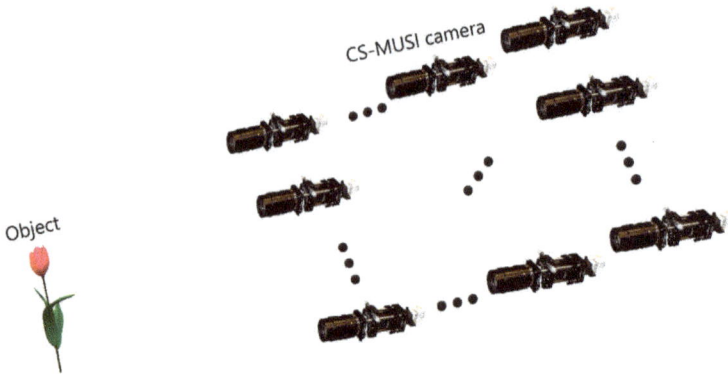

Figure 12. Compressive HS synthetic aperture InIm acquisition setup.

After the acquisition of the data and reconstruction of the spectral information from its compressed version, acquired with the moving CS-MUSI camera, the 3D image for each spectral channel (in terms of focal-stack) can be reconstructed numerically in different ways [49,56–58]. One of the most popular methods is based on back-projection, also known as shift-and-add. In the case of synthetic aperture InIm, the refocusing process can be performed as follows [49]:

$$\tilde{I}(x,y,z,\lambda) = \frac{1}{o(x,y,z,\lambda)} \sum_{j=1}^{P} I_{j,\lambda}\left(x + Sh_{x,z,j}, y + Sh_{y,z,j}\right), \tag{11}$$

where $\tilde{I}(x,y,z,\lambda)$ is the reconstructed data tesseract, $I_{j,\lambda}$ is the EI at wavelength λ showing the perspective from camera j. $o(x,y,z,\lambda)$ is a normalizing matrix that normalizes each pixel in the image $\tilde{I}(x,y,z,\lambda)$ according to the number of EIs that the pixel appears in, and P is the overall number of EIs. $Sh_{x,z,j}$ and $Sh_{y,z,j}$ are the scaled size of the shifts in the horizontal and vertical directions of the CS-MUSI camera [49].

In the 4D imaging application, we acquired spectrally compressed images with our CS-MUSI camera from six perspectives, where in each perspective we captured 29 compressed images. Then, by using the TwIST solver [27], a HS datacube with 261 spectral bands (430–690 nm) was reconstructed. Next, we generated refocused images at different depths by using a shift-and-add algorithm [49]. The data can be ordered as a tesseract, as shown in Figure 13. Figure 13a demonstrates the reconstruction results at three selected depths for four selected wavelengths. The zoom on the HS datacube from the depth of 270 cm illustrates the spectral reconstruction quality, which can be observed from the fact that the laser beam appears clearly only at 635 nm, whereas it is completely filtered out in the other spectral bands (520 nm, 580 nm and 626 nm). From the grayscale images from Figure 13b it can be observed that the closest object was a green alien toy, whose best focus is at 225 cm; the Pinocchio toy's best focus is at 254 cm and the best focus of the different colored shape objects and red laser is at 270 cm.

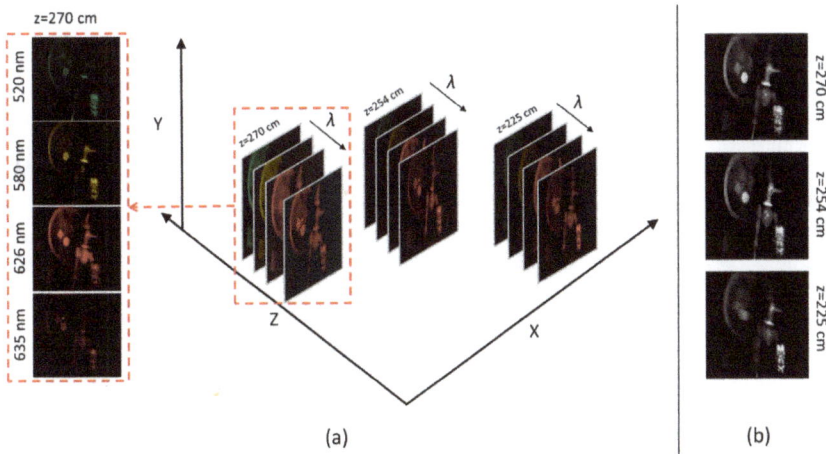

Figure 13. (a) 4D Spectro-Volumetric imaging. (b) Grayscale representation of HS images at three different depths (225 cm, 254 cm and 270 cm).

6. Target Detection

One key usage of spectral imagery is subpixel target detection, when an a priori known spectral signature is sought in each pixel of the spectral datacube. Previous researches dealt with target detection in the reconstructed domain [59], but in the case of the CS-MUSI camera, target detection can be applied in the compressed domain [60,61] since the CS-MUSI camera performs compression only in the spectral domain, without any spatial multiplexing. This yields a significant reduction of processing time and memory storage compared to non-compressing systems, of around an order of magnitude.

In order to test the subpixel target detection performance, we used the match filter (MF) algorithm [62], which can be derived by maximizing the SNR, or even by simply considering two hypotheses, as shown in Equation (12).

$$\begin{aligned} H_0 &: \mathbf{x} = \mathbf{W} \\ H_1 &: \mathbf{x} = \mathbf{S} + \mathbf{W} \end{aligned} \tag{12}$$

where H_0 assumes that no target, $\mathbf{S} = [s_1, s_2, \ldots, s_n]^T$, is present in the pixel and the pixel contains only background, $\mathbf{W} \sim N(0, \sigma^2 I_{n \times n})$. H_1 assumes that both background and target are present in the pixel. For simplicity, both hypotheses are modeled as multidimensional Gaussian distributions.

By applying the log likelihood ratio test for H_0 and H_1 we may derive the MF, which is equal to:

$$\mathbf{MF}(\mathbf{x}) = \mathbf{t}^T \mathbf{\Gamma}^{-1}(\mathbf{x} - \mathbf{m}), \tag{13}$$

where $\mathbf{x}$ is the pixel signature, $\mathbf{t}$ is the target spectral signature and $\mathbf{m}$ is the estimated background. $\mathbf{\Gamma}$ is the covariance matrix, which holds the statistics of the background and can be approximated using:

$$\mathbf{\Gamma} = \frac{1}{L} \sum (x - m)(\mathbf{x} - \mathbf{m})^T, \tag{14}$$

where L is the number of pixels in the datacube. For a pixel that does not include the target, the MF takes the form of Equation (13). However, when the target is present we use an additive, as shown in Equation (15), and the MF takes the form of Equation (16):

$$x\prime = \mathbf{x} + p\mathbf{t}, \tag{15}$$

$$\mathbf{MF(x\prime)} = \mathbf{t}^T \mathbf{\Gamma}^{-1}(\mathbf{x\prime - m}), \tag{16}$$

where x′ is a pixel that contains the target and p is the ratio of the target present in the pixel.

In order to assess the algorithm's performance, we adopt the performance metric that is mentioned in References [60,63]. Finally, we compare the Receiver Operating Characteristics (ROC) curve of the algorithm applied to an original HS datacube and the ROC curve of the same algorithm applied to a simulated compressed CS-MUSI datacube. The curve presents the positive detection vector as a function of false alarm probability, using the calculated value per threshold. The simulation is performed by applying a measured CS-MUSI sensing matrix, $\mathbf{\Phi}$, to each voxel of the HS datacube.

Figure 14 presents results of CS-MUSI camera target detection performance. Figure 14a shows the comparison of ROC curves obtained from conventional HS datacubes (solid lines) to those captured with the CS-MUSI camera (dotted lines). Four pairs of ROC curves are presented for the four images shown in Figure 14b (from [42,64,65]). The compression ratio sets for the compressed datacube varied between 3.5:1 (lowest) and 25:1 (highest). From the ROC curves it can be seen that the performance is not degraded by the compression. Moreover, the detection speed in the compressed HS datacubes is increased due to lower computational complexity.

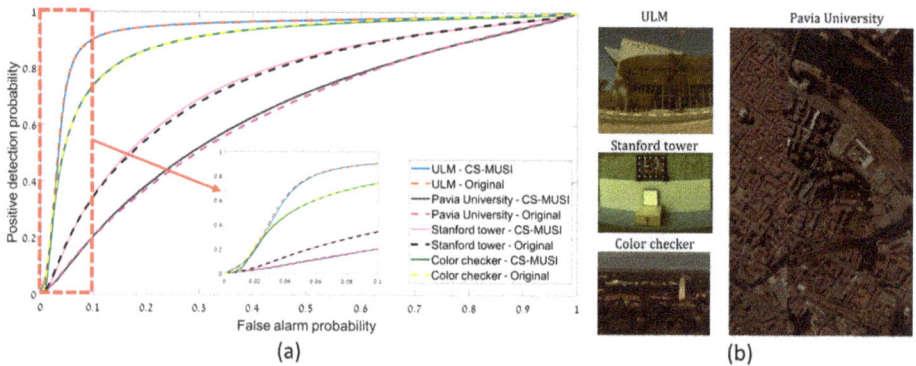

Figure 14. (a) Comparison of ROC curves for target detection in conventional (dotted lines) and CS-MUSI (solid lines) HS datacubes. (b) RGB representation of the four HS datacubes in the comparison [42,64,65].

7. Discussion

We have overviewed an evaluation of the CS-MUSI camera [28] together with its different applications. We demonstrated reconstruction of HS images in the case where the camera and scene are stationary and for the case where the camera moves in the along-track direction. Furthermore, we demonstrated the ability to use the CS-MUSI camera for 4D spectral-volumetric imaging. Experiments in these scenarios and applications have demonstrated compressibility of at least an order of magnitude. Moreover, the results provide a spectral uncertainty of less than one nanometer, e.g., in Reference [28] we demonstrated an example that has a spectral localization accuracy of 0.44 ± 0.04 nm.

Additionally, we presented a remarkable property of the CS-MUSI camera showing that the target detection algorithm performs similarly with the CS-MUSI camera as with traditional HS systems, despite the fact that the CS-MUSI data is up to an order of magnitude less than that in conventional HS datacubes. Another important advantage of the CS-MUSI camera is its high optical throughput, due to the Fellgett's multiplex advantage [66]. Furthermore, the CS-MUSI camera (Figure 5a) can be built with a small geometrical form and low weight by fabricating the LC cell to be attached to the sensor array.

It should be mentioned that the reviewed CS-MUSI camera has some limitations. The response time for a thick LC cell is relatively slow (in the order of a few seconds), which limits the acquisition frame rate. This limitation can be reduced by operating the cell in its transition state or with specially designed electronic functions, or by using faster LC structures such as ferroelectric LCs. Another limitation of the camera is the requirement for large computational resources for processing the data. This limitation can be mitigated by using parallel processing using GPU or multi-core CPU systems. Additionally, as with any HS processing algorithm, our reconstruction algorithm demands high memory capacity, since the storage of HS images can require gigabit sized memory. The additional memory requirements associated with the CS implementation are negligible compared to those of the HS data storage. From a theoretical CS point of view, the fact that there is no encoding in the spatial domain can be viewed as a limitation, since the compression obtained with the spectral encoding is lower than could have been theoretically obtained with encoding in all the three spatial-spectral domains [18]. On the other hand, the lack of spatial encoding makes it possible to maintain the full spatial resolution, allows parallel processing and facilitates the spectral imaging of moving objects.

The method of spectral multiplexing used in the CS-MUSI camera was carried out with a LC phase retarder as the spectral modulator. This method can be also realized with other spectral modulators [67]. In References [68,69], we used a modified Fabry-Perot resonator (mFPR) for spectrometry [68] and for imaging [69], which has a much faster response time compared to the LC cell. The method of spectral multiplexing can also be performed in parallel in order to achieve a snapshot HS camera. Lastly, in Reference [70] we presented a snapshot compressive HS camera that uses an array of mFPRs together with a lens array in order to acquire an array of spectrally multiplexed modulated sub-images.

8. Patents

In reference to the work presented here, a patent with the patent number US10036667 has been granted.

Supplementary Materials: The following are available online at http://www.mdpi.com/2313-433X/5/1/3/s1, Video S1: Along-track scanning measurement. Each frame represents a multiplexed intensity measurement (total of 300 shots).

Author Contributions: Conceptualization, I.A. and A.S.; methodology, I.A. and Y.O.; software, I.A., Y.O., V.F. (Section 6) and D.G. (Section 7); formal analysis, I.A. and Y.O.; investigation, I.A., Y.O., V.F. (Section 6) and D.G. (Section 7); data curation, I.A. and Y.O.; writing—original draft preparation, Y.O. and A.S.; writing—review and editing, I.A., V.F. and D.G.; supervision, A.S.; project administration, A.S.; funding acquisition, A.S.

Funding: This research was funded by Ministry of Science, Technology and Space, Israel, grant number 3-18410 and 3-13351.

Acknowledgments: We wishes to thank Ibrahim Abdulhalim's research group (Department of Electro-Optical Engineering and The Ilse Katz Institute for Nanoscale Science and Technology, Ben-Gurion University) for providing the liquid crystal cell.

Conflicts of Interest: The authors declare no conflict of interest.

References

1. Schott, J.R. *Remote Sensing: The Image Chain Approach*; Oxford University Press: Oxford, UK, 2007.
2. Borengasser, M.; Hungate, W.S.; Watkins, R. *Hyperspectral Remote Sensing: Principles and Applications*; CRC Press: Boca Raton, FL, USA, 2007.
3. Eismann, M.T. *Hyperspectral Remote Sensing*; SPIE PRESS: Bellingham, WA, USA, 2012.
4. Bioucas-Dias, J.M.; Plaza, A.; Camps-Valls, G.; Scheunders, P.; Nasrabadi, N.; Chanussot, J. Hyperspectral remote sensing data analysis and future challenges. *IEEE Geosci. Remote Sens. Mag.* **2013**, *1*, 6–36. [CrossRef]
5. Akbari, H.; Halig, L.; Schuster, D.M.; Fei, B.; Osunkoya, A.; Master, V.; Nieh, P.; Chen, G. Hyperspectral imaging and quantitative analysis for prostate cancer detection. *J. Biomed. Opt.* **2012**, *17*, 076005. [CrossRef] [PubMed]

6. Lu, G.; Fei, B. Medical hyperspectral imaging: A review. *J. Biomed. Opt.* **2014**, *19*, 010901. [CrossRef] [PubMed]
7. Calin, M.A.; Parasca, S.V.; Savastru, D.; Manea, D. Hyperspectral imaging in the medical field: Present and future. *Appl. Spectrosc. Rev.* **2014**, *49*, 435–447. [CrossRef]
8. Sun, D.W. *Hyperspectral Imaging for Food Quality Analysis and Control*; Academic Press/Elsevier: San Diego, CA, USA, 2010.
9. Kamruzzaman, M.; ElMasry, G.; Sun, D.; Allen, P. Non-destructive prediction and visualization of chemical composition in lamb meat using NIR hyperspectral imaging and multivariate regression. *Innov. Food Sci. Emerg. Technol.* **2012**, *16*, 218–226. [CrossRef]
10. ElMasry, G.; Kamruzzaman, M.; Sun, D.; Allen, P. Principles and applications of hyperspectral imaging in quality evaluation of agro-food products: A review. *Crit. Rev. Food Sci. Nutr.* **2012**, *52*, 999–1023. [CrossRef] [PubMed]
11. Li, B.; Beveridge, P.; O'Hare, W.T.; Islam, M. The application of visible wavelength reflectance hyperspectral imaging for the detection and identification of blood stains. *Sci. Justice* **2014**, *54*, 432–438. [CrossRef] [PubMed]
12. Yang, J.; Messinger, D.W.; Dube, R.R. Bloodstain detection and discrimination impacted by spectral shift when using an interference filter-based visible and near-infrared multispectral crime scene imaging system. *Opt. Eng.* **2018**, *57*, 033101. [CrossRef]
13. Brook, A.; Ben-Dor, E. A spatial/spectral protocol for quality assurance of decompressed hyperspectral data for practical applications. In Proceedings of the 2010 2nd Workshop on Hyperspectral Image and Signal Processing: Evolution in Remote Sensing (WHISPERS), Reykjavik, Iceland, 14–16 June 2010; pp. 1–4. [CrossRef]
14. Li, C.; Sun, T.; Kelly, K.F.; Zhang, Y. A compressive sensing and unmixing scheme for hyperspectral data processing. *IEEE Trans. Image Process.* **2012**, *21*, 1200–1210. [CrossRef] [PubMed]
15. August, Y.; Vachman, C.; Stern, A. Spatial versus spectral compression ratio in compressive sensing of hyperspectral imaging. In *Compressive Sensing II, Proceedings of the SPIE Defense, Security, and Sensing 2013, Baltimore, MD, USA, 29 April–3 May 2013*; SPIE: Bellingham, WA, USA, 2013; Volume 8717.
16. Willett, R.M.; Duarte, M.F.; Davenport, M.; Baraniuk, R.G. Sparsity and structure in hyperspectral imaging: Sensing, reconstruction, and target detection. *IEEE Signal Process Mag.* **2014**, *31*, 116–126. [CrossRef]
17. Parkinnen, J.; Hallikainen, J.; Jaaskelainen, T. Characteristic spectra of surface Munsell colors. *J. Opt. Soc. Am. A* **1989**, *6*, 318–322. [CrossRef]
18. August, Y.; Vachman, C.; Rivenson, Y.; Stern, A. Compressive hyperspectral imaging by random separable projections in both the spatial and the spectral domains. *Appl. Opt.* **2013**, *52*, D46–D54. [CrossRef] [PubMed]
19. Stern, A.; Yitzhak, A.; Farber, V.; Oiknine, Y.; Rivenson, Y. Hyperspectral Compressive Imaging. In Proceedings of the 2013 12th Workshop on Information Optics (WIO), Puerto de la Cruz, Spain, 15–19 July 2013; pp. 1–3. [CrossRef]
20. Lin, X.; Wetzstein, G.; Liu, Y.; Dai, Q. Dual-coded compressive hyperspectral imaging. *Opt. Lett.* **2014**, *39*, 2044–2047. [CrossRef] [PubMed]
21. Arce, G.R.; Brady, D.J.; Carin, L.; Arguello, H.; Kittle, D.S. Compressive coded aperture spectral imaging: An introduction. *IEEE Signal Process Mag.* **2014**, *31*, 105–115. [CrossRef]
22. Stern, A. *Optical Compressive Imaging*; CRC Press: Boca Raton, FL, USA, 2016.
23. Golub, M.A.; Averbuch, A.; Nathan, M.; Zheludev, V.A.; Hauser, J.; Gurevitch, S.; Malinsky, R.; Kagan, A. Compressed sensing snapshot spectral imaging by a regular digital camera with an added optical diffuser. *Appl. Opt.* **2016**, *55*, 432–443. [CrossRef] [PubMed]
24. Arce, G.R.; Rueda, H.; Correa, C.V.; Ramirez, A.; Arguello, H. Snapshot compressive multispectral cameras. In *Wiley Encyclopedia of Electrical and Electronics Engineering*; John Wiley & Sons, Inc.: Hoboken, NJ, USA, 2017; pp. 1–22. [CrossRef]
25. Saragadam, V.; Wang, J.; Li, X.; Sankaranarayanan, A.C. Compressive spectral anomaly detection. In Proceedings of the 2017 IEEE International Conference on Computational Photography (ICCP), Stanford, CA, USA, 12–14 May 2017; pp. 1–9. [CrossRef]
26. Wang, X.; Zhang, Y.; Ma, X.; Xu, T.; Arce, G.R. Compressive spectral imaging system based on liquid crystal tunable filter. *Opt. Express* **2018**, *26*, 25226–25243. [CrossRef]

27. August, Y.; Stern, A. Compressive sensing spectrometry based on liquid crystal devices. *Opt. Lett.* **2013**, *38*, 4996–4999. [CrossRef]

28. August, I.; Oikinine, Y.; AbuLeil, M.; Abdulhalim, I.; Stern, A. Miniature Compressive Ultra-spectral Imaging System Utilizing a Single Liquid Crystal Phase Retarder. *Sci. Rep.* **2016**, *6*, 23524. [CrossRef]

29. Yariv, A.; Yeh, P. *Optical Waves in Crystals*; Wiley: New York, NY, USA, 1984.

30. Candès, E.J.; Romberg, J.; Tao, T. Robust uncertainty principles: Exact signal reconstruction from highly incomplete frequency information. *IEEE Trans. Inf. Theory* **2006**, *52*, 489–509. [CrossRef]

31. Donoho, D.L. Compressed sensing. *IEEE Trans. Inf. Theory* **2006**, *52*, 1289–1306. [CrossRef]

32. Eldar, Y.C.; Kutyniok, G. *Compressed Sensing: Theory and Applications*; Cambridge University Press: Cambridge, UK, 2012; ISBN 9781107005587.

33. Bioucas-Dias, J.M.; Figueiredo, M.A. A new TwIST: Two-step iterative shrinkage/thresholding algorithms for image restoration. *IEEE Trans. Image Process.* **2007**, *16*, 2992–3004. [CrossRef] [PubMed]

34. Figueiredo, M.A.; Nowak, R.D.; Wright, S.J. Gradient projection for sparse reconstruction: Application to compressed sensing and other inverse problems. *IEEE J. Sel. Top. Signal Process.* **2007**, *1*, 586–597. [CrossRef]

35. Wright, S.J.; Nowak, R.D.; Figueiredo, M.A. Sparse reconstruction by separable approximation. *IEEE Trans. Signal Process.* **2009**, *57*, 2479–2493. [CrossRef]

36. Li, C.; Yin, W.; Zhang, Y. User's guide for TVAL3: TV minimization by augmented lagrangian and alternating direction algorithms. *CAAM Rep.* **2009**, *20*, 46–47.

37. Elad, M. *Sparse and Redundant Representations: From Theory to Applications in Signal and Image Processing*; Springer Science & Business Media: New York, NY, USA, 2010.

38. Aharon, M.; Elad, M.; Bruckstein, A. K-SVD: An algorithm for designing overcomplete dictionaries for sparse representation. *IEEE Trans. Signal Process.* **2006**, *54*, 4311–4322. [CrossRef]

39. Chakrabarti, A.; Zickler, T. Statistics of real-world hyperspectral images. In Proceedings of the 2011 IEEE Conference on Computer Vision and Pattern Recognition (CVPR), Colorado Springs, CO, USA, 20–25 June 2011; pp. 193–200. [CrossRef]

40. Oikinine, Y.; Arad, B.; August, I.; Ben-Shahar, O.; Stern, A. Dictionary based hyperspectral image reconstruction captured with CS-MUSI. In Proceedings of the 2018 9nd Workshop on Hyperspectral Image and Signal Processing: Evolution in Remote Sensing (WHISPERS), Amsterdam, The Netherlands, 23–26 September 2018.

41. Pudil, P.; Novovičová, J.; Kittler, J. Floating search methods in feature selection. *Pattern Recognit. Lett.* **1994**, *15*, 1119–1125. [CrossRef]

42. Arad, B.; Ben-Shahar, O. Sparse Recovery of Hyperspectral Signal from Natural RGB Images. In Proceedings of the European Conference on Computer Vision, Amsterdam, The Netherlands, 8–16 October 2016; pp. 19–34. [CrossRef]

43. Kokaly, R.F.; Clark, R.N.; Swayze, G.A.; Livo, K.E.; Hoefen, T.M.; Pearson, N.C.; Wise, R.A.; Benzel, W.M.; Lowers, H.A.; Driscoll, R.L. USGS Spectral Library Version 7. *USGS* **2017**, *1035*, 61. [CrossRef]

44. Oikinine, Y.; August, I.; Stern, A. Along-track scanning using a liquid crystal compressive hyperspectral imager. *Opt. Express* **2016**, *24*, 8446–8457. [CrossRef]

45. Reddy, B.S.; Chatterji, B.N. An FFT-based technique for translation, rotation, and scale-invariant image registration. *IEEE Trans. Image Process.* **1996**, *5*, 1266–1271. [CrossRef]

46. Stern, A.; Kopeika, N.S. Motion-distorted composite-frame restoration. *Appl. Opt.* **1999**, *38*, 757–765. [CrossRef]

47. Usama, S.; Montaser, M.; Ahmed, O. A complexity and quality evaluation of block based motion estimation algorithms. *Acta Polytech.* **2005**, *45*, 29–41.

48. Oikinine, Y.; August, Y.I.; Revah, L.; Stern, A. Comparison between various patch wise strategies for reconstruction of ultra-spectral cubes captured with a compressive sensing system. In *Compressive Sensing V: From Diverse Modalities to Big Data Analytics, Proceedings of the SPIE Commercial + Scientific Sensing and Imaging 2016, Baltimore, MD, USA, 17–21 April 2016*; SPIE: Bellingham, WA, USA, 2016; Volume 985705. [CrossRef]

49. Farber, V.; Oikinine, Y.; August, I.; Stern, A. Compressive 4D spectro-volumetric imaging. *Opt. Lett.* **2016**, *41*, 5174–5177. [CrossRef] [PubMed]

50. Stern, A.; Farber, V.; Oikinine, Y.; August, I. Compressive hyperspectral synthetic aperture integral imaging. In *3D Image Acquisition and Display: Technology, Perception and Applications*; Paper DW1F. 1; Optical Society of America (OSA): Washington, DC, USA, 2017.

51. Farber, V.; Oiknine, Y.; August, I.; Stern, A. 3D reconstructions from spectral light fields. In *Three-Dimensional Imaging, Visualization, and Display 2018, Proceedings of the SPIE Commercial + Scientific Sensing and Imaging 2018, Orlando, Florida, USA, 15–19 April 2018*; SPIE: Bellingham, WA, USA, 2018; Volume 10666. [CrossRef]
52. Farber, V.; Oiknine, Y.; August, I.; Stern, A. Spectral light fields for improved three-dimensional profilometry. *Opt. Eng.* **2018**, *57*, 061609. [CrossRef]
53. Lippmann, G. Epreuves reversibles Photographies integrals. *C. R. Acad. Sci* **1908**, *146*, 446–451.
54. Arimoto, H.; Javidi, B. Integral three-dimensional imaging with digital reconstruction. *Opt. Lett.* **2001**, *26*, 157–159. [CrossRef] [PubMed]
55. Stern, A.; Javidi, B. Three-dimensional image sensing, visualization, and processing using integral imaging. *Proc. IEEE* **2006**, *94*, 591–607. [CrossRef]
56. Hong, S.; Jang, J.; Javidi, B. Three-dimensional volumetric object reconstruction using computational integral imaging. *Opt. Express* **2004**, *12*, 483–491. [CrossRef]
57. Aloni, D.; Stern, A.; Javidi, B. Three-dimensional photon counting integral imaging reconstruction using penalized maximum likelihood expectation maximization. *Opt. Express* **2011**, *19*, 19681–19687. [CrossRef]
58. Llavador, A.; Sánchez-Ortiga, E.; Saavedra, G.; Javidi, B.; Martínez-Corral, M. Free-depths reconstruction with synthetic impulse response in integral imaging. *Opt. Express* **2015**, *23*, 30127–30135. [CrossRef]
59. Busuioceanu, M.; Messinger, D.W.; Greer, J.B.; Flake, J.C. Evaluation of the CASSI-DD hyperspectral compressive sensing imaging system. In *Algorithms and Technologies for Multispectral, Hyperspectral, and Ultraspectral Imagery XIX, Proceedings of the SPIE Defense, Security, and Sensing 2013, Baltimore, MD, USA, 29 April–3 May 2013*; SPIE: Bellingham, WA, USA, 2013; Volume 8743. [CrossRef]
60. Gedalin, D.; Oiknine, Y.; August, I.; Blumberg, D.G.; Rotman, S.R.; Stern, A. Performance of target detection algorithm in compressive sensing miniature ultraspectral imaging compressed sensing system. *Opt. Eng.* **2017**, *56*, 041312. [CrossRef]
61. Oiknine, Y.; Gedalin, D.; August, I.; Blumberg, D.G.; Rotman, S.R.; Stern, A. Target detection with compressive sensing hyperspectral images. In *Image and Signal Processing for Remote Sensing XXIII, Proceedings of the SPIE Remote Sensing, 2017, Warsaw, Poland, 11–14 September 2017*; SPIE: Bellingham, WA, USA, 2017; Volume 10427.
62. Caefer, C.E.; Stefanou, M.S.; Nielsen, E.D.; Rizzuto, A.P.; Raviv, O.; Rotman, S.R. Analysis of false alarm distributions in the development and evaluation of hyperspectral point target detection algorithms. *Opt. Eng.* **2007**, *46*, 076402. [CrossRef]
63. Bar-Tal, M.; Rotman, S.R. Performance measurement in point source target detection. In Proceedings of the Eighteenth Convention of Electrical and Electronics Engineers in Israel, Tel Aviv, Israel, 7–8 March 1995; pp. 3.4.6/1–3.4.6/5. [CrossRef]
64. Skauli, T.; Farrell, J. A collection of hyperspectral images for imaging systems research. In *Digital Photography IX, Proceedings of the IS&T/SPIE Electronic Imaging, Burlingame, CA, USA, 3–7 February 2013*; SPIE: Bellingham, WA, USA, 2013; Volume 8660. [CrossRef]
65. Hyperspectral Remote Sensing Scenes. Available online: http://www.ehu.eus/ccwintco/index.php?title= Hyperspectral_Remote_Sensing_Scenes (accessed on 26 October 2018).
66. Fellgett, P. The Multiplex Advantage. Ph.D. Thesis, University of Cambridge, Cambridge, UK, 1951.
67. Oiknine, Y.; August, I.; Stern, A. Compressive spectroscopy by spectral modulation. In *Optical Sensors 2017, Proceedings of the SPIE Optics + Optoelectronics, 2017, Prague, Czech Republic, 24–27 April 2017*; SPIE: Bellingham, WA, USA, 2017; Volume 10231. [CrossRef]
68. Oiknine, Y.; August, I.; Blumberg, D.G.; Stern, A. Compressive sensing resonator spectroscopy. *Opt. Lett.* **2017**, *42*, 25–28. [CrossRef] [PubMed]
69. Oiknine, Y.; August, I.; Blumberg, D.G.; Stern, A. NIR hyperspectral compressive imager based on a modified Fabry–Perot resonator. *J. Opt.* **2018**, *20*, 044011. [CrossRef]
70. Oiknine, Y.; August, I.; Stern, A. Multi-aperture snapshot compressive hyperspectral camera. *Opt. Lett.* **2018**, *43*, 5042–5045. [CrossRef]

Journal of
Imaging

MDPI

Review

Recent Trends in Compressive Raman Spectroscopy Using DMD-Based Binary Detection

Derya Cebeci [1,*], **Bharat R. Mankani** [2] **and Dor Ben-Amotz** [3]

1   PortMera Corp., Stony Brook, NY 11790, USA
2   MarqMetrix Inc., Seattle, WA 98103, USA; bmanks@gmail.com
3   Department of Chemistry, Purdue University, West Lafayette, IN 47907, USA; bendor@purdue.edu
*   Correspondence: derya.cebeci@portmera.com; Tel.: +1-631-572-8535

Received: 21 November 2018; Accepted: 13 December 2018; Published: 21 December 2018

Abstract: The collection of high-dimensional hyperspectral data is often the slowest step in the process of hyperspectral Raman imaging. With the conventional array-based Raman spectroscopy acquiring of chemical images could take hours to even days. To increase the Raman collection speeds, a number of compressive detection (CD) strategies, which simultaneously sense and compress the spectral signal, have recently been demonstrated. As opposed to conventional hyperspectral imaging, where full spectra are measured prior to post-processing and imaging CD increases the speed of data collection by making measurements in a low-dimensional space containing only the information of interest, thus enabling real-time imaging. The use of single channel detectors gives the key advantage to CD strategy using optical filter functions to obtain component intensities. In other words, the filter functions are simply the optimized patterns of wavelength combinations characteristic of component in the sample, and the intensity transmitted through each filter represents a direct measure of the associated score values. Essentially, compressive hyperspectral images consist of 'score' pixels (instead of 'spectral' pixels). This paper presents an overview of recent advances in compressive Raman detection designs and performance validations using a DMD based binary detection strategy.

Keywords: Raman spectroscopy; chemical imaging; compressive detection; spatial light modulators (SLM); digital micromirror device (DMD); digital light processor (DLP); optimal binary filters; Chemometrics; multivariate data analysis

1. Introduction

Raman spectroscopy may be used to study the chemical composition and construction of spectral images of various compounds, and has proven to be a useful tool for a variety of scientific fields. However, Raman scattering has intrinsically low cross-section, yielding low signals. Furthermore, conventional Raman spectroscopy with a multichannel array detector, such as the charged-couple device (CCD) camera, is limited by the inherent read noise of the detector electronics. These array-based spectrometers disperse different wavelengths of light onto separate detector pixels (or wavelength channels). Because of the inherent read-noise associated with CCD measurements, array-based Raman spectrometers have a major drawback in the low-signal regime. For example, if 100 photons are distributed over 100 pixels of a CCD detector, then the resulting signal in each pixel would be well below the typical CCD read-noise (of a few counts per channel), essentially rendering the Raman photons undetectable. As a result, Raman measurements often require collection times on the orders of hundreds of milliseconds or longer to obtain a spectrum with decent signal-to-noise ratio. This limitation hinders the use of Raman spectroscopy for hyperspectral imaging applications, where thousands to millions of different spatial points are measured. For example, the collection of a one-megapixel image would take 12 days with a typical 1 s per spectrum acquisition rate.

Recent advances in spatial light modulators (SLM) bring about a paradigm shift in the way that Raman imaging is performed. SLMs provide a means of producing variable programmable filter functions. Raman light is modulated by the filter functions loaded on SLMs, sent to the single-channel detector and recorded. The introduction of compressive data collection strategies enabled by SLMs allow fast Raman measurements by multiplexing Raman photons from different wavelengths onto a low-noise single channel detector [1–10]. Therefore, the scanning speed is basically only defined by the limits of the count rate of the detector. These recent studies have led to the conclusion that digital (binary) SLM optical filters created using the Digital Micromirror Device (DMD, Texas Instruments, Dallas, TX, USA) can be as effective in data compression as analogue SLMs, with a simpler and more robust optical design [1,3,4,6–10]. Moreover, two detectors can be used to detect all the collected Raman photons, and thus increase compression speed, with appropriately optimized digital filters [7].

Hyperspectral Raman imaging is the combination of two technologies: spectroscopy and imaging, whereas compressive hyperspectral imaging combines three technologies: spectroscopy, imaging, and signal processing. In the hyperspectral imaging mode, an image is acquired by recording the full spectrum at each x, y point on the sample, which produces 'data cubes', where signal intensity is measured as a function of the x and y spatial dimensions and a spectral (wavenumber, cm^{-1}) dimension. More generally, in hyperspectral imaging each pixel contains two-dimensional spatial information (x and y), and a third dimension of spectral information (e.g., wavenumbers). Subsequently, to derive the sought information from this large, multidimensional data set, in order to obtain a chemical image, it is necessary to transform it to lower-dimensional space through multivariate data analysis algorithms, such as principal component analysis (PCA), partial least squares (PLS), multivariate curve resolution (MCR), or total least-squares (TLS) [11,12], or in some cases through univariate analysis if a component of interest has a unique spectral peak (that has not overlapped with peaks due to other components) [13]. Generally, in a univariate analysis a property of interest is calculated based on a single value, which is correlated to the property to which that peak corresponds, such as the amount of that component at each image pixel location. In other words, the area (intensity) of the peak of interest is used as a measure of the amount of that component at each spatial location in the image. When an isolated peak of the investigated component does not exist, then more advanced multivariate data analysis techniques are typically used to extract the desired information. In fact, even when one or more components do have an isolated peak it can still be advantageous to perform a multivariate data analysis, as this will make use of the information contained in the entire measured spectrum.

Compressive Raman spectroscopy is similar to conventional hyperspectral Raman spectroscopy, except that the array detector is replaced by a spatial light modulator (SLM), and one or two single channel detectors, such as photon counting amplified photodiodes or photomultiplier tube detectors. In the compressive detection (CD) mode, instead of recording the full spectrum at each pixel, the spectral response (a score value) for each SLM filter function is recorded. In other words, CD differs from the conventional Raman detection in that the scores are directly detected using the hardware, rather than by obtaining scores after post-processing the full spectra. The total number of photons transmitted through each filter on the SLM is counted using a photon-counting detector, to obtain a direct measure of the score value. More specifically, CD effectively measures the dot-product of the filter vectors and the spectral vector coming from the sample. Chemometric techniques may be used to create optical filter functions, which are trained using full-spectral reference spectra. Once the filters are generated, photon counting is performed in each pixel (instead of full-spectral acquisition as in conventional systems). Because all the light transmitted by SLMs is measured by a single-channel detector rather than being separately detected by multiple channels (often >1000 channels), CD benefits from Felgett's (or multiplex) signal-to-noise ratio (SNR) advantage. Thus, for example, a Raman spectrum with a total number of 100 photons would have a SNR of ~10 on a single-channel detector while, as previously noted, the same 100 photons would have been practically undetectable on a multi-channel CCD detector.

Here we review the latest advances in the design and performance of compressive hyperspectral Raman strategies, which facilitate the rapid collection of chemical images by directly applying programmable optical filters optimized to distinguish compounds of interest using spatial light modulators and single channel detectors.

2. Compressive Detection (CD) Strategies

Two design strategies were recently described for compressive Raman detection, in which filter functions are applied to either a digital micromirror array device (DMD) [1,3,14–16] or analog-based liquid crystal [2,6,17] spatial light modulators (LC-SLM). DMDs provide binary states, as each mirror pixel can be programmed to be either "on" or "off", corresponding to a mirror tilt of ±12°. LC-SLMs, on the other hand, use light polarization to produce either phase or amplitude modulated variable analog filters [18]. Each pixel on LC-SLMs is a separately addressable optical phase modulator, which is used to rotate the polarization of the detected light between 0 (*p*-polarized) and 90 (*s*-polarized). The filter functions on liquid crystal cell control the degree to which, for example, the input *p*-polarized signal is rotated to s-polarization and thus reflected into the detection optical path. In this way, the spectral component that became s-polarized by means of the liquid crystal cell is entirely transmitted to the single channel detector while no *p*-polarized light reaches the detector. Earlier applications using transmissive LC-SLMs in compressive spectroscopy suffered from low light throughput of ~20% [6]. Recent developments in reflectance LC-SLM with higher light throughput (~80%) and fill factor [19] made possible for better performing spectrometers [2,20].

The data obtained in CD technology is fundamentally photon counts, which essentially corresponds to the dot product of filter functions and the spectra vectors. The filter functions are simply the combination of wavelengths specifically designed to regenerate the eigenvectors (often referred to principal component) obtained from chemometric algorithms. Multivariate techniques like PLS and PCA are appropriate to use to generate the optimal eigenvectors for a given experiment when the components are known [21,22]. These techniques use pure component samples as a built-in calibration set [23–25]. However, if components are not known then techniques like MCR may be more valuable to extract component information [26]. The amplitude of the measured signal is proportional to the amplitude of the eigenvectors, and thus to the amount of the corresponding compound.

Filter functions ensure that only the photons with certain wavelengths that are the most effective in discriminating the components of interest are detected. Irrelevant photons are disposed. To create filters for a given application, one must first obtain a high SNR training spectrum of each component of interest. Both LC-SLM and DMD-based CD systems can also function as a general Raman spectrometer to obtain full Raman spectra by notch scanning the SLM arrays one array column at a time. In DMD-based systems notch scanning could be performed by sequentially directing one mirror column towards the detector while all others are directed away, and count the number of photons at each notch position. In LC-SLM systems, SLM is used to produce band-pass filters with variable center wavelengths. However, the efficiency of band-pass scanning (or notch-scanning) is quite low since most Raman photons are discarded. An advantageous alternative to notch-scanning may be to use Hadamard [27] filter functions to obtain full spectra at higher SNR [2,17,23,24,28]. The efficiency of Hadamard strategy comes from that half of the Raman photons is always detected by each Hadamard filter.

The present review is focused primarily on recent developments in DMD-based CD systems.

Digital Micromirror Device (DMD)-Based Compressive Raman Detection

DMD is a micro-electronic mechanical system (MEMS) which consists of thousands of individually addressable moving micromirrors controlled by underlying electronics. DMD is also an SLM as the mirrors are highly reflective and are used to modulate light; to rotate the light to either a +12 degree or −12 degree position relative to the flat state of the array depending on the binary state of the cell below each pixel. These two positions determine the direction that light is deflected. Each tiltable mirror-pixel can be moved to reflect light to, or away from an intended target [29–32].

In DMD-based Raman detection systems the micromirrors on DMDs are horizontally binned (x mirrors/pixels) and vertically fully binned. That is, all mirrors in each column of the array are set to the same angle of either −12 or +12 and mirrors in each row are divided into adjacent groupings. Bins are defined by bands of photon energy, then groups of x adjacent columns are set in unison. The filter functions on DMD, then tells which columns of pixels are turned "on", sending those selected photons to the detector and which columns are turned "off", directing those photons away from the detector. More specifically, while photons with certain energy levels corresponding to "on" columns are collected in a single-channel detector and recorded, photons with wavelengths reaching "off" columns are disposed.

The first reported use of DMD SLMs in spectrometry dates back to 1995 by Wagner et al. [33]. In this early work the contrast ratio of the DMD was only about 60:1; today it goes as high as 2000:1 with a higher fill factor, allowing the design of better performing compressive Raman systems. Two approaches were reported recently to construct binary filters for DMD based-Raman systems [15,34]. The filter design developed by Scotte et al. is based on maximizing the precision of the components proportion estimates [15,35] using a new Cramer-Rao lower bound based algorithm. Buzzard and Lucier's approach was to minimize the error in estimating photon emission rates of the chemical species investigated [34,36]. Both approaches based their theory on the fact that the photons transmitted through filter functions are modeled by Poisson random variables when the measurements are photon-noise limited [1,7,15,34–37]. Here, optimized binary compressive detection (OBCD) procedures based on Buzzard and Lucier's approach is overviewed.

OBCD design: The recently-developed optimized binary compressive detection (OBCD) method relies on binary filters, which provides optimal measurement settings. Input data to be modeled to generate filter functions are photon counts, modeled by Poisson random variables whose variances equal to their means. Photon emission rates are correlated to the concentration of components of interest. In other words, concentrations are not directly measured, rather photon emission rates of each compound are estimated and the concentrations are calculated from this estimation. Objective is to minimize the mean square error between estimated and true emission rates.

OBCD design has been shown to enable high-speed chemical classification, quantitation, imaging [1,36,37], as well as facilitating Raman classification in the presence of fluorescence background [8]. The design of an OBCD Raman spectrometer with 785 nm laser excitation whose schematic is shown in Figure 1A is described in detail in reference [1]. This design is configured to collect backscattered Raman photons with the same objective lens used to focus the excitation laser onto the sample. After separating Rayleigh photons using dichroic and notch filters, then Raman light is directed to the spectrometer module. It is then dispersed onto the DMD ((Texas Instruments, DLP D4000, 1920 × 1080 aluminum mirror array with 10.8 μm mirror pitch) after passing through volume holographic grating (VHG). In this design, 15 columns of adjacent mirrors are binned to yield a total of 128 bins, each bin corresponding to ~30 cm^{-1} and the whole spectral window being ~200–1700 cm^{-1}. The Raman light transmitted by the "on" mirrors (corresponding to +12 degree tilt of mirrors) is then sent to the low-noise photon-counting avalanche photodiode (APD) module (dark count rate of ~200 photons/s and no read noise). The input binary optical filters tell which mirrors will point toward (assigned value of one) or point away (assigned value of zero) the detector. Authors have demonstrated that the OBCD with 785 nm excitation can be used to rapidly quantify binary and tertiary liquid mixtures with known components, and also to generate chemical images of mixed powders as well as generating filter functions using the MCR algorithm to facilitate high speed chemical imaging of samples for which pure components spectra are not available. [37]. They reported that with the OBCD strategy, a mixture of glucose and fructose is discriminated with as low as ~10 photons per pixel, corresponding to pixel dwell time of ~ 30 μs.

In order to demonstrate the accuracy of the OBCD detection mode, pairs of liquid mixtures with various degrees of spectral overlap were tested. Classification error was found to vary both with the degree of overlap and acquisition time. Low to moderately overlapping spectra (benzene/acetone with a correlation coefficient of 0.12, and n-hexane/methylcyclohexane with a correlation coefficient of 0.71) were accurately classified with as few as 10–25 photons per measurement in tens to hundreds of microseconds. The highly overlapped case of n-heptane/n-octane mixture with a correlation coefficient of 0.99, correct classification was achieved with ~200 photons in a few milliseconds. These acquisition times obtained using OBCD strategy were not accessible using comparative CCD-based Raman spectroscopy.

Another OBCD Raman spectrometer prototype with 514 nm laser excitation with similar design to the 785 nm system mentioned above was also prototyped in Ben-Amotz's lab [8]. For this design, a DMD chip of 608×684 mirror array with 10.8 µm mirror pitch was used. Two columns of adjacent mirrors were binned to give a total of 342 bins with each bin corresponding to 12 cm^{-1} and yielding a spectrometer with a ~200–4100 cm^{-1} spectral window. As a single channel detector, a photomultiplier tube (PMT) with a dark count rate of ~500 photons/s was used in this design. In this work [8] the feasibility of the OBCD strategy for Raman imaging of moderately fluorescing samples was demonstrated. A strategy for fitting a fluorescence background to the third-degree Bernstein polynomials was adopted to train OBCD filters, which were then used to quantitatively separate Raman signals from the fluorescence background, facilitating Raman imaging of chemicals in the presence of a fluorescence background.

OBCD2 design: In the OBCD detection strategy only a fraction of Raman photons, which were transmitted by "on" (+12 degrees) stage of micromirrors, were read by the detector. Raman light reaching to "off" (−12 degrees) micromirrors on DMD was disposed. A new strategy, termed as OBCD2, was proposed to increase the efficiency of Raman detection, wherein binary filters were generated in pairs [7]. Two detectors were used to count all Raman photons transmitted by two complimentary OB filters. OBCD2 is considered a derivative of OBCD, accordingly many of the assumption made in formulating the OBCD strategy [1,7,36] remain valid for OBCD2 strategy, as well.

A schematic of this technique is shown in Figure 1B. In the OBCD2 strategy, when one OBCD filter is generated corresponding to the "on" mirrors on DMD, the exact complement of that filter is also generated for implementation to "off" mirrors. To describe a system with n components a minimum of $2(n-1)$ filters, which constitutes to $n-1$ pairs of complementary filters, are required. Photons of different wavelengths are selectively reflected by micromirrors either positive 12 degrees or negative 12 degrees to the surface of the DMD and are directed to either one or the other PMT detector (dark count of ~500 photons/s) shown in Figure 1B. With OBCD2 filtering strategy all Raman photons are detected. As a result, Raman scattering rates recovered using OBCD2 filters have lower variance than those using OBCD filters [7]. In order to quantify the performance advantage of the OBCD2 over OBCD strategy, a ternary system of benzene, hexane, and methylcyclohexane were analyzed in [7]. For this system there were three OBCD filters and $2 \times (3-1) = 4$ OBCD2 filters (or two complementary pairs). The standard deviations of the estimated recovered Raman scattering rates are shown to improve ~63%, ~23%, and ~24% for benzene, hexane, and methylcyclohexane, respectively.

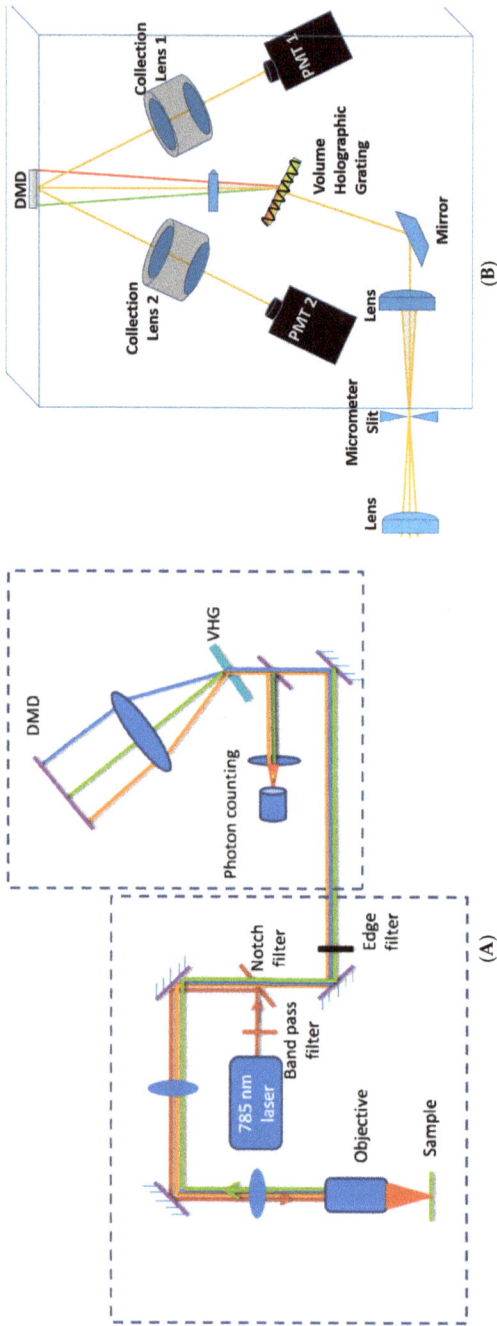

Figure 1. (A) Schematic of optimized binary compressive detection (OBCD) Raman system with 785 nm laser source (Reprinted from Reference [1] with permission from Elsevier); **(B)** schematic of Dual Detector CD (OBCD2) (detection module is shown only). Collimated beam of light is focused on a slit at the entrance of the spectrometer. The beam is recollimated and dispersed using a volume holographic grating (VHG). The dispersed beam is focused onto the DMD. The DMD can send parts of all the light towards either detector.

3. Assessment of DMD-Based Raman vs. CCD-Based Raman Detection

The performance of compressive Raman detection has been assessed compared to conventional Raman measurements by Scotte et al. in a recent paper [15], and by Ben-Amotz et al. in a forthcoming publication. Scotte et al. evaluated the performance of a custom-built DMD-based system with two commercially available spectrometers with different detectors (CCD and EMCCD) for the detection of low concentrations of biologically relevant components (microcalcification powders relevant to human breast cancer) [15]. They reported that in the high signal regime CD technology outperformed Raman imaging of the biological system they studied with a ×100 to ×10 speed improvements compared with CCD and EMCCD-based Raman imaging, respectively. In the low signal regime where noise is the limiting factor, they chose to compare the systems' limits of detection (LOD). The LOD is defined as the minimum Raman light needed to reach on the detector to be able to correctly estimate component proportions. LOD for a compressive system is found to be similar to EMCCD and up to ×100 higher than CCD system. At equal signal-to-noise ratio CD is still faster than hyperspectral imaging. However, it is important to note that the custom-built DMD-Raman system in this work can further be improved as PMT used only has 40% quantum efficiency (QE) while two cameras used in commercial systems reach 90 to 95% QE.

Figure 2 shows the classification of acetone and benzene using full spectral acquisition using a CCD detector. Figure 3 corresponds to optimized binary compressive detection (OBCD) measurements, and Figure 4 shows the results for OBCD2 strategy performed under otherwise identical conditions to those used to obtain the results in Figure 2. In all three figures (Figures 2–4), panel A shows training spectra obtained using 30 mW of laser power and an integration time of 1 s, while panel B shows classification results obtained using lower laser powers and integration times. More specifically, Panel A in Figures 2–4 shows the normalized training spectra of acetone (red, top left) and benzene (blue, top right). Training spectra for the OBCD and OBCD2 measurements were obtained by measuring counts through Hadamard filters applied on the DMD for 1 s each (again with 30 mW of laser power). The counts were then invers Hadamard transformed to produce the high SNR spectra shown in Panels A of Figures 3 and 4.

Panel B in Figures 2–4 are each subdivided into three sub panels a, b, and c corresponding to the classification of acetone and benzene in 1 ms using 30 mW, 3 mW, and 1 mW of laser power at the sample, respectively. The normalized spectrum of acetone (red) and benzene (blue) at each laser power are shown on the top and the right of the two-dimensional classification plot in each sub panel. The ellipses represent the 95% confidence interval.

Note that the CCD cannot collect spectra faster than 1 ms, which is why the laser power is chosen to be reduced rather than reducing integration. However, the CD measurements can be performed using the PMT detectors with integration times as short as ~3µs—this is a key advantage of CD-based (OBCD or OBCD2) measurements as opposed to conventional full spectral CCD-based measurements.

In the assessment of CCD based-Raman in Panel B of Figure 2, each cloud consists 1000 independently measured spectra classified by post processing the spectra using least squares to extract the lower dimension concentration information. The bottom of panel B shows the dimension reduced linear discriminant analysis (LDA) histograms, which make it visually clear that the classification error increases with the reduction in signal (Raman scattered photons). Note Panel B of Figure 2 also includes the full spectra obtained in 1 ms of integration using the three laser powers. These spectra clearly reveal that no spectral peaks are evident at laser powers of 3 mW and below. However, the LDA histogram reveal that it is still possible to accurately discriminate the acetone and benzene spectra at a power of 3 mW using such extremely noisy spectra.

Figure 3 shows the results obtained using OBCD filtering strategies for the same classification problem. OBCD filters are shaded in gray in panel A of Figure 3. These shaded wavelengths are the OB wavelengths, that were applied programmatically onto a DMD, then multiplexed onto one PMT. Each cloud consists of 1000 independently measured intensities (photon counts) obtained using two sequentially applied filters onto one PMT detector. OBCD2 filters are shaded in red and blue in panel A of Figure 4 All Raman photons reflected by the DMD are detected using two PMTs in OBCD2 strategy. The photons from both PMTs are used to transform counts to concentration space and are shown in the classification plots in panel B of Figure 4 Panel B shows OBCD2 classification plots and LDA histograms of acetone and benzene. Comparison of these LDA histogram results with those in Figure 2 clearly reveals the greater discriminating power of CD-based as opposed to CCD-based measurements, as well as the fact that OBCD2 outperforms OBCD.

Table 1 compares the resolution of LDA histograms of acetone and benzene produced by different classification strategies and using different laser powers. The resolution (R) was calculated using Equation (1) below. It is defined by the ratio of the absolute separation in the mean values (μ) of the clouds obtained from the acetone and benzene histograms divided by the sum of the standard deviation (σ) of each of the histograms. The higher the resolution, the greater the separation between the histograms and the better the classification between the two chemicals.

$$R = \frac{|\mu_1 - \mu_2|}{\sigma_1 + \sigma_2} \tag{1}$$

The best resolution is achieved for the highest laser power at the sample. This makes sense because higher laser power generates more Raman photons yielding a higher signal-to-noise. The resolution for the three classification methods are comparable at the higher laser power. However, for lower laser powers, the resolution of classification of acetone and benzene is greatest for the OBCD2 classification, followed by OBCD and CCD classifications.

Figure 5 compares the relative standard deviations (RSD, Equation (2)) in the classification using CCD and optimal binary CD measurements. The higher the RSD the worse the classification. The *x*-axis in Figure 5 represents the total photon counts, which were measured by adjusting both the laser power and the integration times. The same Raman photons were sent to the CD spectrometer as the CCD spectrometer by using a flip mirror to direct the light either towards the CCD or OBCD detection systems.

$$RSD = \frac{\sigma}{\mu} \times 100 \tag{2}$$

In the high intensity (photon count) limit, the *RSD* for all three methods were comparable. However, below about 1000 photons the OBCD classification has lower *RSD*. Below about 2000 photons the OBCD2 classification has lower *RSD*. Thus, although full hyperspectral CCD measurements are always a bit better than OBCD and OBCD2 measurements in the high intensity limit, in the low signal regime both OBCD and OBCD2 far outperform CCD-based measurements, and OBCD2 is slightly better than OBCD. Additional measurements (not shown) performed using components that are more highly overlapped Raman spectra have been found to have similar crossover points. Thus, it appears to be generally true that OBCD/OBCD2 measurements outperform conventional full spectral CCD-based measurements when the integrated number of photoelectrons in the spectra is below a few thousand counts.

Table 1. The table compares the resolution of the linear discriminant analysis (LDA) histograms of acetone and benzene produced by different classification strategies and using different laser powers.

Classification Strategy	30 mW, 1 ms	3 mW, 1 ms	1 mW, 1 ms
CCD	26	3	1
OBCD	22	7	4
OBCD2	32	13	6

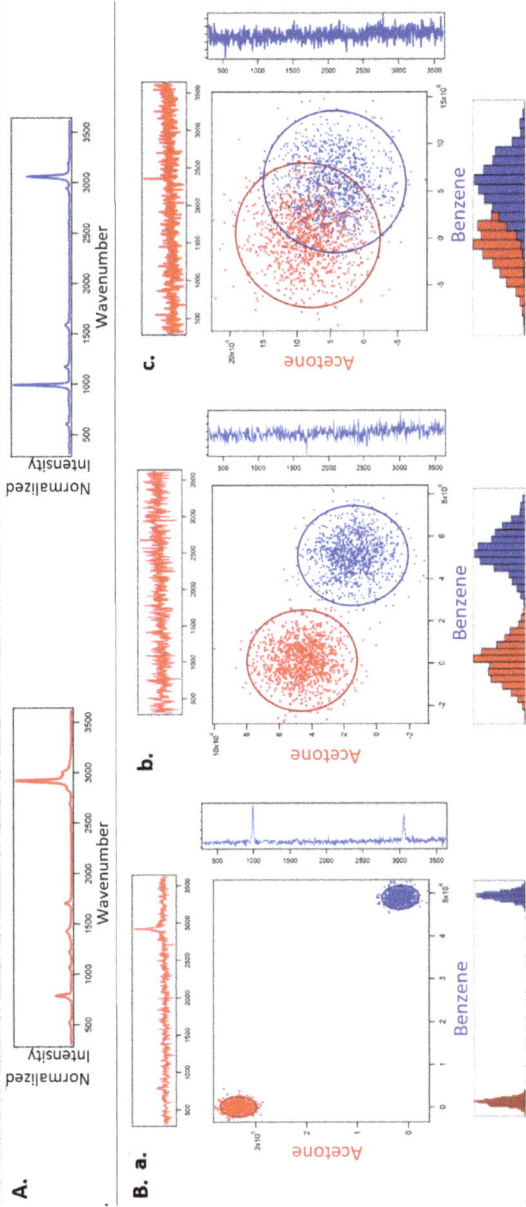

Figure 2. The figure shows the classification of acetone and benzene using a charge coupled device (CCD)-based Raman spectroscopy. Panel (**A**) shows the normalized spectra of acetone (red, top left) and benzene (blue, top right) collected using 30 mW laser power at the sample, and 1 s total spectral acquisition time to be used as training spectra in least square analysis. Subpanels (**a**–**c**) in Panel (**B**) correspond to the 1000 full spectral least squares classification of acetone and benzene in 1 ms using 30 mW, 3 mW, and 1 mW laser power at the sample, respectively.

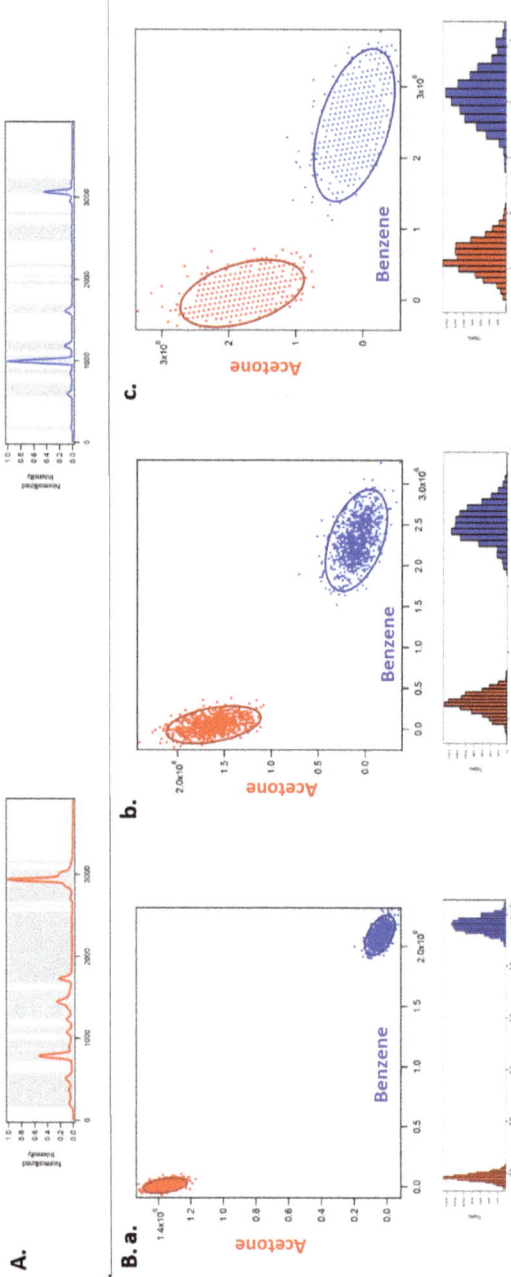

Figure 3. The figure shows the classification of acetone and benzene using optimized binary compressive detection (OBCD) filtering. Panel (**A**) shows the normalized spectra of acetone (red, top left) and benzene (blue, top right) collected to be used as training spectra to generate OBCD filters. This panel also shows OBCD filters in gray, i.e., the OB wavelengths that are multiplexed onto one PMT. Sub panels (**a**–**c**) in Panel (**B**) correspond to 1000 OBCD classification of acetone and benzene in 1 ms using 30 mW, 3 mW, and 1 mW laser power at the sample, respectively.

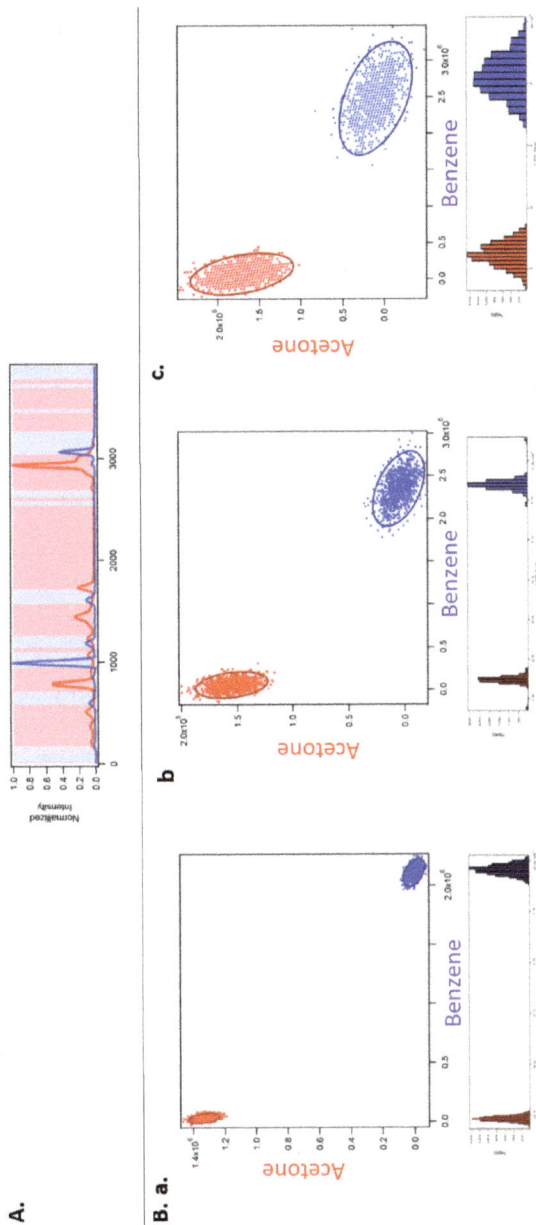

Figure 4. The figure shows the classification of acetone and benzene with OBCD2 filtering (using complementary OBCD filters). Panel (**A**) shows the normalized spectra of acetone (red, top left) and benzene (blue, top right) collected to be used as training spectra to generate OBCD2 filters. This panel also shows OBCD2 filters (shaded red and blue), i.e., the OB wavelengths that are multiplexed onto two PMTs. Sub panels (**a–c**) in Panel (**B**) correspond to 1000 OBCD classification of acetone and benzene in 1 ms using 30 mW, 3 mW, and 1 mW laser power at the sample, respectively.

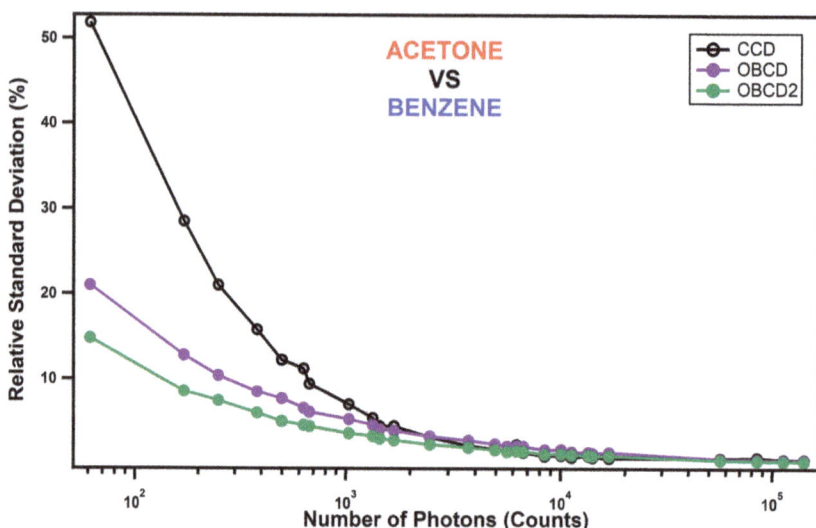

Figure 5. The figure compares the relative standard deviation in the classification using CCD, OBCD and OBCD2. The instrument was set to laser powers 30, 3, and 1 mW and counts were measured at nine integration times; 100, 80, 60,40, 10, 8, 6, 4, and 1 ms at each power.

4. Discussion

A key bottleneck to fast Raman analysis, including real-time monitoring and hyperspectral imaging, is the time required to acquire multivariate hyperspectral data and post-processing the full spectra. Multichannel detectors (e.g., CCD) are generally more expensive and less sensitive than single channel detectors and also require cooling when long integration time and low dark counts are needed. A CCD-based Raman spectrometer cannot operate fast enough to be applicable for dynamic system measurements. A compressive spectrometer, which incorporates SLM technology and a single channel detector, offers not only higher sensitivity and speed, but also a potentially lower-cost alternative to CCD-based Raman imaging. Most importantly, compressive detection can be used to obtain chemical imaging information in the very low signal limit at which conventional CCD-based Raman spectroscopy is completely impossible.

Here we overviewed the latest advances in compressive Raman detection with focus on the optimized binary compressive detection (OBCD) strategy. At the heart of OBCD strategy, is the widely adopted reflective light modulator DMD used in standard computer projection systems (manufactured by Texas Instruments), whose switch speed, contrast ratio, and broad spectral capability outperforms analog-based SLMs. DMD is a semiconductor-based "light switch" array of hundreds of thousands of individually switchable mirror pixels. The light switching speeds in the order of kHz at which each mirror can modulate between "on" and "off" states enable CD measurements at kHz frequencies. DMDs have faster modulation rates [38]. Analog-based SLMs, on the other hand, do not have the speed and precision capability which make the DMD more attractive for use in Raman spectroscopy. Also, they have slower pixel response and have to operate on linearly polarized light. DMDs maintain higher throughput due to polarization insensitivity.

Rapid evaluation of chemical species in complex chemical matrices is of high importance in a diverse array of applications. The projected advances brought about by DMD use in Raman spectroscopy will make real-time measurement possible, for example real-time imaging systems for medical and scientific communities, in-line quality inspections to evaluate chemical compositions, etc. DMD-based Raman systems are shown to effectively suppress laser-induced fluorescence backgrounds, which makes fast Raman mapping of samples with large background possible [8,9]. In a very recent

J. Imaging **2019**, *5*, 1

publication by Scotte et al., the performance of DMD-based compressive Raman technology is assessed on a biologically relevant sample mimicking microcalcification in breast cancer [15]. For this study, four micromirrors on DMD were binned horizontally, giving a spectrometer with spectral resolution of 40 cm^{-1} and the spatial resolution of 1.4 μm with 532 nm laser excitation. At these parameters, compressive Raman spectroscopy is reported to be a very useful technique correlating the state of the cancer to the chemical composition of microcalcification.

In addition, the studies in references [23,24] showed the potential of compressive Raman detection as a process analytical technology (PAT) tool for pharma industry. Non-invasive, real-time measurement systems for qualitative and quantitative analysis of raw, in-process, and finished products in continuous pharmaceutical manufacturing settings are quite critical in the success of PAT program initiated by US FDA in 2004 [39]. CD-Raman spectroscopy speeds up the collection of Raman data, which makes it attractive as a PAT tool for real-time measurement applications in continuous manufacturing.

CD-Raman systems can reproduce the functionality of conventional array-based Raman spectroscopy to collect full spectral information by raster-scanning each array column. However, it is important to emphasize that the full speed advantage is only realized when it is used in a compressive detection mode with filter functions. Furthermore, compressive Raman detection is most advantageous when it is used in low signal regime or high-speed conditions. A CCD cannot acquire at high speeds, but a single channel detector such as PMT can. Full spectral measurements may be preferred under high-signal conditions where dark and read-noise do not affect SNR of spectra or in cases where the number of filter functions may be too many to be practical. With CCD measurements, there is no loss of data due to information compression and full spectral data can be further investigated in the future. However, in low photon budget OBCD is advantageous because multiplexing on a single channel detector increases the SNR dramatically.

Funding: The work was supported in part by the Office of Naval Research (Contract N0001413-039).

Conflicts of Interest: The authors declare no conflict of interest.

References

1. Wilcox, D.S.; Buzzard, G.T.; Lucier, B.J.; Wang, P.; Ben-Amotz, D. Photon level chemical classification using digital compressive detection. *Anal. Chim. Acta* **2012**, *755*, 17–27. [CrossRef] [PubMed]
2. Davis, B.M.; Hemphill, A.J.; Cebeci Maltas, D.; Zipper, M.A.; Wang, P.; Ben-Amotz, D. Multivariate Hyperspectral Raman Imaging Using Compressive Detection. *Anal. Chem.* **2011**, *83*, 5086–5092. [CrossRef]
3. Quyen, N.T.; Da Silva, E.; Dao, N.Q.; Jouan, M.D. New Raman Spectrometer Using a Digital Micromirror Device and a Photomultiplier Tube Detector for Rapid On-Line Industrial Analysis. Part I: Description of the Prototype and Preliminary Results. *Appl. Spectrosc.* **2008**, *62*, 273–278. [CrossRef] [PubMed]
4. Smith, Z.J.; Strombom, S.; Wachsmann-Hogiu, S. Multivariate optical computing using a digital micromirror device for fluorescence and Raman spectroscopy. *Opt. Express* **2011**, *19*, 16950–16962. [CrossRef] [PubMed]
5. Nelson, M.P.; Aust, J.F.; Dobrowolski, J.A.; Verly, P.G.; Myrick, M.L. Multivariate Optical Computation for Predictive Spectroscopy. *Anal. Chem.* **1998**, *70*, 73–82. [CrossRef] [PubMed]
6. Uzunbajakava, N.; de Peinder, P.; Hooft, G.; van Gogh, A. Low-cost spectroscopy with a variable multivariate optical element. *Anal. Chem.* **2006**, *78*, 7302–7308. [CrossRef] [PubMed]
7. Rehrauer, O.G.; Dinh, V.C.; Mankani, B.R.; Buzzard, G.T.; Lucier, B.J.; Ben-Amotz, D. Binary Complementary Filters for Compressive Raman Spectroscopy. *Appl. Spectrosc.* **2018**, *72*, 69–78. [CrossRef]
8. Rehrauer, O.G.; Mankani, B.R.; Buzzard, G.T.; Lucier, B.J.; Ben-Amotz, D. Fluorescence modeling for optimized-binary compressive detection Raman spectroscopy. *Opt. Express* **2015**, *23*, 23935–23951. [CrossRef]
9. Corden, C.J.; Shipp, D.W.; Matousek, P.; Notingher, I. Fast Raman spectral mapping of highly fluorescing samples by time-gated spectral multiplexed detection. *Opt. Lett.* **2018**, *43*, 5733–5736. [CrossRef]
10. Sturm, B.; Soldevila, F.; Tajahuerce, E.; Gigan, S.; Rigneault, H.; De Aguiar, H.B. High-sensitivity high-speed compressive spectrometer for Raman imaging. *arXiv*, 2018; arXiv:1811.06954.

11. Zhang, L.; Henson, M.J.; Sekulic, S.S. Multivariate data analysis for Raman imaging of a model pharmaceutical tablet. *Anal. Chim. Acta* **2005**, *545*, 262–278. [CrossRef]
12. Rajalahti, T.; Kvalheim, O.M. Multivariate data analysis in pharmaceutics: A tutorial review. *Int. J. Pharm.* **2011**, *417*, 280–290. [CrossRef] [PubMed]
13. Sasic, S.; Clark, D.A.; Mitchell, J.C.; Snowden, M.J. A comparison of Raman chemical images produced by univariate and multivariate data processing-a simulation with an example from pharmaceutical practice. *Analyst* **2004**, *129*, 1001–1007. [CrossRef]
14. Duarte, M.F.; Davenport, M.A.; Takhar, D.; Laska, J.N.; Sun, T.; Kelly, K.F.; Baraniuk, R.G. Single-pixel imaging via compressive sampling. *IEEE Signal Process. Mag.* **2008**, *25*, 83–91. [CrossRef]
15. Scotté, C.; de Aguiar, H.B.; Marguet, D.; Green, E.M.; Bouzy, P.; Vergnole, S.; Winlove, C.P.; Stone, N.; Rigneault, H. Assessment of Compressive Raman versus Hyperspectral Raman for Microcalcification Chemical Imaging. *Anal. Chem.* **2018**. [CrossRef] [PubMed]
16. Mankani, B.R. *Advances in Raman Hyperspectral Compressive Detection Instrumentation for Fast Label Free Classificiation, Quantification and Imaging*; Purdue University: West Lafayette, IN, USA, 2016.
17. Vornehm, J.E.; Dong, A.J.; Boyd, R.W.; Shi, Z. Multiple-output multivariate optical computing for spectrum recognition. *Opt. Express* **2014**, *22*, 25005–25014. [CrossRef] [PubMed]
18. Armitage, D.; Thackara, J.I.; Clark, N.A.; Handschy, M.A. Ferroelectric Liquid Crystal Spatial Light Modulator. *Mol. Cryst. Liq. Cryst.* **1987**, *144*, 309–316. [CrossRef]
19. Harriman, J.; Serati, S.; Stockley, J. Comparison of transmissive and reflective spatial light modulators for optical manipulation applications. In Proceedings of the Optics and Photonics, San Diego, CA, USA, 26 August 2005; p. 10.
20. Van Beek, M.C.; Schuurmans, F.J.P.; Bakker, L.P. *Optical Analysis System Using Multivariate Optical Elements*; Koninklijke Philips Electronics N.V.: Eindhoven, The Netherlands, 2010.
21. Mehmood, T.; Ahmed, B. The diversity in the applications of partial least squares: An overview. *J. Chemom.* **2016**, *30*, 4–17. [CrossRef]
22. Geladi, P.; Kowalski, B.R. Partial least-squares regression: A tutorial. *Anal. Chim. Acta* **1986**, *185*, 1–17. [CrossRef]
23. Cebeci-Maltaş, D.; McCann, R.; Wang, P.; Pinal, R.; Romañach, R.; Ben-Amotz, D. Pharmaceutical Application of Fast Raman Hyperspectral Imaging with Compressive Detection Strategy. *J. Pharm. Innov.* **2013**. [CrossRef]
24. Cebeci Maltaş, D.; Kwok, K.; Wang, P.; Taylor, L.S.; Ben-Amotz, D. Rapid classification of pharmaceutical ingredients with Raman spectroscopy using compressive detection strategy with PLS-DA multivariate filters. *J. Pharm. Biomed. Anal.* **2013**, *80*, 63–68. [CrossRef]
25. Brereton, R.G. *Applied Chemometrics for Scientists*; John Wiley and Sons: West Sussex, UK, 2007.
26. Ruckebusch, C.; Blanchet, L. Multivariate curve resolution: A review of advanced and tailored applications and challenges. *Anal. Chim. Acta* **2013**. [CrossRef] [PubMed]
27. Treado, P.J.; Morris, M.D. A thousand points of light: The Hadamard transform in chemical analysis and instrumentation. *Anal. Chem.* **1989**, *61*, 723A–734A. [CrossRef] [PubMed]
28. Corcoran, T.C. Compressive Detection of Highly Overlapped Spectra Using Walsh–Hadamard-Based Filter Functions. *Appl. Spectrosc.* **2018**, *72*, 392–403. [CrossRef] [PubMed]
29. ApplicationNote. *DMD 101: Introduction to Digital Micromirror Device (DMD) Technology*; Texas Instruments: Dallas, TX, USA, 2008.
30. Nelson, P. *Texas Instruments DLP® Technology for Spectroscopy*; Texas Instruments: Dallas, TX, USA, 2014.
31. Dudley, D.; Duncan, W.M.; Slaughter, J. Emerging digital micromirror device (DMD) applications. In Proceedings of the Micromachining and Microfabrication, San Jose, CA, USA, 20 January 2003; p. 12.
32. Hornbeck, L.J. Digital Light Processing for high-brightness high-resolution applications. In Proceedings of the Electronic Imaging, San Jose, CA, USA, 8 May 1997; p. 14.
33. Wagner, E.P.; Smith, B.W.; Madden, S.; Winefordner, J.D.; Mignardi, M. Construction and Evaluation of a Visible Spectrometer Using Digital Micromirror Spatial Light Modulation. *Appl. Spectrosc.* **1995**, *49*, 1715–1719. [CrossRef]
34. Buzzard, G.T.; Lucier, B.J. Optimal filters for high-speed compressive detection in spectroscopy. In Proceedings of the IS&T/SPIE Electronic Imaging, Burlingame, CA, USA, 14 February 2013; p. 10.
35. Réfrégier, P.; Scotté, C.; de Aguiar, H.B.; Rigneault, H.; Galland, F. Precision of proportion estimation with binary compressed Raman spectrum. *J. Opt. Soc. Am. A* **2018**, *35*, 125–134. [CrossRef] [PubMed]

36. Ben-Amotz, D.; Lucier, B.J.; Buzzard, G.T.; Wilcox, D.S.; Wang, P.; Mankani, B.R. Optical Chemical Classification. U.S. Patent 9476824 B2, 2016.

37. Wilcox, D.S.; Buzzard, G.T.; Lucier, B.J.; Rehrauer, O.G.; Wang, P.; Ben-Amotz, D. Digital compressive chemical quantitation and hyperspectral imaging. *Analyst* **2013**, *138*, 4982–4990. [CrossRef] [PubMed]

38. Turtaev, S.; Leite, I.T.; Mitchell, K.J.; Padgett, M.J.; Phillips, D.B.; Čižmár, T. Comparison of nematic liquid-crystal and DMD based spatial light modulation in complex photonics. *Opt. Express* **2017**, *25*, 29874–29884. [CrossRef] [PubMed]

39. USFDA. *PAT-A Framework for Innovative Pharmaceutical Development, Manufacturing, and Quality Assurance*; USFDA: Silver Spring, MD, USA, 2004.

Journal of
Imaging

MDPI

Review

Deep Learning Meets Hyperspectral Image Analysis: A Multidisciplinary Review

Alberto Signoroni *, Mattia Savardi, Annalisa Baronio and Sergio Benini

Information Engineering Department, University of Brescia, I25123 Brescia, Italy; m.savardi001@unibs.it (M.S.); a.baronio005@studenti.unibs.it (A.B.); sergio.benini@unibs.it (S.B.)
* Correspondence: alberto.signoroni@unibs.it; Tel.: +39-030-371-5432

Received: 9 April 2019 ; Accepted: 2 May 2019 ; Published: 8 May 2019

Abstract: Modern hyperspectral imaging systems produce huge datasets potentially conveying a great abundance of information; such a resource, however, poses many challenges in the analysis and interpretation of these data. Deep learning approaches certainly offer a great variety of opportunities for solving classical imaging tasks and also for approaching new stimulating problems in the spatial–spectral domain. This is fundamental in the driving sector of Remote Sensing where hyperspectral technology was born and has mostly developed, but it is perhaps even more true in the multitude of current and evolving application sectors that involve these imaging technologies. The present review develops on two fronts: on the one hand, it is aimed at domain professionals who want to have an updated overview on how hyperspectral acquisition techniques can combine with deep learning architectures to solve specific tasks in different application fields. On the other hand, we want to target the machine learning and computer vision experts by giving them a picture of how deep learning technologies are applied to hyperspectral data from a multidisciplinary perspective. The presence of these two viewpoints and the inclusion of application fields other than Remote Sensing are the original contributions of this review, which also highlights some potentialities and critical issues related to the observed development trends.

Keywords: deep learning; hyperspectral imaging; neural networks; machine learning; image processing

---

1. Introduction

In the last few decades, hyperspectral imaging (HSI) has gained importance and a central role in many fields of visual data analysis. The concept of spectroscopy combined with imaging was first introduced in the late 1970s in the Remote Sensing (RS) field [1]. Since then HSI has found applications in an increasing number of fields for a variety of specific tasks, and nowadays it is also largely used, other than in RS [2], in biomedicine [3], food quality [4], agriculture [5,6] and cultural heritage [7], among others [8].

Hyperspectral images are able to convey much more spectral information than RGB or other multispectral data: each pixel is in fact a high-dimensional vector typically containing reflectance measurements from hundreds of contiguous narrow band spectral channels (full width at half maximum, FWHM between 2 and 20 nm) covering one or more relatively wide spectral intervals (typically, but not exclusively, in the 400–2500 nm wavelength range) [9]. Current HSI acquisition technologies are able to provide high spectral resolution while guaranteeing enough spatial resolution and data throughput for advanced visual data analysis [10] in a variety of quality demanding application contexts [8].

However, the great richness of HSI come with some data handling issues that, if not correctly addressed, limits its exploitation. The main problem for the computational interpretation of hyperspectral data is the well-known *curse of dimensionality*, related to the great number of channels and to the fact that data distribution becomes sparse and difficult to model as soon as the space dimensionality increases. Nevertheless, the presence of data redundancy (due to the fine spectral resolution and, in some cases, to fairly high spatial resolution) enables the adoption of dimensionality reduction strategies. Doing this while preserving the rich information content is not a simple task, since the spectral–spatial nature of the hyperspectral data is complex, as it can also be observed in terms of inter- and intra-class variability of spectral signatures arising in non-trivial classification problems.

While these difficulties inevitably have repercussions on the performance of traditional machine learning methods, which strongly depend on the quality of (hand-crafted) selected features, relevant solutions to the above issues have been appearing in recent years with the spread of representation learning approaches [11] and their implementation through Deep Learning (DL) architectures.

1.1. Hyperspectral Data Analysis Meets Deep Learning

Traditional learning-based approaches to HSI data interpretation rely on the extraction of hand-crafted features on which to hinge a classifier. Starting early on with simple and interpretable low-level features followed by a linear classifier, subsequently both the feature set and the classifiers started becoming more complex. This is the case, for instance, of Scale-Invariant Feature Transform (SIFT) [12], Histogram of Oriented Gradients (HOG) [13] or Local Binary Patterns [14], in conjunction with kernel-based Support Vector Machines (SVM) [15], Random Forests [16] or statistical learning methods [17]. It is interesting to look at the new trend of DL as something whose clues were already embedded in the pathway of Computer Vision and Digital Signal Processing [11,18]. For example, Neural Networks (NN) can approximate what a traditional bag-of-local-features does with convolutional filters [19] very well and SVM can be seen as a single layer NN with a hinge loss. At the same time DL solutions cannot be seen as the ultimate solution for the fundamental questions Computer Vision is called to answer [20].

The advantages introduced with DL solutions lie in the automatic and hierarchical learning process from data itself (or spatial–spectral portions of it) which is able to build a model with increasingly higher semantic layers until a representation suitable to the task at hand (e.g., classification, regression, segmentation, detection, etc.) is reached. Despite these potentials, some attention is needed when DL is applied to hyperspectral data. Most importantly, given the large amount of parameters of DL models (typically of the order of tens of millions), a sufficiently large dataset is needed to avoid overfitting. Hereinafter, large datasets are meant to be composed of hundreds of thousands examples (where a typical example can consist of a spectral signature associated to a pixel or to a small area or a HSI sub-volume). Conversely, a dataset composed of hundreds of examples can be considered small. The very limited availability, where not complete lacking, of public (labeled) datasets is the most evident shortcoming in the current "DL meets HSI" scenario. Due to the curse of dimensionality, the effects of the shortage of labeled training data is amplified by the high data dimensionality and may lead to effects spanning from the so-called Hughes phenomena (classification performance sub-optimalities) to the models' complete inability to generalize (severe overfitting). Other pitfalls hidden behind limited availability of data for research purposes are limitations in terms of breadth of the studied solutions that may be limited to the scope of the dataset itself. This also leads to the necessity to work with unsupervised algorithms to partially overcome the lack of labeled data. Data augmentation techniques (such as in [21,22]) in conjunction with the use of some specific DL architectures (such as Convolutional Neural Networks and Autoencoders) also play an important role in handling the above data-driven issues.

1.2. Purpose and Relations with Other Surveys

The purpose of this survey is to give an overview of the application of DL in the context of hyperspectral data processing and to describe the state-of-the-art in this context. While this review is not meant to gain further insight into technical aspects of specific application fields and instrumentation, its objective is to be at the intersection of these two important trends: DL, driver of disruptive innovation, especially in computer vision and natural language processing, and exploitation of HSI technologies and data analysis, which is expected to have a high growth even beyond the RS field. This two trends meet up in a field where data is at the same time a challenge (for its dimensionality) and a precious resource (for its informative wealth).

Highly informative reviews about DL methods in the RS field have been produced [23–25] where there are several references or sections dedicated to HSI data. Conversely, recent work dedicated to reviewing HSI data analysis comprises DL methods [10,26–29] but their scope is strictly limited to the RS field. With the present work we want to provide an overview of the main principles and advances related to the use of DL in HSI, not only in RS (from airborne or spaceborne platform), but also in other relevant small-scale (from micro to macro ground based acquisitions) applications of HSI data, where DL is already finding fertile terrain for its exploitation. The aim is to define a complete framework to which even non-RS professionals can refer. With this aim in mind, this review has been conceived (and schematized in Figure 1) to be accessible to different categories of readers while maintaining a single and coherent logical flow.

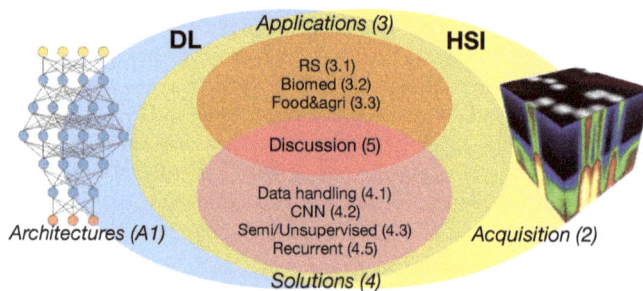

Figure 1. Graphical structure of the article.

In order to create the context for what follows, in Section 2 we provide a concise overview about the main ways to acquire HSI datasets. This also gives the opportunity to evidence the possibility of exploiting DL solutions in the creation of HSI data from undersampled spectral representations. In Section 3, we adopt the point of view of "what" has been done until now by using DL approaches on HSI data in different application fields. This part is meant to be more accessible to domain expert readers. On the other hand, Machine learning and Computer Vision experts could be more interested in Section 4, which aims to review "how" different DL architectures and their configurations are used on HSI data for different analysis and interpretation tasks. With the final discussion in Section 5, we also want to draw conclusive remarks aimed at pointing out some residual issues and trying to envisage the future developments and challenges to address from the joint exploitation of HSI and DL technologies. Finally, a basic introduction to DL architectures, in particular those mentioned in this work, is provided in Appendix A in order to give additional context and references, especially to domain expert readers.

2. HSI Acquisition Systems

In this section we give a concise review of the most diffused approaches that can be exploited for the formation of HSI datasets. Interestingly, we also include a review of recent DL-based solutions conceived for the production of HSI volumes starting from RGB or other sparse spectral representations.

2.1. HSI Formation Methods

Hyperspectral imaging (HSI) refers to imaging methods also able to acquire, other than 2D spatial information xy, a densely sampled spectral information λ. The prefix *hyper* is used when the acquired contiguous spectral bands are of the order of 10^2 to 10^3, as opposed to Multispectral imaging (MSI) aimed at the acquisition of order of dozens of bands (with typical FWHM of 100–200 nm), not necessarily contiguous/isometric. Thus, HSI makes it possible to finely capture absorption features, facilitating the identification of the presence of specific substances; while with MSI (and even worse with RGB imaging) physico-chemical absorption features are spread over the channel bandwidth and become much less detectable. Available HSI devices are able to acquire the 3D $xy\lambda$ volumes by means of 2D sensors ij by converting in time, or arranging in space, the spectral dimension. There are various ways to acquire HSI volumes in practice. Here we review the main and most widespread, each one involving physical limitations requiring a balance between key parameters, such as spectral and spatial resolution, acquisition time (or temporal resolution), device compactness, computational complexity among the main ones.

Relative motion between the HSI sensor and the sample are exploited in whiskbroom (area raster scan) and pushbroom (linear) scanners to respectively acquire the spectrum λ of a single point $x_i y_j$ (at time t_{ij}) or of a line xy_j (at time t_j) of the sample. This is typically done by means of a prism or a diffraction grating able to disperse the incoming light. For whiskbroom mode, the temporal resolution is highly penalized especially if one wants to obtain decent spatial resolution and this prevents, in most cases, the practical use of point-wise spectrography for HSI production. In Figure 2a a pushbroom acquisition is depicted which is far more interesting and widespread since high spatial and spectral resolution can be obtained at the cost of the time needed for the linear scanning over the sample. Commercial pushbroom HSI cameras are currently able to offer easy balancing between frame-rate and spectral resolution (See, for example http://www.specim.fi/fx/ (last visit March 2019)).

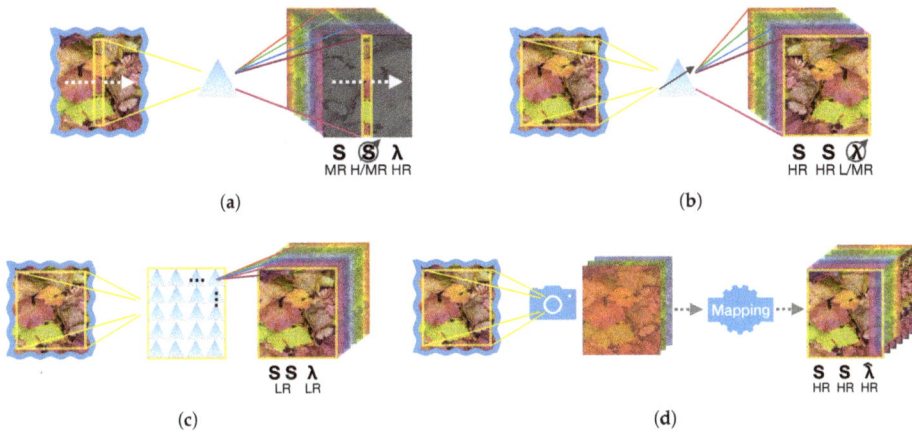

Figure 2. Basic schemes of HSI formation methods. H/M/LR: High/Medium/low Resolution. S: space, either x or y. λ: spectral dimension. (**a**) Pushbroom linear scanner. (**b**) Spectral selective acquisition. (**c**) Spectrally resolved detector array (snapshot). (**d**) HSI from RGB images.

Selective spectral acquisition in time is at the basis of another acquisition mode that requires the incoming images to be filtered to produce a $xy\lambda_k$ image at time t_k (see Figure 2b). The main trade-off here is between spectral and temporal resolution, where spectral filtering can be done with mechanical filter wheels (typically limited to MSI) or by means of acusto-optical or liquid-crystal tunable filters (enabling HSI at a higher cost).

The possibility of obtaining a spectral image by just taking a snapshot is highly attractive for time-constrained applications and this has driven a lot of research [30]. In these cases, physical limitations due to the simultaneous use of spatial and spectral divisions, severely limit both resolutions. Relatively economic systems have been commercialized recently by exploiting a technology able to deposit filter mosaics directly onto the image acquisition chip (See, for example https://www.imecint.com/en/hyperspectral-imaging (last visit March 2019)). Figure 2c depicts this idea of spectrally resolved detector array, while we refer to [31] for a complete and up-to-date review.

An alternative way to rapidly obtain a HSI dataset from single shots is to derive a pixelwise estimation of $\hat{\lambda}$ by means of an inverse mapping starting from highly subsampled (snapshot) spectral measures, such as RGB images taken by commercial digital cameras. This idea, pioneered in [32,33], has attracted some research interest in the CV community especially toward systems able to simulate the production of HSI images in a very cheap and effective way starting from single RGB images (see Figure 2d). Since in many cases this involved the exploitation of Deep Learning solutions we provide a review of this domain in the next subsection.

2.2. HSI from RGB

The possibility to use deep learning approaches to generate hyperspectral images just starting from RGB images, or other sparse spectral representations, has been investigated recently [34,35] and generated a certain interest, especially in the Computer Vision community. The intent is to find alternative solutions to the cost issues and spatial resolution limitations of HSI acquisition devices, by introducing learned inverse mappings from a highly subsampled space to a dense spectral representation.

Different DL solutions (CNN [36,37], 3D CNN [38], Dense and Residual Networks [39], Dirichlet networks [40], Generative Adversarial Networks [41]) have been proposed to improve the mapping and the spectral reconstruction by leveraging spatial context. Following results in [42], which show a non negligible dependency of the spectral reconstruction quality to the colour spectral sensitivity (CSS) functions of the camera, some approaches include the CSS functions to either jointly learn optimal CSS and spectral recovery maps [43], or to produce CSS estimates directly from the RGB images in unknown settings, to better condition the spectral reconstruction [44], or even to learn an optimal filter to construct an optimized multispectral camera for snapshot HSI [45]. A review of recent initiatives in this field can be also found in the report of the first challenge on spectral reconstruction from single RGB images (NITRE 2018 workshop [46]). In a recent work, exploiting properties of computational snapshot multispectral cameras [47], Wang et al. [48] proposed a DL-based HSI volume reconstruction from single 2D compressive images by jointly optimizing the coded aperture pattern and the reconstruction method.

Of course, while these approaches produce interesting results for some applications, their validity is actually limited to the visible spectrum. In fact, to our knowledge no DL-based MSI-to-HSI spectral upsampling has been proposed in the NIR-SWIR spectrum (750–3000 nm) where, because of technological reasons related to currently available detectors, both cost-based and spatial-resolution conditions change and do not lead to the same convenience considerations.

3. HSI Applications Meet DL Solutions

In this section we present an overview of DL applications to HSI data subdivided into the main working fields. There is still an imbalance between the number of RS related papers with respect to other application fields. This is due to many factors, including the origins of the development of HSI technologies, the dimension of the RS research community, and the existence of specialized venues. Despite the greater variety and average maturity of works related to RS, in our multidisciplinary review we try to give the greatest value even to exploratory works in other fields being aware that, as it frequently happens, some works done in one domain may inspire other works in another sector.

3.1. Remote Sensing

The main purposes of HSI data analysis for RS focus on image processing (comprising calibration and radiometric corrections), feature extraction, classification, target recognition and scene understanding. All these steps are a breeding ground for the exploitation of DL approaches, especially for the potential advantages they bring in terms of data management and feature extraction with

J. Imaging **2019**, *5*, 52

a consequent performance boost. Past and future missions (for an updated overview see [49] (Ch. 1)) will feed application professionals with an increasing number of HSI data and big interpretation challenges to address (starting from proper handling of the volume of generated data). Conversely, most of the technological breakthroughs coming from representation learning studies and DL architectures have been quite rapidly tested in RS applications, and RS-related HSI does not represent an exception to this.

3.1.1. Classification

Many DL approaches in the literature include *classification* as a final goal, while land cover classification is one of the main task in RS. The main classes are related to crops (corn, grass, soybean, ...) or urban areas (asphalt, trees, bricks, ...) and, according to available labels in the benchmark datasets, a combination of those classes is considered in the majority of RS-HSI classification works that exploit DL methods.

DL classification architectures have *feature extraction* capability by design. Conversely, classical techniques consider classification on top of a separate hand-crafted feature extraction and remains critical for the representativeness and robustness of the selected features with respect to the task at hand. HSI-DL classification and feature extraction solutions have been recently explored using very different approaches in terms of feature extraction and exploitation. HSI data offer different opportunities to approach the analysis using a pure spectral or a joint spectral–spatial approach. In this section, few works are usually selected as representative of the main paradigms, while in Section 4 many other works are considered according to technological and methodological criteria.Pixel classification can be based on the exploitation of the spectral features thanks to their richness and abundance. Representative studies adopting this approach are [50–53]. Another kind of classification is based on spatial features, since RS data have a contiguity in space so that classification can exploit the similarities and patterns of neighbouring pixels as in [54,55]. Moreover, jointly considering spectral and spatial features has been proven to enhance the classification, as described for example in [56–59]. Moreover, the introduction of multiscale spatial features could improve the performance slightly more as demonstrated in [60–62]. Yang et al. in [63] tested four DL models ranging from 2D-CNN up to a 3D recurrent CNN model, producing a near-perfect classification result.

Labeled and publicly available HSI datasets (for training and benchmarking) are very few and also quite outdated. The ones considered in the majority of RS land cover classification works are Salinas, Pavia, Indian Pines, and Kennedy Space Center (Information about these datasets can be found at http://www.ehu.eus/ccwintco/index.php/Hyperspectral_Remote_Sensing_Scenes (last visit March 2019)). Moreover, this problem is exacerbated by the current practice in the remote sensing community which carries out training and testing on the same image due to limited available datasets, possibly introducing a bias in the evaluation. Therefore, when this practice is used, this makes fair comparison difficult, since improved accuracy does not always necessarily mean a better approach. As a side effect, this soon leads to accuracy performance that has already compressed and tending to an asymptotic optimal value, and what can generate confusion is that this has happened with very different DL approaches in terms, for example, of number of levels, weights and hyper-parameters to learn. Therefore, even if benchmarking is always valuable, near-perfect results (even obtained taking care of overfitting issues) should not be interpreted as if all land cover classification issues can be considered solved. To reduce the bias deriving from indirect influence of training data on test data when they are taken from the same image (even when random sampling is adopted), a spatially constrained random sampling strategy has been proposed in [64], which can be used in case of limited available labeled HSI volumes.

3.1.2. Segmentation

DL approaches have also been used in RS-HSI for *segmentation* purposes. Hypercube segmentation can be exploited in several ways, and it is useful to better handle a subsequent image classification in several situations. In [65], Alam et al. presented a technique that operates on a superpixel partitioning based on both spectral and spatial properties; in [66] the segmentation of the image was used as a preliminary step to focus the subsequent classification on meaningful and well circumscribed regions.

3.1.3. Target Detection and Anomaly Detection

In RS *target detection and recognition* is receiving increasing interest. In [67,68], parallelized and multiscale approaches were respectively proposed for vehicle detection from satellite images. In [69] Zhang et al. described an oil tank detection system, while in [70] a building detection method was presented.

Target detection could be treated in an unsupervised way as well. In this case, it can be seen, depending on the objective, as *anomaly detection* and usually, it does not need prior information about target objects. These approaches are especially useful, for instance, in the case of forest fire, oil spills in the sea or more in general to detect targets with low probabilities or significant changes that have occurred with respect to a previous acquisition in a certain image scene. Elective areas of application for these methods include, for example, disaster monitoring and defense applications, as well as food processing and various manufacturing related quality controls. Approaches to anomaly detection were found in [71] taking advantage of stacked autoencoders and in [72] where Deep Belief Networks were employed. In [72,73] two different approaches to perform real-time and classical anomaly detection were proposed. Similar to them, in [74], a method exploiting change detection was described. In [75] instead, a DL solution based on Deep Belief Networks and a wavelet texture extraction technology outperformed many baseline models on two HSI datasets.

3.1.4. Data Enhancement: Denoising, Spatial Super-Resolution and Fusion

The physical limitations that characterize the HSI acquisition phase (see Section 2) can relate to issues affecting the quality of the acquired data. This can be partially addressed with data enhancement solutions aimed to increase the practical value or the possibility to exploit the data. A recent example of DL-based solutions in this field is described for *restoration and denoising* in [76], where authors use encoding-decoding architectures as intrinsic image priors to effectively acting as an HSI restoration algorithm with no training needed. With this set-up, they demonstrated the superior capability of 2D priors compared to 3D-convolutional ones, outperforming single-image algorithms and obtaining performance comparable to trained CNNs. A denoising technique powered by CNN is also presented in [77] and related advancements [78,79], where improved noise removal has been obtained with concurrent spectral profile preservation and reduced computational time.

Another popular enhancement task for HSI is (spatial) *super-resolution*. This is aimed to overcome resolution limitations so that, starting from a lower resolved HSI data, high resolution hyperspectral images are produced by exploiting high spatial resolution information coming from another imaging source. This is similar to what happens with *pan-sharpening* [80] where panchromatic images are used to enhance the spatial resolution of satellite MSI data (DL methods have also been applied in this field [81,82]). In general HSI super-resolution comes from the exploitation of RGB or other high-spatial low-spectral images at least in a training phase. To this end, in [83], a simple transfer-learning approach was applied, while in [76,84,85] complete end-to-end architectures were presented. In [86] an end-to-end approach based on 3D convolutions was suggested instead. Within the scope of this work the term end-to-end refers to network architectures that take the HSI volume as input and produce the target data without using separate pre- or post- processing stages. Other approaches are composed of multiple stages in which CNNs are applied extensively as in [87,88] or, more interestingly, without requiring auxiliary images, as in [89].

In certain applications the information provided by HSI alone is not sufficient or, in general, the presence of different and complementary data sources can be exploited to improve results or to enable the accomplishment of specific tasks. This is the case in multi-branch DL solutions conceived to enable *data fusion*, especially involving Lidar and HSI images as in [90–94]. Similarly, in [95] data fusion was carried out on three different data sources, with the inclusion of RGB images as well.

3.2. Biomedical Applications

The synergy between HSI and DL can also be exploited in the biomedical sector. For example, the possibility to extract and analyze spectral signatures, spatial maps and joined spatial–spectral representations from specimens in a wide variety of specific application fields (e.g., clinical microbiology, histopathology, dermatology, to name a few) allows the development of (supportive) diagnostic tools in either invasive or non-invasive (or reduced invasiveness) settings. Likewise for RS, where the cover-type classification task is the prominent application, classification operated on the surface of different kinds of specimens, acquired through HSI systems at various scales (from micro to macro), is gaining high interest [3]. Concurrently, the adoption of DL solutions is rapidly becoming the first choice when approaching the majority of medical image analysis tasks [96]. However, despite the high potential, the number of studies able to fully take advantage of both HSI and DL technologies is still relatively low. This may be due to the fact that HSI acquisitions in many biomedical fields are still experimental and unconventional, other than leading to a high amount of data that may be difficult to handle. There are also cost factors and other experimental challenges in terms of infrastructure and experimental setup that, despite the conceptual non-invasiveness of HSI acquisitions, still interfere with a wider usage of HSI systems. However, the interest in HSI and modern DL-based handling of the produced data can grow towards well integrated, safe and effective investigation procedures, and the emerging studies we examine below are proof of this.

3.2.1. Tissue Imaging

The discrimination between normal and abnormal conditions was pursued in an exploratory study [97] to assess the presence of corneal epithelium diseases by means of CNN. In [98,99] different 2D-CNN solutions were considered to classify head and neck cancer from surgical resections and animal models, respectively. Other studies further investigated the possibility of delineating tumor margins on excised tissues [58] and to demonstrate a richer "optical biopsy" classification of normal tissue areas into sub-categories like epithelium, muscle, mucosa [100], also by means of deeper CNN architectures and full spatial–spectral patches. In an interesting study, where a dual-mode endoscopic probe was developed for both 3D reconstruction and hyperspectral acquisitions [101], a CNN-based system was proposed to obtain super-resolved HSI data from dense RGB images and sparse HSI snapshot acquisitions. The latter were obtained by exploiting linear unbundling of a circular optical fiber bundle.

3.2.2. Histology

The task of cell classification is another conceptually similar discrimination that was explored in [102,103] to recognize white blood cells in microscopy images, where different bands were acquired by exploiting Liquid Crystal Tunable Filters (LCTFs). Conversely, in [104], an two-channel global-local feature end-to-end architecture was proposed for blood cell segmentation and classification. Increased spectral information at pixel level can also be exploited as a sample-preserving alternative to invasive chemical procedures, such as in [105], where a virtual staining network was tested to possibly avoid chemical staining of histopathological samples.

3.2.3. Digital Microbiology

In the field of clinical microbiology, multi-class classifications, based on CNN and softmax output, were used for the recognition of bacteria species over VNIR (visible near-infrared, 400–1400 nm) HSI acquisitions of bacteria culture plates where spectral signatures was extracted from single bacterial colonies [106,107]. Interestingly, the exploitation of spectral signatures at a colony level can be seen as an alternative to another form of chemical staining taking place when so called chromogenic culturing plates (filled with agar media enriched with species-selective pigmentation agents) are used to introduce some colour differentiation among bacteria species. This is also significant in recent years since clinical microbiology laboratories are interested by an epochal change in terms of automation and digitization of the whole culturing processes [108]. As a side issue of possible massive usage of HSI data one should consider data conservation needs, typically arising in biomedical domains, which can lead to data handling (storage and transfer) problems especially for high spatial–spectral resolution HSI volumes, each one typically occupying hundreds of MB in raw format. Therefore studying adequate compression techniques and strategies capable of guaranteeing the preservation of the classification and discrimination performance is of high interest, especially in contexts characterized by a high data throughput, such as digital microbiology, where bacteria culturing is massively performed for routine exams and a great volume of digital data is created on a daily basis [109].

3.2.4. Vibrational Spectroscopic Imaging

Despite our focus on HSI, it is worth observing that, especially in the biomedical field, vibrational spectral imaging techniques [110,111] have also recently started to benefit from the possibility offered by representation learning approaches to directly analyze raw spectra (avoiding pre-processing and/or manual-tuning), even improving performance with respect to more classical machine learning solutions [112]. In [113], automatic differentiation of normal and cancerous lung tissues was obtained by a deep CNN model operating on coherent anti-Stokes Raman scattering (CARS) images [114]. In the context of histological applications of Fourier Transform Infrared (FTIR) spectroscopic imaging [115], CNN-based approaches have been introduced to leverage both spatial and spectral information for the classification of cellular constituents [116] and to accomplish cellular-level digital staining to the micrometer scale [117].

3.3. Food and Agriculture

HSI techniques are widely recognized for their added value in the agricultural field for a variety of monitoring, modeling, quantification and analysis tasks [6], while in the food industry sector, noninvasive and nondestructive food quality testing can be carried out on the production and distribution chain by means of HSI-based inspection [118]. Examples of HSI-DL techniques were used to assess the freshness of shrimps [119,120] and to prevent meat adulteration [121]. In agriculture either pre- or post-harvesting controls can be conducted. In the first case nutrient inspection [122] or early pathogenic diagnosis [123] were tested, while the possibility of post-harvesting controls were investigated with the assessment of fruit ripening indicators [124], to help segregate damaged fruits [125] and to detect the presence of plant diseases [126]. The main rationale of adopting DL-based data analysis and interpretation combined with HSI is the need to fully exploit the richness of spectral (frequently linked to chemometric principles in the NIR range) and spatial (usually related to the complexity and non-uniformity of the samples) information, contrasting the complexity of hand-crafted feature extraction by relying on representation learning and DL abstraction hierarchies. Additional complexity can also derive from environmental variables that interfere in case of acquisition in the open field, as in [123]. Discrimination among different (plant) species is another salient application that was trialled in the case of cereal [127] or flower [128] varieties.

3.4. Other Applications

HSI-DL works in other application fields are still very rare. The authors of a recent review of HSI applications [8] proposed a solution for ink analysis based on CNN for automated forgery detection [129] in hyperspectral document analysis [130]. Interesting developments can be expected within the scope of historical and artistic document analysis (manuscripts, paintings, archaeological artifacts and sites), forensic analysis, anti-counterfeiting and authentication domains, surveillance and homeland security, to name a few.

4. Deep Learning Approaches to HSI

In recent years, a variety of DL approaches and architectures have been proposed to address the HSI analysis task described in the previous section. We will mainly focus on Convolutional Neural Networks (CNN) in different configurations (spectral, spatial, spectral–spatial) which have primarily been employed with the aim of feature extraction and classification. In doing so, we will introduce various methods, from classical networks to the integration with multiscale and fusion strategies, as in [131]. Other significant architectures we consider are Autoencoders, Deep Belief Networks, Generative Adversarial Networks and Recurrent Neural networks (all concisely revised in Appendix A). These architectures are flexible and adaptable to different data analysis tasks and suit HSI analysis as well. Dataset augmentation, post-processing solutions and an overview about new directions in HSI data handling conclude this section.

4.1. Data Handling

Hyperspectral data can be treated according to different spatial–spectral viewpoints. Most of the early DL methods only exploit data pixel-wise (1-dimensional approaches), working in the spectral direction. This can be done by extracting spectral signatures from single pixels or from groups of them either surrounding a central pixel or belonging to an object area. The latter approach generally needs some a-priori knowledge and a pre-processing phase to detect the object of interest (by segmentation). In [107] a spectral *cosine distance transform* is exploited to identify and weight pixels belonging to objects of interest in a biomedical application.

Dimensionality reduction is used to tackle the spectral information redundancy. Of the different dimensionality reduction techniques, PCA is still a classic way to proceed. Depending on the context, other approaches can be used as well, such as ICA [132] and stacked autoencoders [66].

Otherwise, a 2-dimensional process can be applied. In this case a preliminary dimensionality reduction is usually carried out as well. Spatial processing is exploited to extract spatial features from the whole bands or on 2D patches.

Finally, HSI data can be handled as a whole with the aim of extracting both spatial and spectral features (3-dimensional). Some of these approaches still use a pre-processing stage to condition the data, but often the final goal is to work directly on the "raw" hypercubes. Since this can be a computationally expensive and complex way to proceed, operating on 3D patches (i.e., sub-volumes) is often a preferred method.

4.2. Convolutional Neural Networks

Nowadays CNNs are the most popular DL approach in computer vision, thanks to their ability to include additional meaningful restriction in the learning process, like space-invariant features and robustness to slight rotation and deformation. They can also work with a limited training dataset thanks to new and powerful regularization techniques, which are one of the most important characteristics behind their success. In the following subsections we first consider CNNs when they are mainly used as feature extractors (Section 4.2.1). We then map the remaining CNN-based approaches according to whether they work with only one (spectral or spatial) data characteristic (Section 4.2.2) or if they jointly exploit the spectral–spatial nature of HSI data (Section 4.2.3). Where not otherwise specified, classification objectives are related to pixel labeling according to the land cover classes defined in the benchmark datasets (see Section 3.1.1). In Table 1 the HSI-DL papers reviewed in the current section are subdivided into their application domain categories.

Table 1. HSI-DL studies exploiting CNNs represented by target use (columns) and field—task (raws).

	Feature Extractor	Spectral or Spatial	Spectral–spatial
RS–Classification	[68,133–138]	[50,54,61,139–141]	[57,62,142–164]
RS–Data fusion	[90–92,94,95]		
RS–Detection	[67]		
RS–Image processing		[55,79]	
Biomedical	[97]	[102,103,107]	[58,100,113]
Food-agriculture		[123,127,128]	[121,126]

4.2.1. Cnn as a Feature Extractor

CNNs have often been combined with classical ML methods, especially SVM. In this setup a CNN is used as a way to dynamically learn a feature extractor from data. This approach has the advantage of exploiting the ability to automatically retrieve a good feature set, from the CNN side, and the robustness to overfitting even on small datasets, from the classical machine learning side. In [136] Leng et al. described a hybrid CNN-SVM for hyperspectral land-cover classification, in which a target pixel and the spectral information of its neighbours are organized into a spectral–spatial multi-feature cube without extra modification of the CNN. In [97] a CNN was combined with SVM to perform binary classification (injured vs healthy) on a small ophthalmic dataset. In [67,68], the introduction of a multiscale approach has proved to be important for the extraction of robust features.

More complex architectures were proposed to jointly handle the space and spectral dimensions in order to produce a more complete feature representation. For instance, in [138] a two-channel deep CNN was used to produce spectral–spatial features from hyperspectral images for land cover classification. Wei et al. [137] proposed a hierarchical framework called spectral–spatial Response that jointly learns spectral and spatial features directly from the images.

In order to perform a robust feature extraction which squeezes all information within HSI data, many methods proposed to optimize and join spatial and spectral features in a single setup. The *fusion* may also involve features extracted from multiple sources and at different levels to make full use of HSI and, for instance, Lidar images as in [91,92,94]. Similarly, in [90] Chen et al. proposed a method in which spatial and spectral features are extracted through CNNs from HSI and Lidar images respectively, and then are fused together by means of a fully connected network. Instead, Xu et al. [95] presented a pixel-wise classification method based on a simple two-channel CNN and multi-source feature extraction. In particular, a 2-D CNN is used to focus on spatial feature extraction and a 1-D CNN is used for spectral features. Eventually, a cascade network is used to combine features at different levels from different sources (HSI, Lidar, RGB). In [134] a two-stream CNN was trained with two separate streams that process the PolSAR and hyperspectral data in parallel before fusing them in a final convolutional layer for land cover classification. A recent effort in this field has been made in [135], in which Jiao et al. proposed a framework for hyperspectral image classification that uses a fully-convolutional network based on VGG-16 to predict spatial features starting from multiscale local information and to fuse them with spectral features through a weighted method. Classification is then carried out with a classical method (SVM). A similar approach was taken in [133] with the addition of a new objective function that explicitly embeds a regularization term into SVM training.

4.2.2. Spectral or Spatial Approaches

Supervised 1D-CNN working at pixel level was proposed in different domains [50,123,128,139] to directly exploit the information relative to each spectral signature. This usually leads to better results with respect to classical ML approaches. For instance in [140], authors proposed an ad-hoc model, carefully tuned to avoid overfitting, providing better results with respect to a comprehensive set of shallow methods. However, especially in the RS domain, performance of pixel-wise methods can be affected by noise [50]. To cope with noise, averaged spectra can be extracted by a group of pixels belonging to an object of interest. This approach is particularly suitable in small-scale domains as in the case of segmented rice seeds [127]. In [107], a similar approach was used in a biomedical scenario, where signatures were obtained by a cosine distance weighted average of pixels belonging to segmented bacterial colonies.

Principal Component Analysis (PCA) is a technique highly exploited in RS to handle data dimensionality and it is used to pre-process data in many DL pipelines as well. In [102], CNN classification of pixel patches obtained after PCA reduction was proposed for cell classification. PCA was used also in [103] to pre-process medical HSI data and improved performance was obtained by the combination or modulation of CNN kernels with Gabor kernels in the preliminary network layers, as suggested in [165].

A different approach for spatial feature extraction was presented by Zhao et al. in [54], and its evolution in [61], in which a multiscale CNN was introduced to learn spatial features. With respect to other methods, data are reorganized into a pyramidal structure containing spatial information at multiple scales. In [55], a band selection method based on spatial features was proposed in order to maximize the HSI classification under the small training set constraint. Similarly, in [141], band selection was performed by means of a distance density measure. The produced spectral signature was then fed to a CNN trained on full bands, exploiting the advantage of a rectified linear unit (only activated for non-zero values), in order to test the band combinations without retraining the model.

4.2.3. Spectral–spatial Approaches

Working jointly with both spectral and spatial features generally leads to improved results. In [163], Zhang et al. described a dual-stream CNN that exploits spectral features using a method similar to [50], spatial features with the approach presented in [139], and a softmax regression classifier to combine them. A similar dual-input approach exploiting a concatenation of spectral and spatial features extracted with 1D-CNN and 3D-CNN respectively was proposed in [121], in a food adulteration detection context. A three-dimensional CNN-based approach can be exploited to extract combined features directly from the hyperspectral images to be used in classification, as done in [126] for plant disease identification. In [157], this allowed to obtain important results in the RS domain, also thanks to a combined L_2 regularization to avoid overfitting and the use of sparse constraints. A similar approach was also described in [144,147] where spectral–spatial feature extraction and consequent classification were done directly on hypercubes and without any pre-processing. The work in [146] presented a similar approach, but with a Siamese CNN [166].

In [58,100], Halicek et al. proposed an effective 3-D CNN based on AlexNet, trained with 3-D patches and an extended version with an inception block (i.e., with filters of multiple sizes operating at the same network level). While in [164], Gao et al. introduced a network with an alternate small convolution and a feature reuse module able to improve the rate of the high-dimensional features in the network, thus allowing a better extraction. In the last few years, RS-HSI research has been particularly focused on this kind of architectures. Densenet-like architectures and VGG16 were also exploited in [135,156], respectively, for classification. In [158], Liu et al. described a 3-D CNN trained via deep few-shot learning [167] to learn a metric space that causes the samples of the same class to be close to each other. This approach has proven to be effective in cases of few labeled data.

An interesting improvement to a CNN-based model was introduced by Paoletti et al. in [150], where the redundant information present in hidden layers was used in order to exploit additional connections between them in an efficient way, generally enhancing the learning process. One additional 3-D approach was proposed in [159] and recently in [160]. In the latter case a complex scheme was proposed, in which virtual sample creation and transfer-learning were adopted in order to mitigate data shortage during training.

Other examples of spatial–spectral approaches were found in [148,153], in which CNN pixel classification methods that hierarchically construct high level features were presented. Furthermore, in [145] a sparse representation method was employed to reduce the computational cost and to increase the inter-class discrimination after the feature extraction from CNN while, in [155], this step was followed by a spectral feature reduction method. In [151] an architecture that extracts band specific spectral–spatial features and performs land cover classification was presented. Yang et al. [152] used a two stream spatial–spectral network to perform transfer-learning, by fine-tuning only final layers, producing an improvement with respect to excluding the transfer-learning part. In [143] Lee et al. first tried to use a very deep CNN, proposing a Contextual Deep CNN for classification, which is able to jointly optimize the spectral and spatial information together.

A multiscale-based approach is presented in [154], in which multiscale object features, obtained from an initial SLIC (simple linear iterative clustering) superpixel segmentation [168], were combined with spectral features and used as input to a CNN for classification. Instead, in [57] authors proposed a Diverse-region-based CNN (DR-CNN), which uses a joint representation from diverse regions in the proposed CNN framework, simultaneously taking advantage of both spectral and spatial features. Furthermore, they adopted a multiscale summation module designed to combine multiple scales and different level features from unequal layers.

In [161], Ouyang et al. demonstrated that networks augmented by reconstruction pathways can bring some advantages to feature extraction and classification. The reconstruction is established by the decoding channel with reconstruction loss computation, which is then used jointly with the classification loss as the loss function for network training. Finally, the high-level features from the encoding and decoding channels are combined by a designed control gate. This is somewhat similar to what can be achieved with the deconvolutional network used in [162] aimed at recovering images starting from the intermediate features in order to improve the training.

The introduction of sensor-specific feature learning (a model is trained to learn the separability of a sensor using a specific dataset) leads to architectures able to produce good feature sets for classification purposes. In [149] Mei et al. created a sensor-specific five layer structure integrating both spatial and spectral features. Fang et al. [142] proposed a new architecture that is capable of adaptively selecting meaningful maps for classification produced by a multi-bias module that decouples input patches into multiple response maps.

Recently in [62], 1D, 2D, and 3D multiscale approaches were compared with a new multiscale-convolutional layer, demonstrating the superiority of the proposed 3D approach.

4.3. Autoencoders and Deep Belief Networks

Autoencoders (AEs) and Stacked Autoencoders (SAEs) have been widely used in hyperspectral imagery for different tasks, mainly in RS but also in food-quality applications. This is due, as in Deep Belief Networks (DBN), to the fact that they tackle the problem of small labeled datasets by attempting to exploit an unsupervised or semi-supervised approach before the desired training, thus producing a well initialized architecture that is suited to HSI tasks.

In [59] this approach was used and tested on RS-HSI for the first time by Lin et al. They proposed a framework in which PCA on spectral components is combined with SAEs on the other two dimensions to extract spectral–spatial features for classification. In line with this in [169] Chen et al. presented different architectures where spectral, spatial (flattened to 1-D vector by using PCA) or jointly driven classifications are obtained by a Logistic Regression (LR) layer operating on features computed with SAEs. Similarly, in [170,171] a SAE was used, followed respectively by a SVM and a Multi Layer Perceptron (MLP) for the classification. In the food quality domain, SAE-based approaches were used in combination with regression methods to predict and quantify the presence of chemical indicators of food freshness [119,120,122] or to assess edible quality attributes [124]. In [172], Ma et al. proposed an effective method called Contextual Deep Learning (CDL) which can extract spectral–spatial features directly from HSI. In order to exploit spectral feature extraction in [52] Karalas et al. used sparse AE composed of a single hidden layer, as well as stacked in a greedy layer-wise fashion; in [66] the same goal was reached using a segmented SAE by employing a dimensionality reduction.

An improvement to plain SAE was introduced by Ma et al. [173] in order to deal with parameter instability when a small training set was used. In particular a SAE is modified not only to minimize the classification error as usual, but also to minimize the discrepancy within each class and maximize the difference between different classes. In [174] an improved version with deep SAE was presented. Zhang et al. [71] proposed a stacked autoencoder suitable for hyperspectral anomaly detection.

Multiscale approaches were also introduced to support AE. In [54] Zhao et al. proposed a combination of AEs and LR. In particular they introduced a method that combines PCA to extract spectral features, multiscale convolutional AEs to extract high-level features and LR to classify them. In [175] a mixture between SAEs and CNN was used. In particular SAEs are exploited to generate deep spectral features (1-D) which are then combined with spatial features extracted with a pyramid pool-based CNN able to manage features at different scales. On top of it, a LR classifier is used.

Many works use stacked denoising AEs, which are SAEs trained on noisy input. Liu et al. [176] used them to generate feature maps that are then classified trough a superpixel segmentation approach and majority voting. In [53], Xing et al. presented a pre-trained network using stacked denoising AEs joined with a logistic regression to perform supervised classification. Conversely, in [82] modified sparse denoising AEs were used to train a mapping between low-resolution and high-resolution image patches for pan-sharpening. Inspired by denoising AEs, an unsupervised DL framework, namely Relit Spectral Angle-Stacked AE (RSA-SAE), was employed in [177] to map hyperspectral image pixels to low-dimensional illumination invariant encodings. In Ball et al. [178], a complete classification pipeline was presented, in which a denoising SAE is fed using an augmentation technique, and a final post-processing provides robust image classification. Lan et al. [179] proposed a framework integrating k-sparse denoising AEs and spectral–restricted spatial characteristics for hyperspectral image classification.

Thanks to their dimensionality reduction capabilities DBN can be used to extract features. In [180] DBN were combined with LR classification, similarly to how SAEs were exploited in [169]. In [56] 1-layer-DBN and 2-layer-DBN with spatial–spectral information were both used after a preliminary PCA. Recently, an unsupervised DBN was presented in [72] by Ma et al. to develop a real-time anomaly detection system able to detect interesting local objects. Instead, in [75], DBNs were fed with a 3D discrete wavelet transformation on the input HSI data. Autoencoders also find applications in non-linear spectral unmixing, for endmember extraction and abundance map estimation. In [181] a solution that relies on the given data and does not require supervision is presented, while in [182] an end-to-end learning method called EndNet is introduced based on an AE network exploiting additional layers and a Spectral Angle Distance metric.

4.4. Generative Adversarial Networks

Generative Adversarial Networks (GANs) have gained a lot of interest for their ability to learn to generate samples from data distribution using two competing neural networks, namely a generator and a discriminator. In [183], authors used the discriminator network of a trained GAN to perform classification. This method has proven to be effective when the number of training examples is small. Similarly, [184–186] applied GANs in order to use their discriminator outputs for the final classification phase. In [105] a conditional generative adversarial network (cGAN) was used to build a mapping from PCA reduced HSI data and RGB images of chemically stained tissue samples.

4.5. Recurrent Neural Networks

Other DL approaches worth mentioning are those based on *Recurrent Neural Networks* (RNNs), i.e., neural network architectures specifically designed to handle time dependencies. In this case, hyperspectral data can be treated as if they were video sequences (with spectral bands as video frames) and a RNN can be applied to model the dependencies between different spectral bands, as in [187]. In [51], Mou et al. presented a supervised classification method which focuses on the use of RNN and parametric rectified *tanh* as an activation function. In [146] Liu et al. introduced a bidirectional-convolutional long short term memory (LSTM) network in which a convolution operator across the spatial domain is incorporated into the network to extract the spatial feature, and a bidirectional recurrent connection is proposed to exploit the spectral information. Recently, Shi et al. [188] presented a 3-D RNN able to address the problem of the mixed spectral pixel in order to remove the noise in the classification stage.

4.6. Dataset Augmentation, Transfer-Learning, and Unsupervised Pre-Training

A way to address the lack of availability of labeled pixels is by using different data augmentation strategies. Among them, random pixel-pair features (PPF) was introduced in [21], which exploits the similarity of the pixels of the same class to augment the training data, where a deep CNN with multi layers is then employed to learn these PPF. This approach was improved in [22], in which Ran et al. proposed a spatial pixel-pair feature, SPFF, with a flexible multi-stream CNN-based classification. In [189] Windrim et al. proposed a data augmentation strategy based on relighting for training samples of the CNN which consists of simulating the spectral appearance of a region under different illumination during training. While in [190], Li et al. made an extensive comparison of common augmentation techniques and proposed a new one that helps the CNN to better learn intra-class correspondences.

Another way to handle this data availability problem is to exploit big labeled datasets containing similar data with a *transfer-learning* approach. The reasoning is that usually the first part of a DNN learns generic filters that are reusable in many contexts. In [191], Windrim et al. used this approach by creating a pre-trained CNN on a similar but more complete HSI dataset and then fine-tuning it on the ground-truth dataset. The advantage is that the ground-truth dataset can be now considerably smaller and the training procedure faster. Similarly a transfer-learning approach was employed in [73] to build an anomaly detection system that works on the difference between pixel pairs or in [192] for classification on both homogeneous and heterogeneous HSI data.

As mentioned above, the lack of training sets makes unsupervised and semi-supervised methods increasingly interesting. For example, in [193], Ratle et al. proposed a semi-supervised neural network framework for large scale HSI classification. In [194], Romero et al. presented a layer-wise unsupervised pre-training for CNN, which leads to both performance gains and improved computational efficiency. In [195], Maggiori et al. introduced an end-to-end framework for dense pixel-wise classification with a new initialization method for the CNN. During initialization, a large amount of possibly inaccurate reference data was used, then a refinement step on a small amount of accurately labeled data was performed. In [196], Mou et al. proposed, for the first time in HSI, an end-to-end 2-D fully Convolution-Deconvolution network for unsupervised spectral–spatial feature learning. It is composed of a convolutional sub-network to reduce the dimensionality, and a deconvolutional sub-network to reconstruct the input data.

Advanced training strategies that use semi-supervised schemes were also presented. These made use of abundant unlabeled data, associating pseudo labels in order to work with a limited labeled dataset as in [197], where a deep convolutional recurrent neural networks (CRNN) for hyperspectral image classification was described. Instead, in [93], a ResNets architecture capable of learning from the unlabeled data was presented. It makes use of the complementary cues of the spectral–spatial features to produce a good HSI classification.

4.7. Post-Processing

Conditional Random Fields (CRF) have been used in several works thanks to their ability to refine CNN results for different tasks. In [65], Alam et al. presented a technique that combines CNN and CRF operating on a superpixel partitioning based on both spectral and spatial properties, while in [198], CNNs were combined with Restricted CRF (CNN-RCRF) to perform high-resolution classification, refining the superpixel image into a pixel-based result. Recently, in [199], a decision method based on fuzzy membership rules applied to single-object CNN classification was adopted to increase classification performance with a considerable gain in accuracy.

4.8. New Directions

Finally, we consider other recent solutions that manage HSI data in a more sophisticated way or that can be considered interesting directions deserving further investigation.

Training sample restrictions Specific DL models and training methods have been proposed to improve accuracy when the number of training samples is not abundant. In [200], the inherent spatial–spectral properties of HSI data were exploited to drive the construction of the network model. The use of an edge preserving filter allows us to better explore the contextual structure in a resilient way with respect to noise and small details. An extension of this approach has been proposed in [201] with the introduction of a multi-grain and semi-supervised approach. A self-improving CNN was described in [202] that is able to handle data dimensionality and the lack of training samples by iteratively selecting the most informative bands. In [203] a domain adaptation method was used to exploit the discriminative information of a source image to a neural network for HSI classification.

Active transfer learning is an iterative procedure of selecting the most informative examples from a subset of unlabeled samples and can be used to train deep networks efficiently [204] also with small training sets. Active learning was used in [205] in order to search for salient samples and is able to exploit high-level feature correlations on both training and target domains. Instead, Haut et al. [206] performed spectral–spatial classification using Active Learning coupled with a Bayesan-CNN, where the idea was to add a prior distribution, allowing a probability or likelihood to be defined on the output.

HSI enhancement As discussed in Section 3.1.4, many sources of degradation negatively impinge on the overall quality of HSI. Thus, different solutions has been proposed in order to recover a high-quality HSI both in the spectral and spatial dimensions. In the area of super-resolution, it is worth mentioning the work by Yuan et al. [83] in which a transfer-learning procedure was applied, and the method in [207] that combined both spectral and spatial constraints with a CNN model. Conversely, in [84], a super-resolution network was employed to improve a classification module in an end-to-end fashion. Remarkably, this approach only used a small amount of training data. Instead, Lin et al. [101] proposed a new architecture called SSRNet (super-spectral-resolution network) that is able to estimate dense hypercubes from standard endoscope RGB images and sparse hyperspectral signals from a RGB to HSI base reconstruction and a sparse to dense HSI refinement. Similarly, an image recovery CNN from spectrally undersampled projections was proposed in [35]. Another HSI super-resolution method [208] took inspiration from deep laplacian pyramid networks (LPN). The spatial resolution is enhanced by an LPN and then refined, taking into account the spectral characteristics between the low- and high-resolution with a non-negative dictionary learning. In [79] Xie et al. presented a promising quality enhancement method. It combines the theory of structure tensors with a deep convolutional neural network (CNN) to solve an HSI quality enhancement problem.

Capsule Networks A new kind of approach in the computer vision field that is currently growing is *Capsule Neural Network*. This kind of network has the aim of improving the CNN robustness to geometric transformations using Capsules, a nested set of neural layers that provide the model with a greater ability to generalize. Examples are found in [209–212]. In particular, in [210], Wang et al. proposed a 2-D CapsNet for HSI classification by using both spatial and spectral information, while in [212] Yin et al. introduced a CapsNet architecture with pretraining and initialization stages to improve speed and convergence while avoiding overfitting.

Classification related tradeoffs In real systems, other requirements/limitations, e.g., in terms of data occupancy or power consumption, can conflict with (classification) performance maximization. The high data flow imposed by HSI in quality inspection or high throughput diagnostic procedures is a challenge when mid- or long-term data conservation is a requirement: for example in [109] authors evaluated the combined use of classification and lossy data compression. To this end, after selecting a suitable wavelet-based compression technology, they tested coding strength-driven operating points, looking for configurations likely able to prevent any classification performance degradation. The result showed that it is possible to derive guidelines for using lossy compression to concurrently guarantee the preservation of the classification quality and the highest compression rate. When computational complexity or power consumption restrictions do emerge, it becomes relevant to evaluate classification performance trade-offs with respect to model implementations on low-power consumption architectures [213]. Concerning computational speed, in [214], Paoletti et al. proposed an implementation of 3-D CNN by integrating a mirroring strategy to effectively process the border areas of the image.

5. Discussion and Future Perspectives

An imbalance that clearly emerged from this overview is the one between the number of HSI-DL studies in the scope of RS with respect to the ones in other application fields. This is depicted in more detail in Figure 3 where, on an annual basis, we subdivided HSI-DL works in this survey by application areas, with RS related studies further split into sub-fields. In this count we did our best to include literature works and their subject mapping. In case of large overlaps of content in multiple works only the most representative works were included. The aforementioned disparity derives from multiple factors: historical and technological reasons (hyperspectral imaging started and developed first and foremost in the RS sector); the development of a wide scientific community; the existence of many venues (journals and conferences) dedicated to RS research themes.

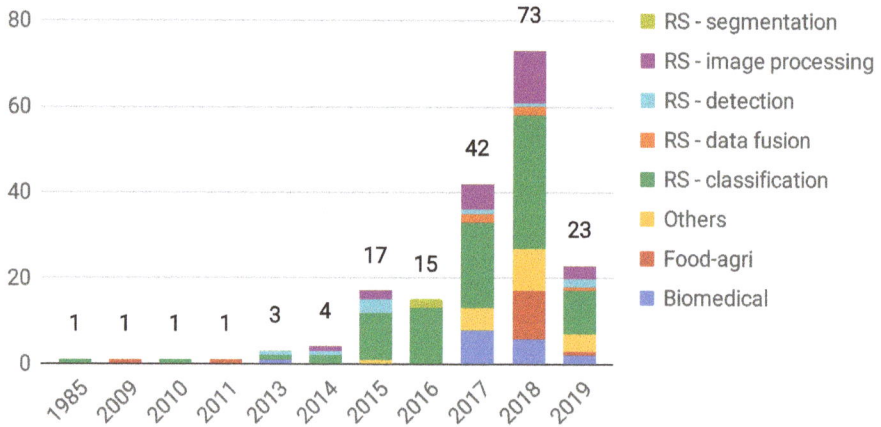

Figure 3. Number of HSI-DL articles per year. The last column comprises published and in-press papers found up to 31 January 2019.

Almost all HSI-DL RS scientific papers, however, still refer to a limited amount of publicly available datasets. While this has proved to be a powerful enabling factor and a stimulus for relevant technological advancements and benchmarking, it can be associated to the risk of incremental and self-referential scientific production as well. Therefore, despite an apparent abundance and exponentially increasing trend (see Figure 3) in the number of RS-related studies (especially for land cover classification), there is still considerable scope and need for the development of workable supervised and unsupervised (or semi-supervised) HSI-DL solutions dedicated to classification studies in specific sub-fields (like soil and geology, water resources and environmental studies, agriculture and vegetation, urban and land development, etc.) as well as vast potential to work on other relevant tasks like change, target, and anomaly detection, analysis of data coming from different sensors (data fusion), spectral unmixing and physico-chemical parameter estimation. Moreover, segmentation is a path not yet well traveled. Architectures like U-net or V-net (for a volumetric approach) can be a good choice to start with, but its formulation in this scenario is yet to be investigated. There is a large variety of HSI classification problems requiring a tailored design or an accurate assessment of existing DL solutions. To comply with the specific application requirements, complexity and computational issues as well as hardware optimization must enter the selection of suitable approaches in addition to pursuing satisfactory accuracy performance. Unfortunately, however, the limited amount of available data also involves difficulties in comparing different methods that lead to similar results, and this again happens for RS image classification studies on benchmark datasets, where near perfect results have been obtained by several, more or less complex, architectures [27,29]. Additional generalization problems arise for data characterized by a relatively high intra-class spectral–spatial variability, not only due to intrinsic target variability but also to atmospheric and daylight conditions. To mitigate these issues, specific data augmentation techniques deserve further investigation, where new generative networks based on GANs can produce very good synthetic data and new and more powerful augmentation techniques. Reinforcement learning could play an interesting role in the near feature in this field as well.

With due proportion, the development in non-RS applications fields seems to be following an important increasing trend as well. This could be the beginning of a successful era in the field of HSI data analysis characterized by a virtuous circle of new industry and professional usages and the development of new acquisition and lighting devices. The market penetration of these systems needs to be backed up by equipment cost reductions, commitment to the generation of representative datasets, the development of advanced DL-based data analysis solutions, and the exploitation of optimized HW/SW computational platforms. This scenario could lead to favourable cost-benefit evaluations and to a greater diffusion of HSI-DL technologies in industrial and professional fields. This could eventually lead to a desirable expansion of dedicated research communities as well. For example, since HSI analysis is still relatively new in many fields related to Computer Vision, there shall be great potential in the future for further investigations in this area from different perspectives, such as 3D modelling and reconstruction, object detection, motion and tracking, multi-sensor data analysis and fusion, etc.

In the professional and industrial fields, datasets are acquired with a precise application purpose and the parameters of the acquisition setup can normally be well controlled by favouring the design of ad-hoc solutions. Although small-scale HSI scenarios can present a high variability, the collection of data is sometimes facilitated as well as the commitment to find resources for data labeling or metadata production by factors such as the race to product development or the mandatory nature of the diagnostic processes. In case of over-abundant availability of data this can be exploited with unsupervised or semi-supervised labeling methods.

Furthermore, for small-scale applications we can identify some peculiarities or aspects that can be addressed differently from what happens in RS. Lighting, for instance, can be controlled and optimized: we think that the exploitation of broadband LED illumination sources in VNIR (400–1400 nm) and SWIR (1400–3000 nm) that are starting to appear on the market (https://www.metaphase-tech.com/hyperspectral-illumination/ (last visit March 2019)) can lead to a further expansion of applications, especially in the biomedical field or where power consumption and temperatures of the halogen lamps can be a problem. This is an interesting development perspective since HSI with LEDs has been often considered unfeasible.

Unlike RS, the problem of data transmission from satellites and the need for on-board compression is not present for small-scale applications. Still, the huge amount of collected data requires compression technologies as well, especially if there are medium- long-term data storage needs arising from statistical needs (e.g., in agricultural studies) or rather from exigencies related to food-traceability or medico-legal regulations. The relationship between compression effects and DL performance demands awareness, experimental validations and methods to guarantee the sought data quality. One pilot example in this direction can be considered the assessment of coding strength aimed at preserving DL-based classification performance in a biomedical application, as proposed in [109].

6. Conclusions

The richness of information contained in HSI data constitutes an indubitable appealing factor especially in sectors that benefit from computer assisted interpretation of visible and invisible (to the human eye) phenomena. However, industrial and professional HSI technologies are subject to cost-benefit evaluations which lead to the need for enabling factors to activate their deployment potentialities. In these years, machine learning technologies are rapidly extending their range and, boosted by the advent of Deep Learning, they are revolutionizing the world of digital data analysis. In this review, we tried to analyze what is currently happening with the meeting of HSI and DL technologies by adopting a multidisciplinary perspective and making our work accessible to both domain experts, machine learning scientists, and practitioners.

Although mitigated by the fact that pixel- and spectral-based analysis tasks can count on an order of thousands training samples for HSI volume, one of the main issues that emerged as an obstacle for quality scientific production is the limited number of publicly available datasets. More in general, the number and quality of acquired data in the various disciplines remains a central issue for the development of sound, effective and broad scope HSI-DL solutions. Rather, the exploration of different DL approaches for the RS field can stimulate efforts and investments in the provision of quality HSI datasets. Moreover, for other application fields where the penetration of HSI technologies is still way behind, the possibility to approach complex visual tasks by means of DL solutions can be seen as an enabling factor and a possible driver for a new era in the deployment of HSI technologies for a wide spectrum of small-scale applications in industry, biology and medicine, cultural heritage and other professional fields.

Author Contributions: Conceptualization, A.S.; Analysis of the literature, A.B., A.S., S.B. and M.S.; Supervision, A.S.; Writing—original draft preparation, A.B., A.S. and M.S.; Writing—review and editing, A.S., M.S. and S.B.

Funding: This research received no external funding.

Acknowledgments: We would like to thank all the anonymous reviewers for the time they spent on the reading of our manuscript and for their useful comments, which allowed to improve this final version.

Conflicts of Interest: The authors declare no conflict of interest.

Appendix A. DL Methods for HSI in Brief

Here, we give a brief introduction to the deep learning world to provide context and references to the core parts of this review. For a more extensive introduction to deep neural networks the reader can refer to [215], while the book [216] is a more comprehensive reference. In a RS perspective, valuable overviews of DL approaches can be found in [23–25].

DL is a branch of representational learning in which models are composed of multiple layers to learn representations from data in an end-to-end fashion. These methods have had a terrific impact to date and are expected to continue revolutionizing the way complex data analysis tasks are approached in domains such as natural language processing, speech recognition, visual object detection and recognition and many others. Together with the classical supervised and unsupervised learning approaches other paradigms have become relevant in the context of DL, where large amounts of data (order of hundreds of thousands) are supposed to be necessary to carry out correct learning of the high number of parameters characterizing a Deep model and to avoid overfitting. In fact, both sufficiently exhaustive data acquisition and labeling (supervision) can be costly or even unfeasible in some contexts. Different data augmentation strategies and techniques can be adopted and are common practice in many cases. Moreover, exploiting the fact that Deep architectures usually build a hierarchical bottom-up representation of the information, in many cases typically the lowest portion of a model trained on somehow related data in a source domain can be transferred to the target domain model, and so called transfer-learning approaches only require a residual estimation of a reduced portion of the parameters or allow a significant reduction of the learning epochs. Other ways to exploit knowledge, this time from the same target domain, belong to the wide family of semi-supervised learning methods. They allow to exploit the typical imbalanced presence of unlabeled data due to the difficulties, also characterizing many HSI domains, to produce large enough and high-quality labelled datasets. Semi-supervised learning can be operated for example by training a classifier with an additional penalty term coming from an Autoencoder (AE) or other unsupervised data embedding methods [217].

Appendix A.1. Fully-Connected

When we refer to *fully-connected*, we are dealing with networks (or layers of a network) in which each node of a layer is connected to all the nodes in the following one without almost any constraints (see Figure A1a). Each neuron acts as a summation node with respect to its inputs. Eventually, a non-linear activation function is applied to the output. Fully-connected is one of the simplest layers and usually is used in the last part of the network for the final classification or regression.

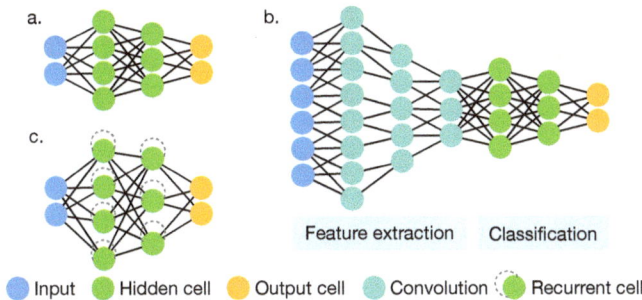

Figure A1. Network architectures. (**a**) Fully-connected; (**b**) Convolutional neural network; (**c**) Recurrent neural network.

Appendix A.2. Convolutional Neural Networks

Convolutional neural networks (CNNs) [218] are particular types of deep feed-forward networks that are simpler to train, and more effective, on sampled data sources (Figure A1b). This is due to the constraints introduced in the hypothesis space that force a structure and reduce the number of parameters. The enforced structure creates features that are spatially invariant and robust to rotation and deformations (up to a certain amount). This is made possible thanks to local connections, shared weights and the use of pooling layers as well. CNNs are designed to process matrices or tensors such as colour images. Many data sources are in the form of multiple arrays: from 1D for sequences and signals, like audio or spectral signatures; 2D for images; and 3D for video or volumetric images.

Notable architectures are: AlexNet [219], which won the ImageNet competition in 2012, outperforming its competitor; GoogleLeNet [220], based on inception blocks which create sub-networks in the main network and increase either depth and width with respect to AlexNet; VGG [221] with its very small (3 × 3) and widely used convolution filter, and a simple and repetitive structure growing in depth; ResNet [222] that builds a very deep structure in which there are skip connections to let the information flow jump over a set or layers, solving the problem of vanishing gradients (i.e., the inability to propagate the error function backwards in very deep networks). This is because it has become too small after a certain point, and thus producing a potential stop of network training). If, instead, skip connections interconnect every following block, the architecture is called DenseNet [223]. Recently, many other networks focusing on low computational devices appear, such as MobileNet [224] and SqueezeNet [225], to name a few.

Appendix A.3. Recurrent Neural Networks

Recurrent neural networks (RNNs) belong to an important branch of the DL family and are mainly designed to handle sequential data (see Figure A1c). A plain RNN is indeed not so powerful and seldom used in works nowadays. Rather, very high performance can be achieved with recurrent hidden units like LSTM (Long Short Term Memory) [226] or GRU (Gate Recurrent Unit) [227]. These units are composed of different internal data paths that can store and release information when needed and are capable of alleviating the vanishing gradient problem.

Appendix A.4. Autoencoders

An autoencoder (AE) [228] is composed of: one visible layer of inputs, one hidden layer of units, one reconstruction layer of units, and an activation function (Figure A2a). During training, it first projects the input to the hidden layer and produces the latent vector. The network corresponding to this step is called the *encoder*. Then, the output of the encoder is mapped by a *decoder* to an output layer that has the same size as the input layer. The power of AEs lies in this form of training that is unsupervised and forces a meaningful compressed representation in its core. During reconstruction, AE only uses the information in hidden layer activity, which is encoded as features from the input. Stacking trained encoders (SAE, see Figure A3) is a way to minimize information loss while preserving abstract semantic information and improving the final model capacity.

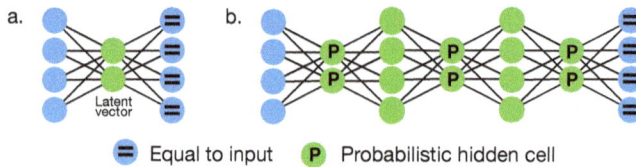

Figure A2. Network architectures. (**a**) Autoencoders; (**b**) Deep belief networks.

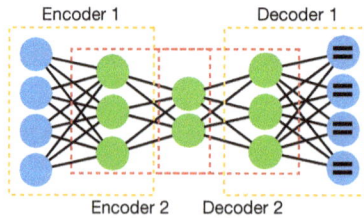

Figure A3. Network architecture of a Stacked Autoencoder

Appendix A.5. Deep Belief Networks

Deep Belief Networks (DBN) can be viewed as a composition of simple, unsupervised networks such as Restricted Boltzmann machines (RBM) [229] or autoencoders [230], in which each sub-network hidden layer serves as the visible layer for the next one (see Figure A2b). If necessary, a feed-forward network is appended for the fine-tune phase.

Appendix A.6. Generative Adversarial Networks

Generative Adversarial Networks (GANs) have recently emerged as a promising approach to constructing and training generative models. In this framework there are two adversarial neural networks that are jointly trained: a *generator G* and a *discriminator D* (see Figure A4). The generator is supposed to learn to generate the samples of a data distribution given random inputs, while *D* tries to discriminate between real data and artificially generated ones. The two networks are trained in a two-player minmax game scheme until the generated data are not distinguishable from the real ones.

J. Imaging **2019**, *5*, 52

After a proper training procedure, *D* can be used as a well trained feature extractor, and applied to a specific problem with the addition of a final block that exploits the needed output (for instance a fully connected layer for classification).

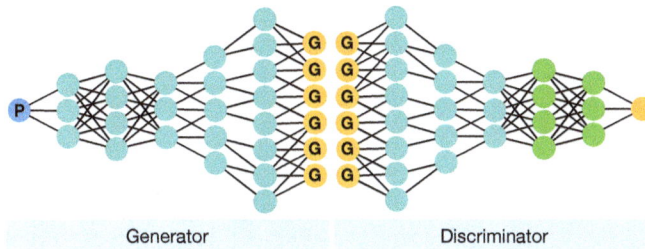

Figure A4. Architecture of Generative adversarial networks.

References

1. Goetz, A.; Vane, G.; Solomon, J.E.; Rock, B. Imaging Spectrometry for Earth Remote Sensing. *Science* **1985**, *228*, 1147–53. [CrossRef]
2. Eismann, M.T. *Hyperspectral Remote Sensing*; SPIE Press: Bellingham, WA, USA, 2012.
3. Lu, G.; Fei, B. Medical hyperspectral imaging: A review. *J. Biomed. Opt.* **2014**, *19*, 010901. [CrossRef]
4. Sun, D.W. *Hyperspectral Imaging for Food Quality Analysis and Control*; Academic Press: Cambridge, MA, USA, 2010.
5. Lowe, A.; Harrison, N.; French, A.P. Hyperspectral image analysis techniques for the detection and classification of the early onset of plant disease and stress. *Plant Methods* **2017**, *13*, 80. [CrossRef] [PubMed]
6. Kamilaris, A.; Prenafeta-Bold, F.X. Deep learning in agriculture: A survey. *Comput. Electron. Agric.* **2018**, *147*, 70–90. [CrossRef]
7. Fischer, C.; Kakoulli, I. Multispectral and hyperspectral imaging technologies in conservation: Current research and potential applications. *Stud. Conserv.* **2006**, *51*, 3–16.
8. Khan, M.J.; Khan, H.S.; Yousaf, A.; Khurshid, K.; Abbas, A. Modern Trends in Hyperspectral Image Analysis: A Review. *IEEE Access* **2018**, *6*, 14118–14129. [CrossRef]
9. Lucas, R.; Rowlands, A.; Niemann, O.; Merton, R. Hyperspectral Sensors and Applications. In *Advanced Image Processing Techniques for Remotely Sensed Hyperspectral Data*; Springer: Berlin/Heidelberg, Germany, 2004; pp. 11–49.
10. Gewali, U.B.; Monteiro, S.T.; Saber, E. Machine learning based hyperspectral image analysis: A survey. *arXiv* **2018**, arXiv:1802.08701.
11. Bengio, Y.; Courville, A.; Vincent, P. Representation Learning: A Review and New Perspectives. *IEEE Trans. Pattern Anal. Mach. Intell.* **2013**, *35*, 1798–1828. [CrossRef] [PubMed]
12. Lowe, D.G. Object recognition from local scale-invariant features. In Proceedings of the Seventh IEEE International Conference on Computer Vision, Kerkyra, Greece, 20–27 September 1999; Volume 2, pp. 1150–1157.
13. Dalal, N.; Triggs, B. Histograms of oriented gradients for human detection. In Proceedings of the International Conference on computer vision & Pattern Recognition, San Diego, CA, USA, 20–25 June 2005; IEEE Computer Society: Washington, DC, USA, 2005; Volume 1, pp. 886–893.
14. Li, W.; Chen, C.; Su, H.; Du, Q. Local Binary Patterns and Extreme Learning Machine for Hyperspectral Imagery Classification. *IEEE Trans. Geosci. Remote Sens.* **2015**, *53*, 3681–3693. [CrossRef]
15. Camps-Valls, G.; Bruzzone, L. Kernel-based methods for hyperspectral image classification. *IEEE Trans. Geosci. Remote Sens.* **2005**, *43*, 1351–1362. [CrossRef]
16. Ham, J.; Chen, Y.; Crawford, M.M.; Ghosh, J. Investigation of the random forest framework for classification of hyperspectral data. *IEEE Trans. Geosci. Remote Sens.* **2005**, *43*, 492–501. [CrossRef]
17. Camps-Valls, G.; Tuia, D.; Bruzzone, L.; Benediktsson, J.A. Advances in Hyperspectral Image Classification: Earth Monitoring with Statistical Learning Methods. *IEEE Signal Process. Mag.* **2014**, *31*, 45–54. [CrossRef]

18. Liu, W.; Wang, Z.; Liu, X.; Zeng, N.; Liu, Y.; Alsaadi, F.E. A survey of deep neural network architectures and their applications. *Neurocomputing* **2017**, *234*, 11–26. [CrossRef]
19. Brendel, W.; Bethge, M. Approximating CNNs with Bag-of-local-Features models works surprisingly well on ImageNet. *arXiv* **2019**, arXiv:1904.00760.
20. Gori, M. What's Wrong with Computer Vision? In Proceedings of the IAPR Workshop on Artificial Neural Networks in Pattern Recognition—LNAI 11081, Siena, Italy, 19–21 September 2018; Springer: Berlin, Germany, 2018; pp. 3–16.
21. Li, W.; Wu, G.; Zhang, F.; Du, Q. Hyperspectral Image Classification Using Deep Pixel-Pair Features. *IEEE Trans. Geosci. Remote Sens.* **2016**, *55*, 844–853. [CrossRef]
22. Ran, L.; Zhang, Y.; Wei, W.; Zhang, Q. A Hyperspectral Image Classification Framework with Spatial Pixel Pair Features. *Sensors* **2017**, *17*, 2421. [CrossRef] [PubMed]
23. Zhang, L.; Zhang, L.; Du, B. Deep Learning for Remote Sensing Data: A Technical Tutorial on the State of the Art. *IEEE Geosci. Remote Sens. Mag.* **2016**, *4*, 22–40. [CrossRef]
24. Ball, J.E.; Anderson, D.T.; Chan, C.S. Comprehensive survey of deep learning in remote sensing: Theories, tools, and challenges for the community. *J. Appl. Remote Sens.* **2017**, *11*, 11–54. [CrossRef]
25. Zhu, X.X.; Tuia, D.; Mou, L.; Xia, G.; Zhang, L.; Xu, F.; Fraundorfer, F. Deep Learning in Remote Sensing: A Comprehensive Review and List of Resources. *IEEE Geosci. Remote Sens. Mag.* **2017**, *5*, 8–36. [CrossRef]
26. Ghamisi, P.; Maggiori, E.; Li, S.; Souza, R.; Tarablaka, Y.; Moser, G.; Giorgi, A.D.; Fang, L.; Chen, Y.; Chi, M.; et al. New Frontiers in Spectral–spatial Hyperspectral Image Classification: The Latest Advances Based on Mathematical Morphology, Markov Random Fields, Segmentation, Sparse Representation, and Deep Learning. *IEEE Geosci. Remote Sens. Mag.* **2018**, *6*, 10–43. [CrossRef]
27. Ghamisi, P.; Yokoya, N.; Li, J.; Liao, W.; Liu, S.; Plaza, J.; Rasti, B.; Plaza, A. Advances in Hyperspectral Image and Signal Processing: A Comprehensive Overview of the State of the Art. *IEEE Geosci. Remote Sens. Mag.* **2017**, *5*, 37–78. [CrossRef]
28. Ghamisi, P.; Plaza, J.; Chen, Y.; Li, J.; Plaza, A.J. Advanced Spectral Classifiers for Hyperspectral Images: A review. *IEEE Geosci. Remote Sens. Mag.* **2017**, *5*, 8–32. [CrossRef]
29. Petersson, H.; Gustafsson, D.; Bergstrom, D. Hyperspectral image analysis using deep learning—A review. In Proceedings of the 2016 Sixth International Conference on Image Processing Theory, Tools and Applications (IPTA), Oulu, Finland, 12–15 December 2016; pp. 1–6.
30. Nathan, A.H.; Kudenov, M.W. Review of snapshot spectral imaging technologies. *Opt. Eng.* **2013**, *52*, 090901.
31. Luthman, A.S. *Spectrally Resolved Detector Arrays for Multiplexed Biomedical Fluorescence Imaging*; Springer: Berlin/Heidelberg, Germany, 2018.
32. Nguyen, R.M.H.; Prasad, D.K.; Brown, M.S. *Training-Based Spectral Reconstruction from a Single RGB Image*; Computer Vision–ECCV 2014; Fleet, D., Pajdla, T., Schiele, B., Tuytelaars, T., Eds.; Springer International Publishing: Cham, Switzerland, 2014; pp. 186–201.
33. Oh, S.W.; Brown, M.S.; Pollefeys, M.; Kim, S.J. Do It Yourself Hyperspectral Imaging with Everyday Digital Cameras. In Proceedings of the 2016 IEEE Conference on Computer Vision and Pattern Recognition (CVPR), Las Vegas, NV, USA, 27–30 June 2016; pp. 2461–2469.
34. Galliani, S.; Lanaras, C.; Marmanis, D.; Baltsavias, E.; Schindler, K. Learned Spectral Super-Resolution. *arXiv* **2017**, arXiv:1703.09470 .
35. Xiong, Z.; Shi, Z.; Li, H.; Wang, L.; Liu, D.; Wu, F. HSCNN: CNN-Based Hyperspectral Image Recovery from Spectrally Undersampled Projections. In Proceedings of the 2017 IEEE International Conference on Computer Vision Workshops (ICCVW), Venice, Italy, 22–29 October 2017; pp. 518–525.
36. Can, Y.B.; Timofte, R. An efficient CNN for spectral reconstruction from RGB images. *arXiv* **2018**, arXiv:1804.04647.
37. Yan, Y.; Zhang, L.; Li, J.; Wei, W.; Zhang, Y. Accurate Spectral Super-Resolution from Single RGB Image Using Multi-scale CNN. In *Pattern Recognition and Computer Vision*; Lai, J.H., Liu, C.L., Chen, X., Zhou, J., Tan, T., Zheng, N., Zha, H., Eds.; Springer International Publishing: Cham, Switzerland, 2018; pp. 206–217.
38. Koundinya, S.; Sharma, H.; Sharma, M.; Upadhyay, A.; Manekar, R.; Mukhopadhyay, R.; Karmakar, A.; Chaudhury, S. 2D-3D CNN Based Architectures for Spectral Reconstruction from RGB Images. In Proceedings of the 2018 IEEE/CVF Conference on Computer Vision and Pattern Recognition Workshops (CVPRW), Salt Lake City, UT, USA, 18–22 June 2018; pp. 957–9577.

39. Shi, Z.; Chen, C.; Xiong, Z.; Liu, D.; Wu, F. HSCNN+: Advanced CNN-Based Hyperspectral Recovery from RGB Images. In Proceedings of the 2018 IEEE/CVF Conference on Computer Vision and Pattern Recognition Workshops (CVPRW), Salt Lake City, UT, USA, 18–22 June 2018; pp. 1052–10528.
40. Qu, Y.; Qi, H.; Kwan, C. Unsupervised Sparse Dirichlet-Net for Hyperspectral Image Super-Resolution. In Proceedings of the 2018 IEEE/CVF Conference on Computer Vision and Pattern Recognition, Salt Lake City, UT, USA, 18–23 June 2018; pp. 2511–2520.
41. Alvarez-Gila, A.; Weijer, J.; Garrote, E. Adversarial Networks for Spatial Context-Aware Spectral Image Reconstruction from RGB. In Proceedings of the 2017 IEEE International Conference on Computer Vision Workshops (ICCVW), Venice, Italy, 22–29 October 2017; pp. 480–490.
42. Arad, B.; Ben-Shahar, O. Filter Selection for Hyperspectral Estimation. In Proceedings of the 2017 IEEE International Conference on Computer Vision (ICCV), Venice, Italy, 22–29 October 2017; pp. 3172–3180.
43. Fu, Y.; Zhang, T.; Zheng, Y.; Zhang, D.; Huang, H. Joint Camera Spectral Sensitivity Selection and Hyperspectral Image Recovery. In Proceedings of the European Conference on Computer Vision (ECCV), Munich, Germany, 8–14 September 2018; Springer International Publishing: Cham, Switzerland, 2018; pp. 812–828.
44. Kaya, B.; Can, Y.B.; Timofte, R. Towards Spectral Estimation from a Single RGB Image in the Wild. *arXiv* **2018**, arXiv:1812.00805.
45. Nie, S.; Gu, L.; Zheng, Y.; Lam, A.; Ono, N.; Sato, I. Deeply Learned Filter Response Functions for Hyperspectral Reconstruction. In Proceedings of the 2018 IEEE/CVF Conference on Computer Vision and Pattern Recognition, Salt Lake City, UT, USA, 18–23 June 2018; pp. 4767–4776.
46. Arad, B.; Ben-Shahar, O.; Timofte, R.; Van Gool, L.; Zhang, L.; Yang, M. NTIRE 2018 Challenge on Spectral Reconstruction from RGB Images. In Proceedings of the 2018 IEEE/CVF Conference on Computer Vision and Pattern Recognition Workshops (CVPRW), Salt Lake City, UT, USA, 18–22 June 2018; pp. 1042–104209.
47. Cao, X.; Yue, T.; Lin, X.; Lin, S.; Yuan, X.; Dai, Q.; Carin, L.; Brady, D.J. Computational Snapshot Multispectral Cameras: Toward dynamic capture of the spectral world. *IEEE Signal Process. Mag.* **2016**, *33*, 95–108. [CrossRef]
48. Wang, L.; Zhang, T.; Fu, Y.; Huang, H. HyperReconNet: Joint Coded Aperture Optimization and Image Reconstruction for Compressive Hyperspectral Imaging. *IEEE Trans. Image Process.* **2019**, *28*, 2257–2270. [CrossRef]
49. Pu, R. *Hyperspectral Remote Sensing: Fundamentals and Practices*; CRC Press: Boca Raton, FL, USA, 2017.
50. Hu, W.; Huang, Y.; Wei, L.; Zhang, F.; Li, H. Deep Convolutional Neural Networks for Hyperspectral Image Classification. *J. Sens.* **2015**, *2015*, 258619. [CrossRef]
51. Mou, L.; Ghamisi, P.; Zhu, X.X. Deep Recurrent Neural Networks for Hyperspectral Image Classification. *IEEE Trans. Geosci. Remote Sens.* **2017**, *55*, 3639–3655. [CrossRef]
52. Karalas, K.; Tsagkatakis, G.; Zervakis, M.; Tsakalides, P. Deep learning for multi-label land cover classification. In *Image and Signal Processing for Remote Sensing XXI*; International Society for Optics and Photonics: Bellingham, WA, USA, 2015; Volume 9643, p. 96430Q.
53. Xing, C.; Ma, L.; Yang, X. Stacked Denoise Autoencoder Based Feature Extraction and Classification for Hyperspectral Images. *J. Sens.* **2016**, *2016*, 3632943. [CrossRef]
54. Zhao, W.; Guo, Z.; Yue, J.; Zhang, X.; Luo, L. On combining multiscale deep learning features for the classification of hyperspectral remote sensing imagery. *Int. J. Remote Sens.* **2015**, *36*, 3368–3379. [CrossRef]
55. Li, Y.; Xie, W.; Li, H. Hyperspectral image reconstruction by deep convolutional neural network for classification. *Pattern Recognit.* **2017**, *63*, 371–383. [CrossRef]
56. Li, T.; Zhang, J.; Zhang, Y. Classification of hyperspectral image based on deep belief networks. In Proceedings of the 2014 IEEE International Conference on Image Processing (ICIP), Paris, France, 27–30 October 2014; pp. 5132–5136.
57. Zhang, M.; Li, W.; Du, Q. Diverse Region-Based CNN for Hyperspectral Image Classification. *IEEE Trans. Image Process.* **2018**, *27*, 2623–2634. [CrossRef]
58. Halicek, M.; Little, J.V.; Wang, X.; Patel, M.; Griffith, C.C.; El-Deiry, M.W.; Chen, A.Y.; Fei, B. Optical biopsy of head and neck cancer using hyperspectral imaging and convolutional neural networks. In *Optical Imaging, Therapeutics, and Advanced Technology in Head and Neck Surgery and Otolaryngology 2018*; International Society for Optics and Photonics: Bellingham, WA, USA, 2018; Volume 10469, p. 104690X.

59. Lin, Z.; Chen, Y.; Zhao, X.; Wang, G. Spectral–spatial Classification of Hyperspectral Image Using Autoencoders. In Proceedings of the 2013 9th International Conference on Information, Communications Signal Processing, Tainan, Taiwan, 10–13 December 2013; pp. 1–5.

60. Guo, Y.; Cao, H.; Bai, J.; Bai, Y. High Efficient Deep Feature Extraction and Classification of Spectral–spatial Hyperspectral Image Using Cross Domain Convolutional Neural Networks. *IEEE J. Sel. Top. Appl. Earth Obs. Remote Sens.* **2019**, *12*, 345–356. [CrossRef]

61. Zhao, W.; Du, S. Learning multiscale and deep representations for classifying remotely sensed imagery. *ISPRS J. Photogramm. Remote Sens.* **2016**, *113*, 155–165. [CrossRef]

62. Gong, Z.; Zhong, P.; Yu, Y.; Hu, W.; Li, S. A CNN With Multiscale Convolution and Diversified Metric for Hyperspectral Image Classification. *IEEE Trans. Geosci. Remote Sens.* **2019**, 1–20. [CrossRef]

63. Yang, X.; Ye, Y.; Li, X.; Lau, R.Y.K.; Zhang, X.; Huang, X. Hyperspectral Image Classification With Deep Learning Models. *IEEE Trans. Geosci. Remote Sens.* **2018**, *56*, 5408–5423. [CrossRef]

64. Liang, J.; Zhou, J.; Qian, Y.; Wen, L.; Bai, X.; Gao, Y. On the Sampling Strategy for Evaluation of Spectral–spatial Methods in Hyperspectral Image Classification. *IEEE Trans. Geosci. Remote Sens.* **2017**, *55*, 862–880. [CrossRef]

65. Alam, F.I.; Zhou, J.; Liew, A.W.; Jia, X. CRF learning with CNN features for hyperspectral image segmentation. In Proceedings of the 2016 IEEE International Geoscience and Remote Sensing Symposium (IGARSS), Beijing, China, 10–15 July 2016; pp. 6890–6893.

66. Zabalza, A.; Ren, J.; Zheng, J.; Huimin Zhao, C.Q.; Yang, Z.; Marshall, S. Novel Segmented Stacked Auto Encoder for Effective Dimensionality Reduction and Feature Extraction in Hyperspectral Imaging. *Neurocomputing* **2016**, *185*, 1–10. [CrossRef]

67. Chen, X.; Xiang, S.; Liu, C.; Pan, C. Vehicle Detection in Satellite Images by Parallel Deep Convolutional Neural Networks. In Proceedings of the 2013 2nd IAPR Asian Conference on Pattern Recognition, Naha, Japan, 5–8 November 2013; pp. 181–185.

68. Chen, X.; Xiang, S.; Liu, C.; Pan, C. Vehicle Detection in Satellite Images by Hybrid Deep Convolutional Neural Networks. *IEEE Geosci. Remote Sens. Lett.* **2014**, *11*, 1797–1801. [CrossRef]

69. Zhang, L.; Shi, Z.; Wu, J. A Hierarchical Oil Tank Detector With Deep Surrounding Features for High-Resolution Optical Satellite Imagery. *IEEE J. Sel. Top. Appl. Earth Obs. Remote Sens.* **2015**, *8*, 4895–4909. [CrossRef]

70. Vakalopoulou, M.; Karantzalos, K.; Komodakis, N.; Paragios, N. Building detection in very high resolution multispectral data with deep learning features. In Proceedings of the 2015 IEEE International Geoscience and Remote Sensing Symposium (IGARSS), Milan, Italy, 26–31 July 2015; pp. 1873–1876.

71. Zhang, L.; Cheng, B. A stacked autoencoders-based adaptive subspace model for hyperspectral anomaly detection. *Infrared Phys. Technol.* **2019**, *96*, 52–60. [CrossRef]

72. Ma, N.; Peng, Y.; Wang, S.; Leong, P.H.W. An Unsupervised Deep Hyperspectral Anomaly Detector. *Sensors* **2018**, *18*, 693. [CrossRef]

73. Li, W.; Wu, G.; Du, Q. Transferred Deep Learning for Anomaly Detection in Hyperspectral Imagery. *IEEE Geosci. Remote Sens. Lett.* **2017**, *14*, 597–601. [CrossRef]

74. Wang, Q.; Yuan, Z.; Du, Q.; Li, X. GETNET: A General End-to-End 2-D CNN Framework for Hyperspectral Image Change Detection. *IEEE Trans. Geosci. Remote Sens.* **2019**, *57*, 3–13. [CrossRef]

75. Huang, F.; Yu, Y.; Feng, T. Hyperspectral remote sensing image change detection based on tensor and deep learning. *J. Vis. Commun. Image Represent.* **2019**, *58*, 233–244. [CrossRef]

76. Sidorov, O.; Hardeberg, J.Y. Deep Hyperspectral Prior: Denoising, Inpainting, Super-Resolution. *arXiv* **2019**, arXiv:1902.00301

77. Xie, W.; Li, Y.; Jia, X. Deep convolutional networks with residual learning for accurate spectral–spatial denoising. *Neurocomputing* **2018**, *312*, 372–381. [CrossRef]

78. Xie, W.; Li, Y.; Hu, J.; Chen, D.Y. Trainable spectral difference learning with spatial starting for hyperspectral image denoising. *Neural Netw.* **2018**, *108*, 272–286. [CrossRef]

79. Xie, W.; Shi, Y.; Li, Y.; Jia, X.; Lei, J. High-quality spectral–spatial reconstruction using saliency detection and deep feature enhancement. *Pattern Recognit.* **2019**, *88*, 139–152. [CrossRef]

80. Loncan, L.; de Almeida, L.B.; Bioucas-Dias, J.M.; Briottet, X.; Chanussot, J.; Dobigeon, N.; Fabre, S.; Liao, W.; Licciardi, G.A.; Simoes, M.; et al. Hyperspectral Pansharpening: A Review. *IEEE Geosci. Remote Sens. Mag.* **2015**, *3*, 27–46. [CrossRef]

81. Zhang, J.; Zhong, P.; Chen, Y.; Li, S. $L_{1/2}$-Regularized Deconvolution Network for the Representation and Restoration of Optical Remote Sensing Images. *IEEE Trans. Geosci. Remote Sens.* **2014**, *52*, 2617–2627. [CrossRef]

82. Huang, W.; Xiao, L.; Wei, Z.; Liu, H.; Tang, S. A New Pan-Sharpening Method With Deep Neural Networks. *IEEE Geosci. Remote Sens. Lett.* **2015**, *12*, 1037–1041. [CrossRef]

83. Yuan, Y.; Zheng, X.; Lu, X. Hyperspectral Image Superresolution by Transfer Learning. *IEEE J. Sel. Top. Appl. Earth Obs. Remote Sens.* **2017**, *10*, 1963–1974. [CrossRef]

84. Hao, S.; Wang, W.; Ye, Y.; Li, E.; Bruzzone, L. A Deep Network Architecture for Super-Resolution-Aided Hyperspectral Image Classification With Classwise Loss. *IEEE Trans. Geosci. Remote Sens.* **2018**, *56*, 4650–4663. [CrossRef]

85. Zheng, K.; Gao, L.; Ran, Q.; Cui, X.; Zhang, B.; Liao, W.; Jia, S. Separable-spectral convolution and inception network for hyperspectral image super-resolution. *Int. J. Mach. Learn. Cybern.* **2019**. [CrossRef]

86. Mei, S.; Yuan, X.; Ji, J.; Zhang, Y.; Wan, S.; Du, Q. Hyperspectral Image Spatial Super-Resolution via 3D Full Convolutional Neural Network. *Remote Sens.* **2017**, *9*, 1139. [CrossRef]

87. Hu, J.; Li, Y.; Xie, W. Hyperspectral Image Super-Resolution by Spectral Difference Learning and Spatial Error Correction. *IEEE Geosci. Remote Sens. Lett.* **2017**, *14*, 1825–1829. [CrossRef]

88. Yang, J.; Zhao, Y.Q.; Chan, J.C.W. Hyperspectral and Multispectral Image Fusion via Deep Two-Branches Convolutional Neural Network. *Remote Sens.* **2018**, *10*, 800. [CrossRef]

89. Jia, J.; Ji, L.; Zhao, Y.; Geng, X. Hyperspectral image super-resolution with spectral–spatial network. *Int. J. Remote Sens.* **2018**, *39*, 7806–7829. [CrossRef]

90. Chen, Y.; Li, C.; Ghamisi, P.; Jia, X.; Gu, Y. Deep Fusion of Remote Sensing Data for Accurate Classification. *IEEE Geosci. Remote Sens. Lett.* **2017**, *14*, 1253–1257. [CrossRef]

91. Ghamisi, P.; Höfle, B.; Zhu, X.X. Hyperspectral and LiDAR Data Fusion Using Extinction Profiles and Deep Convolutional Neural Network. *IEEE J. Sel. Top. Appl. Earth Obs. Remote Sens.* **2017**, *10*, 3011–3024. [CrossRef]

92. Li, H.; Ghamisi, P.; Soergel, U.; Zhu, X.X. Hyperspectral and LiDAR Fusion Using Deep Three-Stream Convolutional Neural Networks. *Remote Sens.* **2018**, *10*, 1649. [CrossRef]

93. Feng, Q.; Zhu, D.; Yang, J.; Li, B. Multisource Hyperspectral and LiDAR Data Fusion for Urban Land-Use Mapping based on a Modified Two-Branch Convolutional Neural Network. *ISPRS Int. J. Geo-Inf.* **2019**, *8*, 28. [CrossRef]

94. Zhang, M.; Li, W.; Du, Q.; Gao, L.; Zhang, B. Feature Extraction for Classification of Hyperspectral and LiDAR Data Using Patch-to-Patch CNN. *IEEE Trans. Cybern.* **2018**, 1–12. [CrossRef] [PubMed]

95. Xu, X.; Li, W.; Ran, Q.; Du, Q.; Gao, L.; Zhang, B. Multisource remote sensing data classification based on convolutional neural network. *IEEE Trans. Geosci. Remote Sens.* **2018**, *56*, 937–949. [CrossRef]

96. Litjens, G.; Kooi, T.; Bejnordi, B.E.; Setio, A.A.A.; Ciompi, F.; Ghafoorian, M.; van der Laak, J.A.; van Ginneken, B.; Sánchez, C.I. A survey on deep learning in medical image analysis. *Med. Image Anal.* **2017**, *42*, 60–88. [CrossRef]

97. Md Noor, S.S.; Ren, J.; Marshall, S.; Michael, K. Hyperspectral Image Enhancement and Mixture Deep-Learning Classification of Corneal Epithelium Injuries. *Sensors* **2017**, *17*, 2644. [CrossRef]

98. Halicek, M.; Lu, G.; Little, J.V.; Wang, X.; Patel, M.; Griffith, C.C.; El-Deiry, M.W.; Chen, A.Y.; Fei, B. Deep convolutional neural networks for classifying head and neck cancer using hyperspectral imaging. *J. Biomed. Opt.* **2017**, *6*, 60503–60503. [CrossRef] [PubMed]

99. Ma, L.; Lu, G.; Wang, D.; Wang, X.; Chen, Z.G.; Muller, S.; Chen, A.; Fei, B. Deep learning based classification for head and neck cancer detection with hyperspectral imaging in an animal model. *Proc. SPIE* **2017**, *10137*, 101372G.

100. Halicek, M.; Little, J.V.; Xu, W.; Patel, M.; Griffith, C.C.; Chen, A.Y.; Fei, B. Tumor margin classification of head and neck cancer using hyperspectral imaging and convolutional neural networks. In *Medical Imaging 2018: Image-Guided Procedures, Robotic Interventions, and Modeling*; SPIE: Houston, TX, USA, 2018; p. 10576.

101. Lin, J.; Clancy, N.T.; Qi, J.; Hu, Y.; Tatla, T.; Stoyanov, D.; Maier-Hein, L.; Elson, D.S. Dual-modality endoscopic probe for tissue surface shape reconstruction and hyperspectral imaging enabled by deep neural networks. *Med. Image Anal.* **2018**, *48*, 162–176. [CrossRef]

102. Li, X.; Li, W.; Xu, X.; Hu, W. Cell classification using convolutional neural networks in medical hyperspectral imagery. In Proceedings of the 2017 2nd International Conference on Image, Vision and Computing (ICIVC), Chengdu, China, 2–4 June 2017; pp. 501–504.

103. Huang, Q.; Li, W.; Xie, X. Convolutional neural network for medical hyperspectral image classification with kernel fusion. In Proceedings of the BIBE 2018 International Conference on Biological Information and Biomedical Engineering, Shanghai, China, 6–8 July 2018; pp. 1–4.

104. Wei, X.; Li, W.; Zhang, M.; Li, Q. Medical Hyperspectral Image Classification Based on End-to-End Fusion Deep Neural Network. *IEEE Trans. Instrum. Meas.* **2019**, 1–12. [CrossRef]

105. Bayramoglu, N.; Kaakinen, M.; Eklund, L.; Heikkilä, J. Towards Virtual H&E Staining of Hyperspectral Lung Histology Images Using Conditional Generative Adversarial Networks. In Proceedings of the 2017 IEEE International Conference on Computer Vision Workshops (ICCVW), Venice, Italy, 22–29 October 2017; pp. 64–71.

106. Turra, G.; Conti, N.; Signoroni, A. Hyperspectral image acquisition and analysis of cultured bacteria for the discrimination of urinary tract infections. In Proceedings of the 2015 37th Annual International Conference of the IEEE Engineering in Medicine and Biology Society (EMBC), Milan, Italy, 25–29 August 2015; pp. 759–762.

107. Turra, G.; Arrigoni, S.; Signoroni, A. CNN-Based Identification of Hyperspectral Bacterial Signatures for Digital Microbiology. In Proceedings of the International Conference on Image Analysis and Processing, Catania, Italy, 11–15 September 2017; pp. 500–510.

108. Bailey, A.; Ledeboer, N.; Burnham, C.A.D. Clinical Microbiology Is Growing Up: The Total Laboratory Automation Revolution. *Clin. Chem.* **2019**, *65*, 634–643 [CrossRef]

109. Signoroni, A.; Savardi, M.; Pezzoni, M.; Guerrini, F.; Arrigoni, S.; Turra, G. Combining the use of CNN classification and strength-driven compression for the robust identification of bacterial species on hyperspectral culture plate images. *IET Comput. Vis.* **2018**, *12*, 941–949. [CrossRef]

110. Salzer, R.; Siesler, H.W. *Infrared and Raman sPectroscopic Imaging*; John Wiley & Sons: Hoboken, NJ, USA, 2014.

111. Pahlow, S.; Weber, K.; Popp, J.; Bayden, R.W.; Kochan, K.; Rüther, A.; Perez-Guaita, D.; Heraud, P.; Stone, N.; Dudgeon, A.; et al. Application of Vibrational Spectroscopy and Imaging to Point-of-Care Medicine: A Review. *Appl. Spectrosc.* **2018**, *72*, 52–84.

112. Liu, J.; Osadchy, M.; Ashton, L.; Foster, M.; Solomon, C.J.; Gibson, S.J. Deep convolutional neural networks for Raman spectrum recognition: A unified solution. *Analyst* **2017**, *142*, 4067–4074. [CrossRef]

113. Weng, S.; Xu, X.; Li, J.; Wong, S.T. Combining deep learning and coherent anti-Stokes Raman scattering imaging for automated differential diagnosis of lung cancer. *J. Biomed. Opt.* **2017**, *22*, 106017. [CrossRef]

114. Duncan, M.D.; Reintjes, J.; Manuccia, T.J. Imaging Biological Compounds Using The Coherent Anti-Stokes Raman Scattering Microscope. *Opt. Eng.* **1985**, *24*, 242352. [CrossRef]

115. Malek, K.; Wood, B.R.; Bambery, K.R. FTIR Imaging of Tissues: Techniques and Methods of Analysis. In *Optical Spectroscopy and Computational Methods in Biology and Medicine*; Springer: Dordrecht, The Netherlands, 2014; pp. 419–473.

116. Berisha, S.; Lotfollahi, M.; Jahanipour, J.; Gurcan, I.; Walsh, M.; Bhargava, R.; Van Nguyen, H.; Mayerich, D. Deep learning for FTIR histology: Leveraging spatial and spectral features with convolutional neural networks. *Analyst* **2019**, *144*, 1642–1653. [CrossRef] [PubMed]

117. Lotfollahi, M.; Berisha, S.; Daeinejad, D.; Mayerich, D. Digital Staining of High-Definition Fourier Transform Infrared (FT-IR) Images Using Deep Learning. *Appl. Spectrosc.* **2019**, *73*, 556–564 . [CrossRef]

118. Reis, M.M.; Beers, R.V.; Al-Sarayreh, M.; Shorten, P.; Yan, W.Q.; Saeys, W.; Klette, R.; Craigie, C. Chemometrics and hyperspectral imaging applied to assessment of chemical, textural and structural characteristics of meat. *Meat Sci.* **2018**, *144*, 100–109. [CrossRef]

119. Yu, X.; Tang, L.; Wu, X.; Lu, H. Nondestructive Freshness Discriminating of Shrimp Using Visible/Near-Infrared Hyperspectral Imaging Technique and Deep Learning Algorithm. *Food Anal. Methods* **2017**, *11*, 1–13. [CrossRef]

120. Yu, X.; Wang, J.; Wen, S.; Yang, J.; Zhang, F. A deep learning based feature extraction method on hyperspectral images for nondestructive prediction of TVB-N content in Pacific white shrimp (Litopenaeus vannamei). *Biosyst. Eng.* **2019**, *178*, 244–255. [CrossRef]

121. Al-Sarayreh, M.; Reis, M.R.; Yan, W.Q.; Klette, R. Detection of Red-Meat Adulteration by Deep Spectral–spatial Features in Hyperspectral Images. *J. Imaging* **2018**, *4*, 63. [CrossRef]

122. Yu, X.; Lu, H.; Liu, Q. Deep-learning-based regression model and hyperspectral imaging for rapid detection of nitrogen concentration in oilseed rape (*Brassica napus* L.) leaf. *Chemom. Intell. Lab. Syst.* **2018**, *172*, 188–193. [CrossRef]

123. Jin, X.; Jie, L.; Wang, S.; Qi, H.J.; Li, S.W. Classifying Wheat Hyperspectral Pixels of Healthy Heads and Fusarium Head Blight Disease Using a Deep Neural Network in the Wild Field. *Remote Sens.* **2018**, *10*, 395. [CrossRef]

124. Yu, X.; Lu, H.; Wu, D. Development of deep learning method for predicting firmness and soluble solid content of postharvest Korla fragrant pear using Vis/NIR hyperspectral reflectance imaging. *Postharvest Biol. Technol.* **2018**, *141*, 39–49. [CrossRef]

125. Wang, Z.; Hu, M.H.; Zhai, G. Application of Deep Learning Architectures for Accurate and Rapid Detection of Internal Mechanical Damage of Blueberry Using Hyperspectral Transmittance Data. *Sensors* **2018**, *18*, 1126. [CrossRef] [PubMed]

126. Nagasubramanian, K.; Jones, S.; Singh, A.K.; Singh, A.; Ganapathysubramanian, B.; Sarkar, S. Explaining hyperspectral imaging based plant disease identification: 3D CNN and saliency maps. *arXiv* **2018**, arXiv:1804.08831.

127. Qiu, Z.; Chen, J.; Zhao, Y.; Zhu, S.; He, Y.; Zhang, C. Variety Identification of Single Rice Seed Using Hyperspectral Imaging Combined with Convolutional Neural Network. *Appl. Sci.* **2018**, *8*, 212. [CrossRef]

128. Wu, N.; Zhang, C.; Bai, X.; Du, X.; He, Y. Discrimination of Chrysanthemum Varieties Using Hyperspectral Imaging Combined with a Deep Convolutional Neural Network. *Molecules* **2018**, *23*, 2831. [CrossRef]

129. Khan, M.J.; Yousaf, A.; Abbas, A.; Khurshid, K. Deep learning for automated forgery detection in hyperspectral document images. *J. Electron. Imaging* **2018**, *27*, 053001. [CrossRef]

130. Qureshi, R.; Uzair, M.; Khurshid, K.; Yan, H. Hyperspectral document image processing: Applications, challenges and future prospects. *Pattern Recognit.* **2019**, *90*, 12–22. [CrossRef]

131. Song, W.; Li, S.; Fang, L.; Lu, T. Hyperspectral Image Classification With Deep Feature Fusion Network. *IEEE Trans. Geosci. Remote Sens.* **2018**, *56*, 3173–3184. [CrossRef]

132. Robila, S.A. Independent Component Analysis. In *Advanced Image Processing Techniques for Remotely Sensed Hyperspectral Data*; Varshney, P.K., Arora, M.K., Eds.; Springer: Berlin/Heidelberg, Germany, 2004; Chapter 4, pp. 109–132.

133. Cheng, G.; Li, Z.; Han, J.; Yao, X.; Guo, L. Exploring Hierarchical Convolutional Features for Hyperspectral Image Classification. *IEEE Trans. Geosci. Remote Sens.* **2018**, *56*, 6712–6722. [CrossRef]

134. Hu, J.; Mou, L.; Schmitt, A.; Zhu, X.X. FusioNet: A two-stream convolutional neural network for urban scene classification using PolSAR and hyperspectral data. In Proceedings of the 2017 Joint Urban Remote Sensing Event (JURSE), Dubai, UAE, 6–8 March 2017; pp. 1–4.

135. Jiao, L.; Liang, M.; Chen, H.; Yang, S.; Liu, H.; Cao, X. Deep Fully Convolutional Network-Based Spatial Distribution Prediction for Hyperspectral Image Classification. *IEEE Trans. Geosci. Remote Sens.* **2017**, *55*, 5585–5599. [CrossRef]

136. Leng, J.; Li, T.; Bai, G.; Dong, Q.; Dong, H. Cube-CNN-SVM: A Novel Hyperspectral Image Classification Method. In Proceedings of the 2016 IEEE 28th International Conference on Tools with Artificial Intelligence (ICTAI), San Jose, CA, USA, 6–8 November 2016; pp. 1027–1034.

137. Wei, Y.; Zhou, Y.; Li, H. Spectral–spatial Response for Hyperspectral Image Classification. *Remote Sens.* **2017**, *9*, 203. [CrossRef]

138. Yang, J.; Zhao, Y.; Chan, J.C.; Yi, C. Hyperspectral image classification using two-channel deep convolutional neural network. In Proceedings of the 2016 IEEE International Geoscience and Remote Sensing Symposium (IGARSS), Beijing, China, 10–15 July 2016; pp. 5079–5082.

139. Slavkovikj, V.; Verstockt, S.; De Neve, W.; Van Hoecke, S.; Van de Walle, R. Hyperspectral image classification with convolutional neural networks. In Proceedings of the 23rd Annual ACM Conference on Multimedia, Brisbane, Australia, 26–30 October 2015; pp. 1159–1162.

140. Yu, S.; Jia, S.; Xu, C. Convolutional neural networks for hyperspectral image classification. *Neurocomputing* **2017**, *219*, 88–98. [CrossRef]

141. Zhan, Y.; Hu, D.; Xing, H.; Yu, X. Hyperspectral Band Selection Based on Deep Convolutional Neural Network and Distance Density. *IEEE Geosci. Remote Sens. Lett.* **2017**, *14*, 2365–2369. [CrossRef]

142. Fang, L.; Liu, G.; Li, S.; Ghamisi, P.; Benediktsson, J.A. Hyperspectral Image Classification With Squeeze Multibias Network. *IEEE Trans. Geosci. Remote Sens.* **2018**, *57*, 1291–1301. [CrossRef]

143. Lee, H.; Kwon, H. Contextual deep CNN based hyperspectral classification. In Proceedings of the 2016 IEEE International Geoscience and Remote Sensing Symposium (IGARSS), Beijing, China, 10–15 July 2016; pp. 3322–3325.

144. Li, Y.; Zhang, H.; Shen, Q. Spectral–spatial Classification of Hyperspectral Imagery with 3D Convolutional Neural Network. *Remote Sens.* **2017**, *9*, 67. [CrossRef]

145. Heming, L.; Li, Q. Hyperspectral Imagery Classification Using Sparse Representations of Convolutional Neural Network Features. *Remote Sens.* **2015**, *8*, 99.

146. Qingshan, L.; Feng, Z.; Renlong, H.; Xiaotong, Y. Bidirectional-Convolutional LSTM Based Spectral–spatial Feature Learning for Hyperspectral Image Classification. *Remote Sens.* **2017**, *9*, 1330. [CrossRef]

147. Liu, B.; Yu, X.; Zhang, P.; Yu, A.; Fu, Q.; Wei, X. Supervised Deep Feature Extraction for Hyperspectral Image Classification. *IEEE Trans. Geosci. Remote Sens.* **2018**, *56*, 1909–1921. [CrossRef]

148. Makantasis, K.; Karantzalos, K.; Doulamis, A.; Doulamis, N. Deep supervised learning for hyperspectral data classification through convolutional neural networks. In Proceedings of the 2015 IEEE International Geoscience and Remote Sensing Symposium (IGARSS), Milan, Italy, 26–31 July 2015; pp. 4959–4962.

149. Mei, S.; Ji, J.; Hou, J.; Li, X.; Du, Q. Learning Sensor-Specific Spatial–spectral Features of Hyperspectral Images via Convolutional Neural Networks. *IEEE Trans. Geosci. Remote Sens.* **2017**, *55*, 4520–4533. [CrossRef]

150. Paoletti, M.; Haut, J.; Plaza, J.; Plaza, A. Deep&Dense Convolutional Neural Network for Hyperspectral Image Classification. *Remote Sens.* **2018**, *10*, 1454.

151. Santara, A.; Mani, K.; Hatwar, P.; Singh, A.; Garg, A.; Padia, K.; Mitra, P. BASS Net: Band-adaptive spectral–spatial feature learning neural network for hyperspectral image classification. *IEEE Trans. Geosci. Remote Sens.* **2017**, *55*, 5293–5301. [CrossRef]

152. Yang, J.; Zhao, Y.; Chan, J.C. Learning and Transferring Deep Joint Spectral–Spatial Features for Hyperspectral Classification. *IEEE Trans. Geosci. Remote Sens.* **2017**, *55*, 4729–4742. [CrossRef]

153. Yue, J.; Zhao, W.; Mao, S.; Liu, H. Spectral–spatial classification of hyperspectral images using deep convolutional neural networks. *Remote Sens. Lett.* **2015**, *6*, 468–477. [CrossRef]

154. Zhang, M.; Hong, L. Deep Learning Integrated with Multiscale Pixel and Object Features for Hyperspectral Image Classification. In Proceedings of the 2018 10th IAPR Workshop on Pattern Recognition in Remote Sensing (PRRS), Beijing, China, 19–20 August 2018; pp. 1–8.

155. Zhao, W.; Du, S. Spectral–Spatial Feature Extraction for Hyperspectral Image Classification: A Dimension Reduction and Deep Learning Approach. *IEEE Trans. Geosci. Remote Sens.* **2016**, *54*, 4544–4554. [CrossRef]

156. Zhi, L.; Yu, X.; Liu, B.; Wei, X. A dense convolutional neural network for hyperspectral image classification. *Remote Sens. Lett.* **2019**, *10*, 59–66. [CrossRef]

157. Chen, Y.; Jiang, H.; Li, C.; Jia, X.; Ghamisi, P. Deep Feature Extraction and Classification of Hyperspectral Images Based on Convolutional Neural Networks. *IEEE Trans. Geosci. Remote Sens.* **2016**, *54*, 6232–6251. [CrossRef]

158. Liu, B.; Yu, X.; Yu, A.; Zhang, P.; Wan, G.; Wang, R. Deep Few-Shot Learning for Hyperspectral Image Classification. *IEEE Trans. Geosci. Remote Sens.* **2018**, *57*, 2290–2304. [CrossRef]

159. Zhong, Z.; Li, J.; Luo, Z.; Chapman, M. Spectral–Spatial Residual Network for Hyperspectral Image Classification: A 3-D Deep Learning Framework. *IEEE Trans. Geosci. Remote Sens.* **2018**, *56*, 847–858. [CrossRef]

160. Liu, X.; Sun, Q.; Meng, Y.; Fu, M.; Bourennane, S. Hyperspectral Image Classification Based on Parameter-Optimized 3D-CNNs Combined with Transfer Learning and Virtual Samples. *Remote Sens.* **2018**, *10*, 1425. [CrossRef]

161. Ouyang, N.; Zhu, T.; Lin, L. Convolutional Neural Network Trained by Joint Loss for Hyperspectral Image Classification. *IEEE Geosci. Remote Sens. Lett.* **2018**, *16*, 457–461. [CrossRef]

162. Ma, X.; Fu, A.; Wang, J.; Wang, H.; Yin, B. Hyperspectral Image Classification Based on Deep Deconvolution Network With Skip Architecture. *IEEE Trans. Geosci. Remote Sens.* **2018**, *56*, 4781–4791. [CrossRef]

163. Zhang, H.; Li, Y.; Zhang, Y.; Shen, Q. Spectral–spatial classification of hyperspectral imagery using a dual-channel convolutional neural network. *Remote Sens. Lett.* **2017**, *8*, 438–447. [CrossRef]

164. Gao, H.; Yang, Y.; Li, C.; Zhou, H.; Qu, X. Joint Alternate Small Convolution and Feature Reuse for Hyperspectral Image Classification. *ISPRS Int. J. Geo-Inf.* **2018**, *7*, 349. [CrossRef]

165. Luan, S.; Chen, C.; Zhang, B.; Han, J.; Liu, J. Gabor Convolutional Networks. *IEEE Trans. Image Process.* **2018**, *27*, 4357–4366. [CrossRef]

166. Chopra, S.; Hadsell, R.; LeCun, Y. Learning a similarity metric discriminatively, with application to face verification. In Proceedings of the 2005 IEEE Computer Society Conference on Computer Vision and Pattern Recognition (CVPR'05), San Diego, CA, USA, 20–25 June 2005; Volume 1, pp. 539–546.

167. Li, F.F.; Fergus, R.; Perona, P. One-shot learning of object categories. *IEEE Trans. Pattern Anal. Mach. Intell.* **2006**, *28*, 594–611.

168. Achanta, R.; Shaji, A.; Smith, K.; Lucchi, A.; Fua, P.; Süsstrunk, S. SLIC Superpixels Compared to State-of-the-Art Superpixel Methods. *IEEE Trans. Pattern Anal. Mach. Intell.* **2012**, *34*, 2274–2282, doi:10.1109/TPAMI.2012.120. [CrossRef]

169. Chen, Y.; Lin, Z.; Zhao, X.; Wang, G.; Gu, Y. Deep Learning-Based Classification of Hyperspectral Data. *IEEE J. Sel. Top. Appl. Earth Obs. Remote Sens.* **2014**, *7*, 2094–2107. [CrossRef]

170. Tao, C.; Pan, H.; Li, Y.; Zou, Z. Unsupervised spectral–spatial feature learning with stacked sparse autoencoder for hyperspectral imagery classification. *IEEE Geosci. Remote Sens. Lett.* **2015**, *12*, 2438–2442.

171. Kussul, N.; Lavreniuk, M.; Skakun, S.; Shelestov, A. Deep learning classification of land cover and crop types using remote sensing data. *IEEE Geosci. Remote Sens. Lett.* **2017**, *14*, 778–782. [CrossRef]

172. Ma, X.; Geng, J.; Wang, H. Hyperspectral image classification via contextual deep learning. *EURASIP J. Image Video Process.* **2015**, *2015*, 20. [CrossRef]

173. Ma, X.; Wang, H.; Geng, J.; Wang, J. Hyperspectral image classification with small training set by deep network and relative distance prior. In Proceedings of the 2016 IEEE International Geoscience and Remote Sensing Symposium (IGARSS), Beijing, China, 10–15 July 2016; pp. 3282–3285.

174. Ma, X.; Wang, H.; Geng, J. Spectral–Spatial Classification of Hyperspectral Image Based on Deep Auto-Encoder. *IEEE J. Sel. Top. Appl. Earth Obs. Remote Sens.* **2016**, *9*, 4073–4085. [CrossRef]

175. Yue, J.; Mao, S.; Li, M. A deep learning framework for hyperspectral image classification using spatial pyramid pooling. *Remote Sens. Lett.* **2016**, *7*, 875–884. [CrossRef]

176. Liu, Y.; Cao, G.; Sun, Q.; Siegel, M. Hyperspectral classification via deep networks and superpixel segmentation. *Int. J. Remote Sens.* **2015**, *36*, 3459–3482. [CrossRef]

177. Windrim, L.; Ramakrishnan, R.; Melkumyan, A.; Murphy, R.J. A Physics-Based Deep Learning Approach to Shadow Invariant Representations of Hyperspectral Images. *IEEE Trans. Image Process.* **2018**, *27*, 665–677. [CrossRef]

178. Ball, J.E.; Wei, P. Deep Learning Hyperspectral Image Classification using Multiple Class-Based Denoising Autoencoders, Mixed Pixel Training Augmentation, and Morphological Operations. In Proceedings of the IGARSS 2018—2018 IEEE International Geoscience and Remote Sensing Symposium, Valencia, Spain, 22–27 July 2018; pp. 6903–6906.

179. Lan, R.; Li, Z.; Liu, Z.; Gu, T.; Luo, X. Hyperspectral image classification using k-sparse denoising autoencoder and spectral–restricted spatial characteristics. *Appl. Soft Comput.* **2019**, *74*, 693–708. [CrossRef]

180. Chen, Y.; Zhao, X.; Jia, X. Spectral–Spatial Classification of Hyperspectral Data Based on Deep Belief Network. *IEEE J. Sel. Top. Appl. Earth Obs. Remote Sens.* **2015**, *8*, 2381–2392. [CrossRef]

181. Wang, M.; Zhao, M.; Chen, J.; Rahardja, S. Nonlinear Unmixing of Hyperspectral Data via Deep Autoencoder Networks. *IEEE Geosci. Remote Sens. Lett.* **2019**, 1–5. [CrossRef]

182. Ozkan, S.; Kaya, B.; Akar, G.B. EndNet: Sparse AutoEncoder Network for Endmember Extraction and Hyperspectral Unmixing. *IEEE Trans. Geosci. Remote Sens.* **2019**, *57*, 482–496. [CrossRef]

183. He, Z.; Liu, H.; Wang, Y.; Hu, J. Generative Adversarial Networks-Based Semi-Supervised Learning for Hyperspectral Image Classification. *Remote Sens.* **2017**, *9*, 1042. [CrossRef]

184. Zhang, M.; Gong, M.; Mao, Y.; Li, J.; Wu, Y. Unsupervised Feature Extraction in Hyperspectral Images Based on Wasserstein Generative Adversarial Network. *IEEE Trans. Geosci. Remote Sens.* **2018**, *57*, 2669–2688. [CrossRef]

185. Zhan, Y.; Wu, K.; Liu, W.; Qin, J.; Yang, Z.; Medjadba, Y.; Wang, G.; Yu, X. Semi-Supervised Classification of Hyperspectral Data Based on Generative Adversarial Networks and Neighborhood Majority Voting. In Proceedings of the IGARSS 2018—2018 IEEE International Geoscience and Remote Sensing Symposium, Valencia, Spain, 22–27 July 2018; pp. 5756–5759.

186. Bashmal, L.; Bazi, Y.; AlHichri, H.; AlRahhal, M.M.; Ammour, N.; Alajlan, N. Siamese-GAN: Learning Invariant Representations for Aerial Vehicle Image Categorization. *Remote Sens.* **2018**, *10*, 351. [CrossRef]

187. Wu, H.; Prasad, S. Convolutional Recurrent Neural Networks forHyperspectral Data Classification. *Remote Sens.* **2017**, *9*, 298. [CrossRef]

188. Shi, C.; Pun, C.M. Superpixel-based 3D deep neural networks for hyperspectral image classification. *Pattern Recognit.* **2018**, *74*, 600–616. [CrossRef]

189. Windrim, L.; Ramakrishnan, R.; Melkumyan, A.; Murphy, R.J. Hyperspectral CNN Classification with Limited Training Samples. *arXiv* **2016**, arXiv:1611.09007.

190. Li, W.; Chen, C.; Zhang, M.; Li, H.; Du, Q. Data Augmentation for Hyperspectral Image Classification With Deep CNN. *IEEE Geosci. Remote Sens. Lett.* **2019**, *16*, 593–597. [CrossRef]

191. Windrim, L.; Melkumyan, A.; Murphy, R.J.; Chlingaryan, A.; Ramakrishnan, R. Pretraining for Hyperspectral Convolutional Neural Network Classification. *IEEE Trans. Geosci. Remote Sens.* **2018**, *56*, 2798–2810. [CrossRef]

192. Lin, J.; Ward, R.; Wang, Z.J. Deep transfer learning for hyperspectral image classification. In Proceedings of the 2018 IEEE 20th International Workshop on Multimedia Signal Processing (MMSP), Vancouver, BC, Canada, 29–31 August 2018; pp. 1–5.

193. Ratle, F.; Camps-Valls, G.; Weston, J. Semisupervised Neural Networks for Efficient Hyperspectral Image Classification. *IEEE Trans. Geosci. Remote Sens.* **2010**, *48*, 2271–2282. [CrossRef]

194. Romero, A.; Gatta, C.; Camps-Valls, G. Unsupervised Deep Feature Extraction for Remote Sensing Image Classification. *IEEE Trans. Geosci. Remote Sens.* **2016**, *54*, 1349–1362. [CrossRef]

195. Maggiori, E.; Tarabalka, Y.; Charpiat, G.; Alliez, P. Convolutional Neural Networks for Large-Scale Remote-Sensing Image Classification. *IEEE Trans. Geosci. Remote Sens.* **2017**, *55*, 645–657. [CrossRef]

196. Mou, L.; Ghamisi, P.; Zhu, X.X. Unsupervised Spectral–Spatial Feature Learning via Deep Residual Conv–Deconv Network for Hyperspectral Image Classification. *IEEE Trans. Geosci. Remote Sens.* **2018**, *56*, 391–406. [CrossRef]

197. Wu, H.; Prasad, S. Semi-Supervised Deep Learning Using Pseudo Labels for Hyperspectral Image Classification. *IEEE Trans. Image Process.* **2018**, *27*, 1259–1270. [CrossRef]

198. Pan, X.; Zhao, J. High-Resolution Remote Sensing Image Classification Method Based on Convolutional Neural Network and Restricted Conditional Random Field. *Remote Sens.* **2018**, *10*, 920. [CrossRef]

199. Hu, Y.; Zhang, J.; Ma, Y.; An, J.; Ren, G.; Li, X. Hyperspectral Coastal Wetland Classification Based on a Multiobject Convolutional Neural Network Model and Decision Fusion. *IEEE Geosci. Remote Sens. Lett.* **2019**, 1–5. [CrossRef]

200. Pan, B.; Shi, Z.; Xu, X. R-VCANet: A New Deep-Learning-Based Hyperspectral Image Classification Method. *IEEE J. Sel. Top. Appl. Earth Obs. Remote Sens.* **2017**, *10*, 1975–1986. [CrossRef]

201. Pan, B.; Shi, Z.; Xu, X. MugNet: Deep learning for hyperspectral image classification using limited samples. *ISPRS J. Photogramm. Remote Sens.* **2018**, *145*, 108–119. [CrossRef]

202. Ghamisi, P.; Chen, Y.; Zhu, X.X. A Self-Improving Convolution Neural Network for the Classification of Hyperspectral Data. *IEEE Geosci. Remote Sens. Lett.* **2016**, *13*, 1537–1541. [CrossRef]

203. Wang, Z.; Du, B.; Shi, Q.; Tu, W. Domain Adaptation With Discriminative Distribution and Manifold Embedding for Hyperspectral Image Classification. *IEEE Geosci. Remote Sens. Lett.* **2019**, 1–5. [CrossRef]

204. Liu, P.; Zhang, H.; Eom, K.B. Active Deep Learning for Classification of Hyperspectral Images. *IEEE J. Sel. Top. Appl. Earth Obs. Remote Sens.* **2017**, *10*, 712–724. [CrossRef]

205. Lin, J.; Zhao, L.; Li, S.; Ward, R.; Wang, Z.J. Active-Learning-Incorporated Deep Transfer Learning for Hyperspectral Image Classification. *IEEE J. Sel. Top. Appl. Earth Obs. Remote Sens.* **2018**, *11*, 4048–4062. [CrossRef]

206. Haut, J.M.; Paoletti, M.E.; Plaza, J.; Li, J.; Plaza, A. Active Learning With Convolutional Neural Networks for Hyperspectral Image Classification Using a New Bayesian Approach. *IEEE Trans. Geosci. Remote Sens.* **2018**, *56*, 6440–6461. [CrossRef]

207. Li, Y.; Hu, J.; Zhao, X.; Xie, W.; Li, J. Hyperspectral image super-resolution using deep convolutional neural network. *Neurocomputing* **2017**, *266*, 29–41. [CrossRef]

208. He, Z.; Liu, L. Hyperspectral Image Super-Resolution Inspired by Deep Laplacian Pyramid Network. *Remote Sens.* **2018**, *10*, 1939. [CrossRef]

209. Paoletti, M.E.; Haut, J.M.; Fernandez-Beltran, R.; Plaza, J.; Plaza, A.; Li, J.; Pla, F. Capsule Networks for Hyperspectral Image Classification. *IEEE Trans. Geosci. Remote Sens.* **2019**, *57*, 2145–2160. [CrossRef]

210. Wang, W.Y.; Li, H.C.; Pan, L.; Yang, G.; Du, Q. Hyperspectral Image Classification Based on Capsule Network. In Proceedings of the IGARSS 2018–2018 IEEE International Geoscience and Remote Sensing Symposium, Valencia, Spain, 22–27 July 2018; pp. 3571–3574.

211. Zhu, K.; Chen, Y.; Ghamisi, P.; Jia, X.; Benediktsson, J.A. Deep Convolutional Capsule Network for Hyperspectral Image Spectral and Spectral–spatial Classification. *Remote Sens.* **2019**, *11*, 223. [CrossRef]
212. Yin, J.; Li, S.; Zhu, H.; Luo, X. Hyperspectral Image Classification Using CapsNet With Well-Initialized Shallow Layers. *IEEE Geosci. Remote Sens. Lett.* **2019**, 1–5. [CrossRef]
213. Haut, J.M.; Bernabé, S.; Paoletti, M.E.; Fernandez-Beltran, R.; Plaza, A.; Plaza, J. Low-High-Power Consumption Architectures for Deep-Learning Models Applied to Hyperspectral Image Classification. *IEEE Geosci. Remote Sens. Lett.* **2019**, *16*, 776–780. [CrossRef]
214. Paoletti, M.; Haut, J.; Plaza, J.; Plaza, A. A new deep convolutional neural network for fast hyperspectral image classification. *ISPRS J. Photogramm. Remote Sens.* **2018**, *145*, 120–147. [CrossRef]
215. LeCun, Y.; Bengio, Y.; Hinton, G. Deep learning. *Nature* **2015**, *521*, 436. [CrossRef] [PubMed]
216. Goodfellow, I.; Bengio, Y.; Courville, A.; Bengio, Y. *Deep Learning*; MIT Press: Cambridge, MA, USA, 2016; Volume 1.
217. Ranzato, M.A.; Szummer, M. Semi-supervised Learning of Compact Document Representations with Deep Networks. In Proceedings of the 25th International Conference on Machine Learning, Helsinki, Finland, 5–9 July 2008; ACM: New York, NY, USA, 2008; pp. 792–799.
218. LeCun, Y.; Boser, B.E.; Denker, J.S.; Henderson, D.; Howard, R.E.; Hubbard, W.E.; Jackel, L.D. Handwritten digit recognition with a back-propagation network. In *Advances in Neural Information Processing Systems*; Morgan Kaufman: Denver, CO, USA, 1990; pp. 396–404.
219. Krizhevsky, A.; Sutskever, I.; Hinton, G.E. Imagenet classification with deep convolutional neural networks. In *Advances in Neural Information Processing Systems*; Curran Associates, Inc.: Lake Tahoe, NV, USA, 2012; pp. 1097–1105.
220. Szegedy, C.; Liu, W.; Jia, Y.; Sermanet, P.; Reed, S.; Anguelov, D.; Erhan, D.; Vanhoucke, V.; Rabinovich, A. Going deeper with convolutions. In Proceedings of the IEEE Conference on Computer Vision and Pattern Recognition, Boston, MA, USA, 7–12 June 2015; pp. 1–9.
221. Simonyan, K.; Zisserman, A. Very deep convolutional networks for large-scale image recognition. *arXiv* **2014**, arXiv:1409.1556.
222. He, K.; Zhang, X.; Ren, S.; Sun, J. Deep residual learning for image recognition. In Proceedings of the IEEE Conference on Computer Vision and Pattern Recognition, Las Vegas, NV, USA, 27–30 June 2016; pp. 770–778.
223. Iandola, F.; Moskewicz, M.; Karayev, S.; Girshick, R.; Darrell, T.; Keutzer, K. Densenet: Implementing efficient convnet descriptor pyramids. *arXiv* **2014**, arXiv:1404.1869.
224. Howard, A.G.; Zhu, M.; Chen, B.; Kalenichenko, D.; Wang, W.; Weyand, T.; Andreetto, M.; Adam, H. Mobilenets: Efficient convolutional neural networks for mobile vision applications. *arXiv* **2017**, arXiv:1704.04861.
225. Iandola, F.N.; Han, S.; Moskewicz, M.W.; Ashraf, K.; Dally, W.J.; Keutzer, K. SqueezeNet: AlexNet-level accuracy with 50x fewer parameters and <0.5 MB model size. *arXiv* **2016**, arXiv:1602.07360.
226. Hochreiter, S.; Schmidhuber, J. Long short-term memory. *Neural Comput.* **1997**, *9*, 1735–1780. [CrossRef] [PubMed]
227. Chung, J.; Gulcehre, C.; Cho, K.; Bengio, Y. Gated feedback recurrent neural networks. In Proceedings of the International Conference on Machine Learning, Lille, France, 6–11 July 2015; pp. 2067–2075.
228. Bengio, Y.; Lamblin, P.; Popovici, D.; Larochelle, H. Greedy layer-wise training of deep networks. In *Advances in Neural Information Processing Systems*; MIT Press: Vancouver, BC, Canada, 2007; pp. 153–160.
229. Hinton, G.E.; Salakhutdinov, R.R. Reducing the dimensionality of data with neural networks. *Science* **2006**, *313*, 504–507. [CrossRef] [PubMed]
230. Larochelle, H.; Erhan, D.; Courville, A.; Bergstra, J.; Bengio, Y. An empirical evaluation of deep architectures on problems with many factors of variation. In Proceedings of the 24th International Conference on Machine Learning, Corvalis, OR, USA, 20–24 June 2007; ACM: New York, NY, USA, 2007; pp. 473–480.

MDPI

St. Alban-Anlage 66

4052 Basel

Switzerland

Tel. +41 61 683 77 34

Fax +41 61 302 89 18

www.mdpi.com

Journal of Imaging Editorial Office

E-mail: jimaging@mdpi.com

www.mdpi.com/journal/jimaging